T0297706

The Art of Insight
in Science and Engineering

The Art of Insight in Science and Engineering

Mastering Complexity

Sanjoy Mahajan

The MIT Press
Cambridge, Massachusetts
London, England

Typeset by the author in 10.5/13.3 Palatino and Computer Modern Sans using ConTeXt and LuaTeX.

Library of Congress Cataloging-in-Publication Data

    Mahajan, Sanjoy, 1969- author.
    The art of insight in science and engineering : mastering complexity / Sanjoy Mahajan.
        pages cm
    Includes bibliographical references and index.
    ISBN 978-0-262-52654-8 (pbk. : alk. paper) 1. Statistical physics.
    2. Estimation theory. 3. Hypothesis. 4. Problem solving. I. Title.
    QC174.85.E88M34 2014
    501'.9-dc23
                                    2014003652

For my teachers, who showed me the way

Peter Goldreich
Carver Mead
Sterl Phinney

And for my students, one of whom said

I used to be curious, naively curious. Now I am fearlessly curious. I feel ready to attack any problem that comes at me, and at least get a feel for why things happen … roughly.

Brief contents

Contents

Preface

Science and engineering, our modern ways of understanding and altering the world, are said to be about accuracy and precision. Yet we best master the complexity of our world by cultivating insight rather than precision.

We need insight because our minds are but a small part of the world. An insight unifies fragments of knowledge into a compact picture that fits in our minds. But precision can overflow our mental registers, washing away the understanding brought by insight. This book shows you how to build insight and understanding first, so that you do not drown in complexity.

LESS
RIGOR
Therefore, our approach will not be rigorous—for rigor easily becomes *rigor mortis* or paralysis by analysis. Forgoing rigor, we'll study the natural and human-created worlds—the worlds of science and engineering. So you'll need some—but not extensive!—knowledge of physics concepts such as force, power, energy, charge, and field. We'll use as little mathematics as possible—algebra and geometry mostly, trigonometry sometimes, and calculus rarely—so that the mathematics promotes rather than hinders insight, understanding, and flexible problem solving. The goal is to help you master complexity; then no problem can intimidate you.

Like all important parts of our lives, whether spouses or careers, I came to this approach mostly unplanned. As a graduate student, I gave my first scientific talk on the chemical reactions in the retinal rod. I could make sense of the chemical chaos only by approximating. In that same year, my friend Carlos Brody wondered about the distribution of twin primes—prime pairs separated by 2, such as 3 and 5 or 11 and 13. Nobody knows the distribution for sure. As a lazy physicist, I approximately answered Carlos's question with a probabilistic model of being prime [32]. Approximations, I saw again, foster understanding.

As a physics graduate student, I needed to prepare for the graduate qualifying exams. I also became a teaching assistant for the "Order-of-Magnitude Physics" course. In three months, preparing for the qualifying exams and learning the course material to stay a day ahead of the students, I learned more physics than I had in the years of my undergraduate degree. Physics teaching and learning had much room for improvement—and approximation and insight could fill the gap.

DEDI-
CATION

In gratitude to my teachers, I dedicate this book to Carver Mead for irreplaceable guidance and faith; and to Peter Goldreich and Sterl Phinney, who developed the "Order-of-Magnitude Physics" course at Caltech. From them I learned the courage to simplify and gain insight—the courage that I look forward to teaching you.

ORGANI-
ZATION

For many years, at the University of Cambridge and at MIT, I taught a course on the "Art of Approximation" organized by topics in physics and engineering. This organization limited the material's generality: Unless you become a specialist in general relativity, you may not study gravitation again. Yet estimating how much gravity deflects starlight (Section 5.3.1) teaches reasoning tools that you can use far beyond that example. Tools are more general and useful than topics.

Therefore, I redesigned the course around the reasoning tools. This organization, which I have used at MIT and Olin College of Engineering, is reflected in this book—which teaches you one tool per chapter, each selected to help you build insight and master complexity.

There are the two broad ways to master complexity: organize the complexity or discard it. Organizing complexity, the subject of Part I, is taught through two tools: divide-and-conquer reasoning (Chapter 1) and making abstractions (Chapter 2).

Discarding complexity (Parts II and III) illustrates that "the art of being wise is the art of knowing what to overlook" (William James [24, p. 369]). In Part II, complexity is discarded *without* losing information. This part teaches three reasoning tools: symmetry and conservation (Chapter 3), proportional reasoning (Chapter 4), and dimensional analysis (Chapter 5). In Part III, complexity is discarded while losing information. This part teaches our final tools: lumping (Chapter 6), probabilistic reasoning (Chapter 7), easy cases (Chapter 8), and spring models (Chapter 9).

FINDING
MEANING

Using these tools, we will explore the natural and human-made worlds. We will estimate the flight range of birds and planes, the strength of chemical bonds, and the angle that the Sun deflects starlight; understand the physics of pianos, xylophones, and speakers; and explain why skies are blue and sunsets are red. Our tools weave these and many other examples into a tapestry of meaning spanning science and engineering.

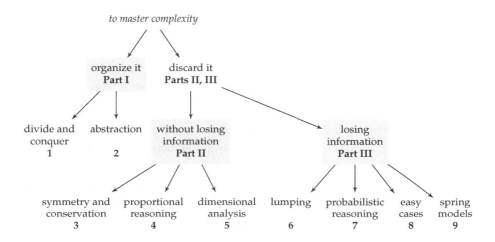

SHARING THIS WORK   Like my earlier *Street-Fighting Mathematics* [33], this book is licensed under a Creative Commons Attribution–Noncommercial–Share Alike license. MIT Press and I hope that you will improve and share the work noncommercially, and we would gladly receive corrections and suggestions.

INTERSPERSED QUESTIONS   The most effective teacher is a skilled tutor [2]. A tutor asks many questions, because questioning, wondering, and discussing promote learning. Questions of two types are interspersed through the book. *Questions marked with a ▶ in the margin*, which a tutor would pose during a tutorial, ask you to develop the next steps of an argument. They are answered in the subsequent text, where you can check your thinking. *Numbered problems*, marked with a shaded background, which a tutor would give you to take home, ask you to practice the tool, to extend an example, to use several tools, and even to resolve an occasional paradox. Merely watching workout videos produces little fitness! So, try many questions of both types.

IMPROVE OUR WORLD   Through your effort, mastery will come—and with a broad benefit. As the physicist Edwin Jaynes said of teaching [25]:

> [T]he goal should be, not to implant in the students' mind every fact that the teacher knows now; but rather to implant a way of thinking that enables the student, in the future, to learn in one year what the teacher learned in two years. Only in that way can we continue to advance from one generation to the next.

May the tools in this book help you advance our world beyond the state in which my generation has left it.

Acknowledgments

In addition to the dedication, I would like to thank the following people and organizations for their generosity.

For encouragement, forbearance, and motivation: my family—Juliet Jacobsen, Else Mahajan, and Sabine Mahajan.

For a sweeping review of the manuscript and improvements to every page: Tadashi Tokieda and David MacKay. Any remaining mistakes were contributed by me subsequently!

For advice on the process of writing: Larry Cohen, Hillary Rettig, Mary Carroll Moore, and Kenneth Atchity (author of *A Writer's Time* [1]).

For editorial guidance over many years: Robert Prior.

For valuable suggestions and discussions: Dap Hartmann, Shehu Abdussalam, Matthew Rush, Jason Manuel, Robin Oswald, David Hogg, John Hopfield, Elisabeth Moyer, R. David Middlebrook, Dennis Freeman, Michael Gottlieb, Edwin Taylor, Mark Warner, and many students throughout the years.

For the free software used for typesetting: Hans Hagen, Taco Hoekwater, and the ConTeXt user community (ConTeXt and LuaTeX); Donald Knuth (TeX); Taco Hoekwater and John Hobby (MetaPost); John Bowman, Andy Hammerlindl, and Tom Prince (Asymptote); Matt Mackall (Mercurial); Richard Stallman (Emacs); and the Debian GNU/Linux project.

For the NB document-annotation system: Sacha Zyto and David Karger.

For being a wonderful place for a graduate student to think, explore, and learn: the California Institute of Technology.

For supporting my work in science and mathematics education: the Whitaker Foundation in Biomedical Engineering; the Hertz Foundation; the Gatsby Charitable Foundation; the Master and Fellows of Corpus Christi College, Cambridge; Olin College of Engineering and its Intellectual Vitality program; and the Office of Digital Learning and the Department of Electrical Engineering and Computer Science at MIT.

Values for backs of envelopes

π	pi	3	
G	Newton's constant	7×10^{-11}	$\mathrm{kg^{-1}\,m^3\,s^{-2}}$
c	speed of light	3×10^8	$\mathrm{m\,s^{-1}}$
$\hbar c$	$\hbar$ shortcut	200	$\mathrm{eV\,nm}$
$m_e c^2$	electron rest energy	0.5	MeV
k_B	Boltzmann's constant	10^{-4}	$\mathrm{eV\,K^{-1}}$
N_A	Avogadro's number	6×10^{23}	$\mathrm{mol^{-1}}$
R	universal gas constant $k_B N_A$	8	$\mathrm{J\,mol^{-1}\,K^{-1}}$
e	electron charge	1.6×10^{-19}	C
$e^2/4\pi\epsilon_0$	electrostatic combination	2.3×10^{-28}	$\mathrm{kg\,m^3\,s^{-2}}$
$(e^2/4\pi\epsilon_0)/\hbar c$	fine-structure constant α	0.7×10^{-2}	
σ	Stefan–Boltzmann constant	6×10^{-8}	$\mathrm{W\,m^{-2}\,K^{-4}}$
M_{Sun}	solar mass	2×10^{30}	kg
m_{Earth}	Earth's mass	6×10^{24}	kg
R_{Earth}	Earth's radius	6×10^6	m
AU	Earth–Sun distance	1.5×10^{11}	m
$\theta_{\mathrm{Moon\ or\ Sun}}$	angular diameter of Moon or Sun	10^{-2}	rad
day	length of a day	10^5	s
year	length of a year	$\pi \times 10^7$	s
t_0	age of the universe	1.4×10^{10}	yr
F	solar constant	1.3	$\mathrm{kW\,m^{-2}}$
p_0	atmospheric pressure at sea level	10^5	Pa
ρ_{air}	air density	1	$\mathrm{kg\,m^{-3}}$
ρ_{rock}	rock density	2.5	$\mathrm{g\,cm^{-3}}$
$L_{\mathrm{vap}}^{\mathrm{water}}$	heat of vaporization of water	2	$\mathrm{MJ\,kg^{-1}}$
γ_{water}	surface tension of water	7×10^{-2}	$\mathrm{N\,m^{-1}}$
P_{basal}	human basal metabolic rate	100	W
a_0	Bohr radius	0.5	$\mathrm{\AA}$
a	typical interatomic spacing	3	$\mathrm{\AA}$
E_{bond}	typical bond energy	4	eV
$\mathcal{E}_{\mathrm{fat}}$	combustion energy density	9	$\mathrm{kcal\,g^{-1}}$
ν_{air}	kinematic viscosity of air	1.5×10^{-5}	$\mathrm{m^2\,s^{-1}}$
ν_{water}	kinematic viscosity of water	10^{-6}	$\mathrm{m^2\,s^{-1}}$
K_{air}	thermal conductivity of air	2×10^{-2}	$\mathrm{W\,m^{-1}\,K^{-1}}$
K	... of nonmetallic solids/liquids	2	$\mathrm{W\,m^{-1}\,K^{-1}}$
K_{metal}	... of metals	2×10^2	$\mathrm{W\,m^{-1}\,K^{-1}}$
c_p^{air}	specific heat of air	1	$\mathrm{J\,g^{-1}\,K^{-1}}$
c_p	... of solids/liquids	25	$\mathrm{J\,mol^{-1}\,K^{-1}}$

Part I

Organizing complexity

We cannot find much insight staring at a mess. We need to organize it. As an everyday example, when I look at my kitchen after a dinner party, I feel overwhelmed. It's late, I'm tired, and I dread that I will not get enough sleep. If I clean up in that scattered state of mind, I pick up a spoon here and a pot there, making little progress. However, when I remember that a large problem can be broken into smaller ones, calm and efficiency return. I begin at one corner of the kitchen, clear its mess, and move to neighboring areas until the project is done. I divide and conquer (Chapter 1).

Once the dishes are clean, I resist the temptation to dump them into one big box. I separate pots from the silverware and, within the silverware, the forks from the spoons. These groupings, or abstractions (Chapter 2), make the kitchen easy to understand and use.

In problem solving, we organize complexity by using divide-and-conquer reasoning and by making abstractions. In Part I, you'll learn how.

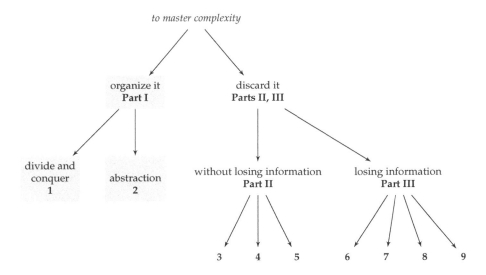

1

Divide and conquer

As imperial rulers knew, you need not conquer all your enemies at once. Instead, conquer them one at a time. Break hard problems into manageable pieces. This process embodies our first reasoning tool: Divide and conquer!

1.1 Warming up

To show how to use divide-and-conquer reasoning, we'll apply it to increasingly complex problems that illustrate its essential features. So we start with an everyday estimate.

▷ *What is, roughly, the volume of a dollar bill?*

Volumes are hard to estimate. However, we should still make a quick guess. Even an inaccurate guess will help us practice courage and, when we compare the guess with a more accurate estimate, will help us calibrate our internal measuring rods. To urge me on, I often imagine a mugger who holds a knife at my ribs, demanding, "Your guess or your life!" Then I judge it likely that the volume of a dollar bill lies between 0.1 and 10 cubic centimeters.

This range is wide, spanning a factor of 100. In contrast, the dollar bill's width probably lies between 10 and 20 centimeters—a range of only a factor of 2. The volume range is wider than the width range because we have no equivalent of a ruler for volume; thus, volumes are less familiar than lengths. Fortunately, the volume of the dollar bill is the product of lengths.

$$\text{volume} = \text{width} \times \text{height} \times \text{thickness}. \qquad (1.1)$$

6 cm	$1 bill
	15 cm

The harder volume estimate becomes three easier length estimates—the benefit of divide-and-conquer reasoning. The width looks like 6 inches, which is roughly 15 centimeters. The height looks like 2 or 3 inches, which is roughly 6 centimeters. But before estimating the thickness, let's talk about unit systems.

▷ *Is it better to use metric or US customary units (such as inches, feet, and miles)?*

Your estimates will be more accurate if you use the units most familiar to you. Raised in the United States, I judge lengths more accurately in inches, feet, and miles than in centimeters, meters, or kilometers. However, for calculations requiring multiplication or division—most calculations—I convert the customary units to metric (and often convert back to customary units at the end). But you may be fortunate enough to think in metric. Then you can estimate and calculate in a single unit system.

The third piece of the divide-and-conquer estimate, the thickness, is difficult to judge. A dollar bill is thin—paper thin.

▷ *But how thin is "paper thin"?*

This thickness is too small to grasp and judge easily. However, a stack of several hundred bills would be graspable. Not having that much cash lying around, I'll use paper. A ream of paper, which has 500 sheets, is roughly 5 centimeters thick. Thus, one sheet of paper is roughly 0.01 centimeters thick. With this estimate for the thickness, the volume is approximately 1 cubic centimeter:

$$\text{volume} \approx \underbrace{15\,\text{cm}}_{\text{width}} \times \underbrace{6\,\text{cm}}_{\text{height}} \times \underbrace{0.01\,\text{cm}}_{\text{thickness}} \approx 1\,\text{cm}^3. \qquad (1.2)$$

Although a more accurate calculation could adjust for the fiber composition of a dollar bill compared to ordinary paper and might consider the roughness of the paper, these details obscure the main result: A dollar bill is 1 cubic centimeter pounded paper thin.

To check this estimate, I folded a dollar bill until my finger strength gave out, getting a roughly cubical packet with sides of approximately 1 centimeter—making a volume of approximately 1 cubic centimeter!

In the preceding analysis, you may have noticed the = and ≈ symbols and their slightly different use. Throughout this book, our goal is insight over accuracy. So we'll use several kinds of equality symbols to describe the accuracy of a relation and what it omits. Here is a table of the equality symbols, in descending order of completeness and often increasing order of usefulness.

≡	equality by definition	read as "is defined to be"
=	equality	"is equal to"
≈	equality except perhaps for a purely numerical factor near 1	"is approximately equal to"
∼	equality except perhaps for a purely numerical factor	"is roughly equal to" or "is comparable to"
∝	equality except perhaps for a factor that may have dimensions	"is proportional to"

As examples of the kinds of equality, for the circle below, $A = \pi r^2$, and $A \approx 4r^2$, and $A \sim r^2$. For the cylinder, $V \sim hr^2$—which implies $V \propto r^2$ and $V \propto h$. In the $V \propto h$ form, the factor hidden in the ∝ symbol has dimensions of length squared.

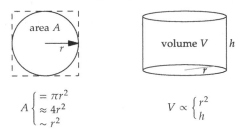

$$A \begin{cases} = \pi r^2 \\ \approx 4r^2 \\ \sim r^2 \end{cases} \qquad V \propto \begin{cases} r^2 \\ h \end{cases}$$

Problem 1.1   Weight of a box of books
How heavy is a small moving-box filled with books?

Problem 1.2   Mass of air in your bedroom
Estimate the mass of air in your bedroom.

Problem 1.3   Suitcase of bills
In the movies, and perhaps in reality, cocaine and elections are bought with a suitcase of $100 bills. Estimate the dollar value in such a suitcase.

Problem 1.4   Gold or bills?
As a bank robber sitting in the vault planning your getaway, do you fill your suit-case with gold bars or $100 bills? Assume first that how much you can carry is a fixed weight. Then redo your analysis assuming that how much you can carry is a fixed volume.

1.2   Rails versus roads

We are now warmed up and ready to use divide-and-conquer reasoning for more substantial estimates. Our next estimate, concerning traffic, comes to mind whenever I drive the congested roads to JFK Airport in New York City. The route goes on the Van Wyck Expressway, which was planned by Robert Moses. As Moses's biographer Robert Caro describes [6, pp. 904ff], when Moses was in charge of building the expressway, the traffic planners recommended that, in order to handle the expected large volume of traffic, the road include a train line to the then-new airport. Alternatively, if build-ing the train track would be too expensive, they recommended that the city, when acquiring the land for the road, still take an extra 50 feet of width and reserve it as a median strip for a train line one day. Moses also rejected the cheaper proposal. Alas, only weeks after its opening, not long after World War Two, the rail-free highway had reached peak capacity.

Let's use our divide-and-conquer tool to compare, for rush-hour commut-ing, the carrying capacities of rail and road. The capacity is the rate at which passengers are transported; it is passengers per time. First we'll estimate the capacity of one lane of highway. We can use the 2-second-following rule taught in many driving courses. You are taught to leave 2 seconds of travel time between you and the car in front. When drivers follow this rule, a sin-gle lane of highway carries one car every 2 seconds. To find the carrying capacity, we also need the occupancy of each car. Even at rush hour, at least in the United States, each car carries roughly one person. (Taxis often have two people including the driver, but only one person is being transported to the destination.) Thus, the capacity is one person every 2 seconds. As an hourly rate, the capacity is 1800 people per hour:

$$\frac{1 \text{ person}}{2 \text{ } \not{s}} \times \frac{3600 \text{ } \not{s}}{1 \text{ hr}} = \frac{1800 \text{ people}}{\text{hr}}. \tag{1.3}$$

The diagonal strike-through lines help us to spot which units cancel and to check that we end up with just the units that we want (people per hour).

This rate, 1800 people per hour, is approximate, because the 2-second following rule is not a law of nature. The average gap might be 4 seconds late at night, 1 second during the day, and may vary from day to day or from highway to highway. But a 2-second gap is a reasonable compromise estimate. Replacing the complex distribution of following times with one time is an application of lumping—the tool discussed in Chapter 6. Organizing complexity almost always reduces detail. If we studied all highways at all times of day, the data, were we so unfortunate as to obtain them, would bury any insight.

> *How does the capacity of a single lane of highway compare with the capacity of a train line?*

For the other half of the comparison, we'll estimate the rush-hour capacity of a train line in an advanced train system, say the French or German system. As when we estimated the volume of a dollar bill (Section 1.1), we divide the estimate into manageable pieces: how often a train runs on the track, how many cars are in each train, and how many passengers are in each car. Here are my armchair estimates for these quantities, kept slightly conservative to avoid overestimating the train-line's capacity. A single train car, when full at rush hour, may carry 150 people. A rush-hour train may consist of 20 cars. And, on a busy train route, a train may run every 10 minutes or six times per hour. Therefore, the train line's capacity is 18 000 people per hour:

$$\frac{150 \text{ people}}{\text{car}} \times \frac{20 \text{ cars}}{\text{train}} \times \frac{6 \text{ trains}}{\text{hr}} = \frac{18\,000 \text{ people}}{\text{hr}}. \tag{1.4}$$

This capacity is ten times the capacity of a single fast-flowing highway lane. And this estimate is probably on the low side; Robert Caro [6, p. 901] gives an estimate of 40 000 to 50 000 people per hour. Using our lower rate, one train track in each direction could replace two highways even if each highway had five lanes in each direction.

1.3 Tree representations

Our estimates for the volume of a dollar bill (Section 1.1) and for the rail and highway capacities (Section 1.2) used the same method: dividing hard problems into smaller ones. However, the structure of the analysis is buried within the sentences, paragraphs, and pages. The sequential presentation hides the structure. Because the structure is hierarchical—big problems

split, or branch, into smaller problems—its most compact representation is
a tree. A tree representation shows us the analysis in one glance.

Here is the tree representation for the capacity
of a train line. Unlike the biological variety, our
trees stand on their head. Their roots, the goals,
sit at the top of the tree. Their leaves, the small
problems into which we have subdivided the
goal, sit at the bottom. The orientation matches
the way that we divide and conquer, filling the
page downward as we subdivide.

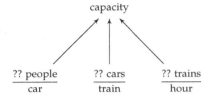

In making this first tree, we haven't estimated
the quantities themselves. We have only identi-
fied the quantities. The question marks remind
us of our next step: to include estimates for the
three leaves. These estimates were 150 people
per car, 20 cars per train, and 6 trains per hour
(giving the tree in the margin).

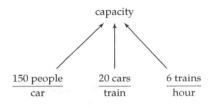

Then we multiplied the leaf values to propagate
the estimates upward from the leaves toward
the root. The result was 18 000 people per hour.
The completed tree shows us the entire estimate
in one glance.

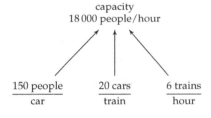

This train-capacity tree had the simplest possible structure with only two layers (the root layer
and, as the second layer, the three leaves). The
next level of complexity is a three-layer tree, which will represent our estimate for the volume of a dollar bill. It started as a two-layer tree with three
leaves.

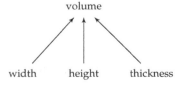

Then it grew, because, unlike the width and height, the thickness was difficult to estimate just by looking at a dollar bill. Therefore, we divided that
leaf into two easier leaves.

The result is the tree in the margin. The thickness leaf, which is the thickness per sheet, has split into (1) the thickness per ream and (2) the number of sheets per ream. The boxed -1 on the line connecting the thickness to the number of sheets per ream is a new and useful notation. The -1 tells us the exponent to apply to that leaf value when we propagate it upward to the root.

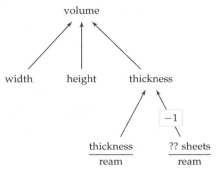

Here is why I write the -1 as a full-sized number rather than a small superscript. Most of our estimates require multiplying several factors. The only question for each factor is, "With what exponent does this factor enter?" The number -1 directly answers this "What exponent?" question. (To avoid cluttering the tree, we don't indicate the most-frequent exponent of 1.)

This new subtree then represents the following equation for the thickness of one sheet:

$$\text{thickness} = \frac{\text{thickness}}{\text{ream}} \times \left(\frac{\text{?? sheets}}{\text{ream}}\right)^{-1}. \tag{1.5}$$

The -1 exponent allows, at the cost of a slight complication in the tree notation, the leaf to represent the number of sheets per ream rather than a less-familiar fraction, the number of reams per sheet.

Now we include our estimates for the leaf values. The width is 15 centimeters. The height is 6 centimeters. The thickness of a ream of paper is 5 centimeters. And a ream contains 500 sheets of paper. The result is the following tree.

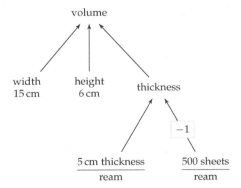

Now we propagate the values to the root. The two bottommost leaves combine to tell us that the thickness of one sheet is 10^{-2} centimeters. This thickness completes the tree's second layer. In the second layer, the three nodes tell us that the volume of a dollar bill—the root—is 1 cubic centimeter.

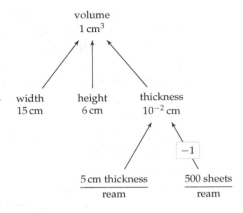

With practice, you can read in this final tree all the steps of the analysis. The three nodes in the second layer show how the difficult volume estimate was subdivided into three easier estimates. That the width and height remained leaves indicates that these two estimates felt reliable enough. In contrast, the two branches sprouting from the thickness indicate that the thickness was still hard to estimate, so we divided that estimate into two more-familiar quantities.

The tree encapsulates many paragraphs of analysis in a compact form, one that our minds can absorb in a single glance. Organizing complexity helps us build insight.

> **Problem 1.5   Tree for the suitcase of bills**
> Make a tree diagram for your estimate in Problem 1.3. Do it in three steps: (1) Draw the tree without any leaf estimates, (2) estimate the leaf values, and (3) propagate the leaf values upward to the root.

1.4 Demand-side estimates

Our analysis of the carrying capacity of highways and railways (Section 1.2) is an example of a frequent application of estimation in the social world—estimating the size of a market. The highway–railway comparison proceeded by estimating the transportation supply. In other problems, a more feasible analysis is based on the complementary idea of estimating the demand. Here is an example.

▷ *How much oil does the United States import (in barrels per year)?*

The volume rate is enormous and therefore hard to picture. Divide-and-conquer reasoning will tame the complexity. Just keep subdividing until the quantities are no longer daunting.

Here, subdivide the demand—the consumption. We consume oil in so many ways; estimating the consumption in each pathway would take a long time without producing much insight. Instead, let's estimate the largest consumption—likely to be cars—then adjust for other uses and for overall consumption versus imports.

$$\text{imports} = \cancel{\text{car usage}} \times \frac{\cancel{\text{all usage}}}{\cancel{\text{car usage}}} \times \frac{\text{imports}}{\cancel{\text{all usage}}}. \tag{1.6}$$

Here is the corresponding tree. The first factor, the most difficult of the three to estimate, will require us to sprout branches and make a subtree. The second and third factors might be possible to estimate without subdividing. Now we must decide how to continue.

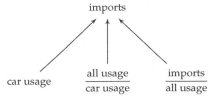

Should we keep subdividing until we've built the entire tree and only then estimate the leaves, or should we try to estimate these leaves and then subdivide what we cannot estimate?

It depends on one's own psychology. I feel anxious in the uncharted waters of a new estimate. Sprouting new branches before making any leaf estimates increases my anxiety. The tree might never stop sprouting branches and leaves, and I'll never estimate them all. Thus, I prefer to harvest my progress right away by estimating the leaves before sprouting new branches. You should experiment to learn your psychology. You are your best problem-solving tool, and it is helpful to know your tools.

Because of my psychology, I'll first estimate a leaf quantity:

$$\frac{\text{all usage}}{\text{car usage}}. \tag{1.7}$$

But don't do this estimate directly. It is more intuitive—that is, easier for our gut—to estimate first the ratio of car usage to other (noncar) usage. The ability to make such comparisons between disjoint sets, at least for physical objects, is hard wired in our brains and independent of the ability to count. Not least, it is not limited to humans. The female lions studied by Karen McComb and her colleagues [35] would judge the relative size of their troop and a group of lions intruding on their territory. The females would approach the intruders only when they outnumbered the intruders by a large-enough ratio, roughly a factor of 2.

Other uses for oil include noncar modes of transport (trucks, trains, and planes), heating and cooling, and hydrocarbon-rich products such as fertilizer, plastics, and pesticides. In judging the relative importance of other uses compared to car usage, two arguments compete: (1) Other uses are so many and so significant, so they are much more important than car usage; and (2) cars are so ubiquitous and such an inefficient mode of transport, so car usage is much larger than other uses. To my gut, both arguments feel comparably plausible. My gut is telling me that the two categories have comparable usages:

$$\frac{\text{other usage}}{\text{car usage}} \approx 1. \tag{1.8}$$

Based on this estimate, all usage (the sum of car and other usage) is roughly double the car usage:

$$\frac{\text{all usage}}{\text{car usage}} \approx 2. \tag{1.9}$$

This estimate is the first leaf. It implicitly assumes that the gasoline fraction in a barrel of oil is high enough to feed the cars. Fortunately, if this assumption were wrong, we would get warning. For if the fraction were too low, we would build our transportation infrastructure around other means of transport—such as trains powered by electricity generated by burning the nongasoline fraction in oil barrels. In this probably less-polluted world, we would estimate how much oil was used by trains.

Returning to our actual world, let's estimate the second leaf:

$$\frac{\text{imports}}{\text{all usage}}. \tag{1.10}$$

This adjustment factor accounts for the fact that only a portion of the oil consumed is imported.

▷ *What does your gut tell you for this fraction?*

Again, don't estimate this fraction directly. Instead, to make a comparison between disjoint sets, first compare (net) imports with domestic production. In estimating this ratio, two arguments compete. On the one hand, the US media report extensively on oil production in other countries, which suggests that oil imports are large. On the other hand, there is also extensive coverage of US production and frequent comparison with countries such as Japan that have almost no domestic oil. My resulting gut feeling is that the categories are comparable and therefore that imports are roughly one-half of all usage:

$$\frac{\text{imports}}{\text{domestic production}} \approx 1 \quad \text{so} \quad \frac{\text{imports}}{\text{all usage}} \approx \frac{1}{2}. \tag{1.11}$$

This leaf, as well as the other adjustment factor, are dimensionless numbers. Such numbers, the main topic of Chapter 5, have special value. Our perceptual system is skilled at estimating dimensionless ratios. Therefore, a leaf node that is a dimensionless ratio probably does not need to be subdivided.

The tree now has three leaves. Having plausible estimates for two of them should give us courage to subdivide the remaining leaf, the total car usage, into easier estimates. That leaf will sprout its own branches and become an internal node.

▷ *How should we subdivide the car usage?*

A reasonable subdivision is into the number of cars N_{cars} and the per-car usage. Both quantities are easier to estimate than the root. The number of cars is related to the US population—a familiar number if you live in the United States. The per-car usage is easier to estimate than is the total usage of all US cars. Our gut can more accurately judge human-scale quantities, such as the per-car usage, than it can judge vast numbers like the total usage of all US cars.

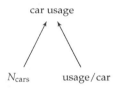

For the same reason, let's not estimate the number of cars directly. Instead, subdivide this leaf into two leaves:

1. the number of people, and

2. the number of cars per person.

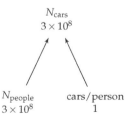

The first leaf is familiar, at least to residents of the United States: $N_{\text{people}} \approx 3 \times 10^8$.

The second leaf, cars per person, is a human-sized quantity. In the United States, car ownership is widespread. Many adults own more than one car, and a cynic would say that even babies seem to own cars. Therefore, a rough and simple estimate might be one car per person—far easier to picture than the total number of cars! Then $N_{\text{cars}} \approx 3 \times 10^8$.

The per-car usage can be subdivided into three easier factors (leaves). Here are my estimates.

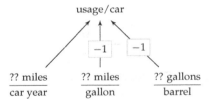

1. *How many miles per car year?* Used cars with 10 000 miles per year are considered low use but are not rare. Thus, for a typical year of driving, let's take a slightly longer distance: say, 20 000 miles or 30 000 kilometers.

2. *How many miles per gallon?* A typical car fuel efficiency is 30 miles per US gallon. In metric units, it is about 100 kilometers per 8 liters.

3. *How many gallons per barrel?* You might have seen barrels of asphalt along the side of the highway during road construction. Following our free-association tradition of equating the thickness of a sheet of paper and of a dollar bill, perhaps barrels of oil are like barrels of asphalt.

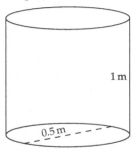

Their volume can be computed by divide-and-conquer reasoning. Just approximate the cylinder as a rectangular prism, estimate its three dimensions, and multiply:

$$\text{volume} \sim \underbrace{1\,\text{m}}_{\text{height}} \times \underbrace{0.5\,\text{m}}_{\text{width}} \times \underbrace{0.5\,\text{m}}_{\text{depth}} = 0.25\,\text{m}^3. \qquad (1.12)$$

A cubic meter is 1000 liters or, using the conversion of roughly 4 liters per US gallon, about 250 gallons. Therefore, 0.25 cubic meters is roughly 60 gallons. (The official volume of a barrel of oil is not too different at 42 gallons.)

Multiplying these estimates, and not forgetting the effect of the two −1 exponents, we get approximately 10 barrels per car per year (also written as barrels per car year):

$$\frac{2 \times 10^4\,\text{miles}}{\text{car year}} \times \frac{1\,\text{gallon}}{30\,\text{miles}} \times \frac{1\,\text{barrel}}{60\,\text{gallons}} \approx \frac{10\,\text{barrels}}{\text{car year}}. \qquad (1.13)$$

In doing this calculation, first evaluate the units. The gallons and miles cancel, leaving barrels per year. Then evaluate the numbers. The 30×60 in the denominator is roughly 2000. The 2×10^4 from the numerator divided by the 2000 from the denominator produces the 10.

This estimate is a subtree in the tree representing total car usage. The car usage then becomes 3 billion barrels per year:

$$3 \times 10^8\,\text{cars} \times \frac{10\,\text{barrels}}{\text{car year}} = \frac{3 \times 10^9\,\text{barrels}}{\text{year}}. \qquad (1.14)$$

This estimate is itself a subtree in the tree representing oil imports. Because the two adjustment factors contribute a factor of 2×0.5, which is just 1, the oil imports are also 3 billion barrels per year.

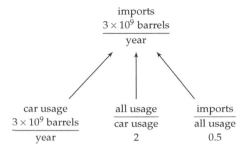

Here is the full tree, which includes the subtree for the total car usage of oil:

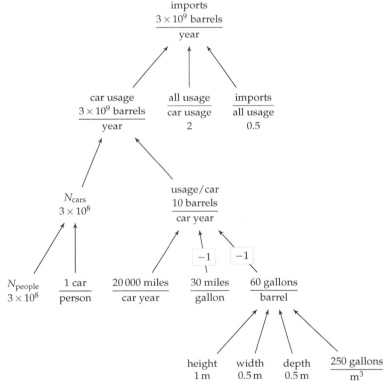

Problem 1.6   Using metric units

As practice with metric units (if you grew up in a nonmetric land) or to make the results more familiar (if you grew up in a metric land), redo the calculation using the metric values for the volume of a barrel, the distance a car is driven per year, and the fuel consumption of a typical car.

How close is our estimate to official values?

For the US oil imports, the US Department of Energy reports 9.163 million barrels per day (for 2010). When I first saw this value, my heart sank twice. The first shock was the 9 in the 9 million. I assumed that it was the number of billions, and wondered how the estimate of 3 billion barrels could be a factor of 3 too small. The second shock was the "million"—how could the estimate be more than a factor of 100 too large? Then the "per day" reassured me. As a yearly rate, 9.163 million barrels per day is 3.34 billion barrels per year—only 10 percent higher than our estimate. Divide and conquer triumphs!

> **Problem 1.7   Fuel efficiency of a 747**
> Based on the cost of a long-distance plane ticket, estimate the following quantities: (a) the fuel efficiency of a 747, in passenger miles per gallon or passenger kilometers per liter; and (b) the volume of its fuel tank. Check your estimates against the technical data for a 747.

1.5 Multiple estimates for the same quantity

After making an estimate, it is natural to wonder about how much confidence to place in it. Perhaps we made an embarrassingly large mistake. The best way to know is to estimate the same quantity using another method. As an everyday example, let's observe how we add a list of numbers.

$$
\begin{array}{r}
12 \\
15 \\
+18 \\
\hline
\end{array}
$$

(1.15)

We often add the numbers first from top to bottom. For 12 + 15 + 18, we calculate, "12 plus 15 is 27; 27 plus 18 is 45." To check the result, we add the numbers in the reverse order, from bottom to top: "18 plus 15 is 33; 33 plus 12 is 45." The two totals agree, so each is probably correct: The calculations are unlikely to contain an error of exactly the same amount. This kind of redundancy catches errors.

In contrast, mindless redundancy offers little protection. If we check the calculation by adding the numbers from top to bottom again, we usually repeat any mistakes. Similarly, rereading written drafts usually means overlooking the same spelling, grammar, or logic faults. Instead, stuff the draft in a drawer for a week, then look at it; or ask a colleague or friend—in both cases, use fresh eyes.

Reliability, in short, comes from *intelligent* redundancy.

This principle helps you make reliable estimates. First, use several methods to estimate the same quantity. Second, make the methods as different from one another as possible—for example, by using unrelated background knowledge. This approach to reliability is another example of divide-and-conquer reasoning: The hard problem of making a reliable estimate becomes several simpler subproblems, one per estimation method.

You saw an example in Section 1.1, where we estimated the volume of a dollar bill. The first method used divide-and-conquer reasoning based on the width, height, and thickness of the bill. The check was a comparison with a folded-up dollar bill. Both methods agreed on a volume of approximately 1 cubic centimeter—giving us confidence in the estimate.

For another example of using multiple methods, return to the estimate of the volume of an oil barrel (Section 1.4). We used a roadside asphalt barrel as a proxy for an oil barrel and estimated the volume of the roadside barrel. The result, 60 gallons, seemed plausible, but maybe oil barrels have a completely different size. One way to catch that kind of error is to use a different method for estimating the volume. For example, we might start with the cost of a barrel of oil—about $100 in 2013—and the cost of a gallon of gasoline—about $2.50 before taxes, or 1/40th of the cost of a barrel. If the markup on gasoline is not significant, then a barrel is roughly 40 gallons. Even with a markup, we can still say that a barrel is at least 40 gallons. Because our two estimates, 60 gallons and > 40 gallons, roughly agree, our confidence in both increases. If they had contradicted each other, one or both would be wrong, and we would look for the mistaken assumption, for the incorrect arithmetic, or for a third method.

1.6 **Talking to your gut**

As you have seen in the preceding examples, divide-and-conquer estimates require reasonable estimates for the leaf quantities. To decide what is reasonable, you have to talk to your gut—what you will learn in this section. Talking to your gut feels strange at first, especially because science and engineering are considered cerebral subjects. Let's therefore discuss how to hold the conversation. The example will be an estimate of the US population based on its area and population density. The divide-and-conquer tree has two leaves. (In Section 6.3.1, you'll see a qualitatively different method, where the two leaves will be the number of US states and the population of a typical state.)

The area is the width times the height, so the area leaf itself splits into two leaves. Estimating the width and height requires only a short dialogue with the gut, at least if you live in the United States. Its width is a 6-hour plane flight at 500 miles per hour, so about 3000 miles; and the height is, as a rough estimate, two-thirds of the width, or 2000 miles. Therefore, the area is 6 million square miles:

$$3000 \text{ miles} \times 2000 \text{ miles} = 6 \times 10^6 \text{ miles}^2. \qquad (1.16)$$

In metric units, it is about 16 million square kilometers.

Estimating the population density requires talking to your gut. If you are like me you have little conscious knowledge of the population density. Your gut might know, but you cannot ask it directly. The gut is connected to the right brain, which doesn't have language. Although the right brain knows a lot about the world, it cannot answer with a value, only with a feeling. To draw on its knowledge, ask it indirectly. Pick a particular population density—say, 100 people per square mile—and ask the gut for its opinion: "O, my intuitive, insightful, introverted right brain: What do you think of 100 people per square mile for the population density?" A response, a gut feeling, will come back. Keep lowering the candidate value until the gut feeling becomes, "No, that value feels way too low."

Here is the dialogue between my left brain (LB) and right brain (RB).

LB: What do you think of 100 people per square mile?

RB: That feels okay based on my experience growing up in the United States.

LB: I can probably support that feeling quantitatively. A square mile with 100 people means each person occupies a square whose side is 1/10th of a mile or 160 meters. Expressed in this form, does the population density feel okay?

RB: Yes, the large open spaces in the western states probably compensate for the denser regions near the coasts.

LB: Now I will lower the estimate by factors of 3 or 10 until you object strongly that the estimate feels too low. [A factor of 3 is roughly one-half of a factor of 10, because $3 \times 3 \approx 10$. A factor of 3 is the next-smallest factor by which to move when a factor of 10 is too large a jump.] In that vein, what about an average population density of 10 people per square mile?

RB: I feel uneasy. The estimate feels a bit low.

LB: I understand where you are coming from. That value may moderately over-estimate the population density of farmland, but it probably greatly underestimates the population density in the cities. Because you are uneasy, let's move more slowly until you object strongly. How about 3 people per square mile?

RB: If the true value were lower than that, I'd feel fairly surprised.

LB: So, for the low end, I'll stop at 3 people per square mile. Now let's navigate to the upper end. You said that 100 people per square mile felt plausible. How do you feel about 300 people per square mile?

RB: I feel quite uneasy. That estimate feels quite high.

LB: I hear you. Your response reminds me that New Jersey and the Netherlands, both very densely populated, are at 1000 people per square mile, although I couldn't swear to this value. I cannot imagine packing the whole United States to a density comparable to New Jersey's. Therefore, let's stop here: Our upper endpoint is 300 people per square mile.

▷ *How do you make your best guess based on these two endpoints?*

A plausible guess is to use their arithmetic mean, which is roughly 150 people per square mile. However, the right method is the geometric mean:

$$\text{best guess} = \sqrt{\text{lower endpoint} \times \text{upper endpoint}}. \tag{1.17}$$

The geometric mean is the midpoint of the lower and upper bounds—but on a ratio or logarithmic scale, which is the scale built into our mental hardware. (For more about how we perceive quantity, see *The Number Sense* [9].) The geometric mean is the correct mean when combining quantities produced by our mental hardware.

Here, the geometric mean is 30 people per square mile: a factor of 10 removed from either endpoint. Using that population density,

$$\text{US population} \sim 6 \times 10^6 \,\cancel{\text{miles}^2} \times \frac{30}{\cancel{\text{miles}^2}} \approx 2 \times 10^8. \tag{1.18}$$

The actual population is roughly 3×10^8. The estimate based almost entirely on gut reasoning is within a factor of 1.5 of the actual population—a pleasantly surprising accuracy.

Problem 1.8 More gut estimates

By asking your gut to help you estimate the lower and upper endpoints, estimate (a) the height of a nearby tall tree that you can see, (b) the mass of a car, and (c) the number of water drops in a bathtub.

1.7 **Physical estimates**

Your gut understands not only the social world but also the physical world. If you trust its feelings, you can tap this vast reservoir of knowledge. For practice, we'll estimate the salinity of seawater (Section 1.7.1), human power output (Section 1.7.2), and the heat of vaporization of water (Section 1.7.3).

1.7.1 **Salinity of seawater**

To estimate the salinity of seawater, which will later help you estimate the conductivity of seawater (Problem 8.10), do not ask your gut directly: "How do you feel about, say, 200 millimolar?" Although that kind of question worked for estimating population density (Section 1.6), here, unless you are a chemist, the answer will be: "I have no clue. What is a millimolar anyway? I have almost no experience of that unit." Instead, offer your gut concrete data—for example, from a home experiment: adding salt to a cup of water until the mixture tastes as salty as the ocean.

This experiment can be a thought or a real experiment—another example of using multiple methods (Section 1.5). As a thought experiment, I ask my gut about various amounts of salt in a cup of water. When I propose adding 2 teaspoons, it responds, "Disgustingly salty!" At the lower end, when I propose adding 0.5 teaspoons, it responds, "Not very salty." I'll use 0.5 and 2 teaspoons as the lower and upper endpoints of the range. Their midpoint, the estimate from the thought experiment, is 1 teaspoon per cup.

I tested this prediction at the kitchen sink. With 1 teaspoon (5 milliliters) of salt, the cup of water indeed had the sharp, metallic taste of seawater that I have gulped after being knocked over by large waves. A cup of water is roughly one-fourth of a liter or 250 cubic centimeters. By mass, the resulting salt concentration is the following product:

$$\frac{1\,\text{tsp salt}}{1\,\text{cup water}} \times \frac{1\,\text{cup water}}{250\,\text{g water}} \times \frac{5\,\text{cm}^3\,\text{salt}}{1\,\text{tsp salt}} \times \underbrace{\frac{2\,\text{g salt}}{1\,\text{cm}^3\,\text{salt}}}_{\rho_{\text{salt}}}. \tag{1.19}$$

The density of 2 grams per cubic centimeter comes from my gut feeling that salt is a light rock, so it should be somewhat denser than water at 1 gram per cubic centimeter, but not too much denser. (For an alternative method, more accurate but more elaborate, try Problem 1.10.) Then doing the arithmetic gives a 4 percent salt-to-water ratio (by mass).

The actual salinity of the Earth's oceans is about 3.5 percent—very close to the estimate of 4 percent. The estimate is close despite the large number of assumptions and approximations—the errors have mostly canceled. Its accuracy should give you courage to perform home experiments whenever you need data for divide-and-conquer estimates.

> **Problem 1.9   Density of water**
> Estimate the density of water by asking your gut to estimate the mass of water in a cup measure (roughly one-quarter of a liter).
>
> **Problem 1.10   Density of salt**
> Estimate the density of salt using the volume and mass of a typical salt container that you find in a grocery store. This value should be more accurate than my gut estimate in Section 1.7.1 (which was 2 grams per cubic centimeter).

1.7.2 Human power output

Our second example of talking to your gut is an estimate of human power output—a power that is useful in many estimates (for example, Problem 1.17). Energies and powers are good candidates for divide-and-conquer estimates, because they are connected by the subdivision shown in the following equation and represented in the tree in the margin:

$$\text{power} = \frac{\text{energy}}{\text{time}}. \tag{1.20}$$

In particular, let's estimate the power that a trained athlete can generate for an extended time (not just during a few-seconds-long, high-power burst). As a proxy for that power, I'll use my own burst power output with two adjustment factors:

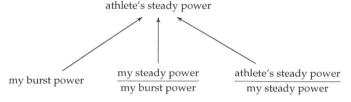

Maintaining a power is harder than producing a quick burst. Therefore, the first adjustment factor, my steady power divided by my burst power, is somewhat smaller than 1—maybe 1/2 or 1/3. In contrast, an athlete's power output will be higher than mine, perhaps by a factor of 2 or 3: Even though I am sometimes known as the street-fighting mathematician [33], I am no athlete. Then the two adjustment factors roughly cancel, so my burst power should be comparable to an athlete's steady power.

To estimate my burst power, I performed a home experiment of running up a flight of stairs as quickly as possible. Determining the power output requires estimating an energy and a time:

$$\text{power} = \frac{\text{energy}}{\text{time}}. \tag{1.21}$$

The energy, which is the change in my gravitational potential energy, itself subdivides into three factors:

$$\text{energy} = \underbrace{\text{mass}}_{m} \times \underbrace{\text{gravity}}_{g} \times \underbrace{\text{height}}_{h}. \tag{1.22}$$

In the academic building at my university, a building with high ceilings and staircases, I bounded up a staircase three stairs at a time. The staircase was about 12 feet or 3.5 meters high. Therefore, my mechanical energy output was roughly 2000 joules:

$$E \sim 65\,\text{kg} \times 10\,\text{m}\,\text{s}^{-2} \times 3.5\,\text{m} \sim 2000\,\text{J}. \tag{1.23}$$

(The units are fine: $1\,\text{J} = 1\,\text{kg}\,\text{m}^2\,\text{s}^{-2}$.)

The remaining leaf is the time: how long the climb took me. I made it in 6 seconds. In contrast, several students made it in 3.9 seconds—the power of youth! My mechanical power output was about 2000 joules per 6 seconds, or about 300 watts. (To check whether the estimate is reasonable, try Problem 1.12, where you estimate the typical human basal metabolism.)

This burst power output should be close to the sustained power output of a trained athlete. And it is. As an example, in the Alpe d'Huez climb in the 1989 Tour de France, the winner—Greg LeMond, a world-class athlete—put out 394 watts (over a 42.5-minute period). The cyclist Lance Armstrong, during the time-trial stage during the Tour de France in 2004, generated even more: 495 watts (roughly 7 watts per kilogram). However, he publicly admitted to blood doping to enhance performance. Indeed, because of widespread doping, many cycling power outputs of the 1990s and 2000s are suspect; as a result, 400 watts stands as a legitimate world-class sustained power output.

> **Problem 1.11   Energy in a 9-volt battery**
> Estimate the energy in a 9-volt battery. Is it enough to launch the battery into orbit?
>
> **Problem 1.12   Basal metabolism**
> Based on our daily caloric consumption, estimate the human basal metabolism.
>
> **Problem 1.13   Energy measured in person flights of stairs**
> How many flights of stairs can you climb using the energy in a stick (100 grams) of butter?

1.7.3   Heat of vaporization of water

Our final physical estimate concerns the most important liquid on Earth.

> *What is the heat of vaporization of water?*

Because water covers so much of the Earth and is such an important part of the atmosphere (clouds!), its heat of vaporization strongly affects our climate—whether through rainfall (Section 3.4.3) or air temperatures.

Heat of vaporization is defined as a ratio:

$$\frac{\text{energy to evaporate a substance}}{\text{amount of the substance}}, \quad (1.24)$$

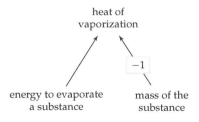

where the amount of substance can be measured in moles, by volume, or (most commonly) by mass. The definition provides the structure of the tree and of the estimate based on divide-and-conquer reasoning.

For the mass of the substance, choose an amount of water that is easy to imagine—ideally, an amount familiar from everyday life. Then your gut can help you make estimates. Because I often boil a few cups of water at a time, and each cup is few hundred milliliters, I'll imagine 1 liter or 1 kilogram of water.

The other leaf, the required energy, requires more thought. There is a common confusion about this energy that is worth discussing.

> *Is it the energy required to bring the water to a boil?*

No: The energy has nothing to do with the energy required to bring the water to a boil! That energy is related to water's specific heat c_p. The heat of vaporization depends on the energy needed to evaporate—boil *away*—the water, once it is boiling. (You compare these energies in Problem 5.61.)

Energy subdivides into power times time (as when we estimated human power output in Section 1.7.2). Here, the power could be the power output of one burner; the time is the time to boil away the liter of water. To estimate these leaves, let's hold a gut conference.

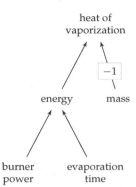

For the time, my dialogue is as follows.

> LB: How does 1 minute sound as a lower bound?
>
> RB: Way too short—you've left boiling water on the stove unattended for longer without its boiling away!
>
> LB: How about 3 minutes?
>
> RB: That's on the low side. Maybe that's the lower bound.
>
> LB: Okay. For the upper bound, how about 100 minutes?
>
> RB: That time feels way too long. Haven't we boiled away pots of water in far less time?
>
> LB: What about 30 minutes?
>
> RB: That's long, but I wouldn't be shocked, only fairly surprised, if it took that long. It feels like the upper bound.

My range is therefore 3...30 minutes. Its midpoint—the geometric mean of the endpoints—is about 10 minutes or 600 seconds.

For variety, let's directly estimate the burner power, without estimating lower and upper bounds.

> LB: How does 100 watts feel?
>
> RB: Way too low: That's a lightbulb! If a lightbulb could boil away water so quickly, our energy troubles would be solved.
>
> LB (feeling chastened): How about 1000 watts (1 kilowatt)?
>
> RB: That's a bit low. A small appliance, such as a clothes iron, is already 1 kilowatt.
>
> LB (raising the guess more slowly): What about 3 kilowatts?
>
> RB: That burner power feels plausible.

Let's check this power estimate by subdividing power into two factors, voltage and current:

$$\text{power} = \text{voltage} \times \text{current}. \tag{1.25}$$

An electric stove requires a line voltage of 220 volts, even in the United States where most other appliances require only 110 volts. A standard fuse is about 15 amperes, which gives us an idea of a large current. If a burner corresponds to a standard fuse, a burner supplies roughly 3 kilowatts:

$$220\,\text{V} \times 15\,\text{A} \approx 3000\,\text{W}. \qquad (1.26)$$

This estimate agrees with the gut estimate, so both methods gain plausibility—which should give you confidence to use both methods for your own estimates. As a check, I looked at the circuit breaker connected to my range, and it is rated for 50 amperes. The range has four burners and an oven, so 15 amperes for one burner (at least, for the large burner) is plausible.

We now have values for all the leaf nodes. Propagating the values toward the root gives the heat of vaporization (L_{vap}) as roughly 2 megajoules per kilogram:

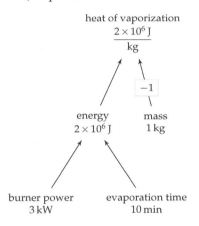

$$L_{\text{vap}} \sim \frac{\overset{\text{power}}{3\,\text{kW}} \times \overset{\text{time}}{600\,\text{s}}}{\underset{\text{mass}}{1\,\text{kg}}} \qquad (1.27)$$

$$\approx 2 \times 10^6\,\text{J}\,\text{kg}^{-1}.$$

The true value is about 2.2×10^6 joules per kilogram. This value is one of the highest heats of vaporization of any liquid. As water evaporates, it carries away significant amounts of energy, making it an excellent coolant (Problem 1.17).

1.8 Summary and further problems

The main lesson that you should take away is courage: No problem is too difficult. We just use divide-and-conquer reasoning to dissolve difficult problems into smaller pieces. (For extensive practice, see the varied examples in the *Guesstimation* books [47 and 48].) This tool is a universal solvent for problems social and scientific.

> **Problem 1.14   Per-capita land area**
> Estimate the land area per person for the world, for your home country, and for your home state or province.

Problem 1.15   Mass of the Earth
Estimate the mass of the Earth. Then look it up (p. xvii) to check your estimate.

Problem 1.16   Billion
How long would it take to count to a billion (10^9)?

Problem 1.17   Sweating
Estimate how much water you need to drink to replace water lost to evaporation, if you ride a bicycle vigorously for 1 hour. Represent your estimate as a divide-and-conquer tree. Hint: Humans are only about 25 percent efficient in generating mechanical work.

Problem 1.18   Pencil line
How long a line can you write with a pencil?

Problem 1.19   Pine needles
Estimate the number of needles on a pine tree.

Problem 1.20   Hairs
How many hairs are on your head?

2
Abstraction

Divide-and-conquer reasoning, the tool introduced in Chapter 1, is powerful, but it is not enough by itself to organize the complexity of the world. Try, for example, to manage the millions of files on a computer—even my laptop says that it has almost 3 million files. Without any organization, with all the files in one monster directory or folder, you could never find information that you need. However, simply using divide and conquer by dividing the files into groups—the first 100 files by date, the second 100 files by date, and so on—does not disperse the chaos. A better solution is to organize the millions of files into a hierarchy: as a tree of folders and subfolders. The elements in this hierarchy get names—for example, "photos of the children" or "files for typesetting this book"—and these names guide us to the needed information.

Naming—or, more technically, abstraction—is our other tool for organizing complexity. A name or an abstraction gets its power from its reusability. Without reusable ideas, the world would become unmanageably complicated. We might ask, "Could you, without tipping it over, move the wooden board glued to four thick sticks toward the large white plastic circle?" instead of, "Could you slide the chair toward the table?" The abstractions "chair," "slide," and "table" compactly represent complex ideas and physical structures. (And even the complex question itself uses abstractions.)

Similarly, without good abstractions we could hardly calculate, and modern science and technology would be impossible. As an illustration, imagine the pain of the following calculation:

$$\text{XXVII} \times \text{XXXVI}, \tag{2.1}$$

which is 27×36 in Roman numerals. The problem is not that the notation is unfamiliar, but rather that it is not based on abstractions useful for calculation. Not least, it does not lend itself to divide-and-conquer reasoning; for example, even though V (5) is a part of XXVII, V×XXXVI has no obvious answer. In contrast, our modern number system, based on the abstractions of place value and zero, makes the whole multiplication simple. Notations are abstractions, and good abstractions amplify our intelligence. In this chapter, we will practice making abstractions, discuss their high-level purpose, and continue to practice.

2.1 Energy from burning hydrocarbons

Our understanding of the world is built on layers of abstractions. Consider the idea of a fluid. At the bottom of the abstraction hierarchy are the actors of particle physics: quarks and electrons. Quarks combine to build protons and neutrons. Protons, neutrons, and electrons combine to build atoms. Atoms combine to build molecules. And large collections of molecules act, under many conditions, like a fluid.

The idea of a fluid is a new unit of thought. It helps us understand diverse phenomena, without our having to calculate or even know how quarks and electrons interact to produce fluid behavior. As one consequence, we can describe the behavior of air and water using the same equations (the Navier–Stokes equations of fluid mechanics); we need only to use different values for the density and viscosity. Then atmospheric cyclones and water vortices, although they result from widely differing sets of quarks and electrons and their interactions, can be understood as the same phenomenon.

A similarly powerful abstraction is a chemical bond. We'll use this abstraction to estimate a quantity essential to our bodies and to modern society: the energy released by burning chains made of hydrogen and carbon atoms (hydrocarbons). A hydrocarbon can be abstracted as a chain of CH_2 units:

fluid

↑

molecules

↑

atoms

electrons   protons, neutrons

↑

quarks

Burning a CH_2 unit requires oxygen (O_2) and releases carbon dioxide (CO_2), water, and energy:

$$CH_2 + \frac{3}{2}O_2 \longrightarrow CO_2 + H_2O + \text{energy.} \tag{2.2}$$

For a hydrocarbon with eight carbons—such as octane, a prime component of motor fuel—simply multiply this reaction by 8:

$$(CH_2)_8 + 12\,O_2 \longrightarrow 8\,CO_2 + 8\,H_2O + \text{lots of energy.} \tag{2.3}$$

(The two additional hydrogens at the left and right ends of octane are not worth worrying about.)

▷ *How much energy is released by burning one CH_2 unit?*

To make this estimate, use the table of bond energies. It gives the energy required to break (not make) a chemical bond—for example, between carbon and hydrogen. However, there is no unique carbon–hydrogen (C–H) bond. The carbon–hydrogen bonds in methane are different from the carbon–hydrogen bonds in ethane. To make a reusable idea, we neglect those differences—placing them below our abstraction barrier—and make an abstraction called the carbon–hydrogen bond. So the table, already in its first column, is built on an abstraction.

The second gives the bond energy in kilocalories per mole of bonds. A kilocalorie is roughly 4000 joules, and a mole is Avogadro's

	bond energy		
	$\left(\dfrac{\text{kcal}}{\text{mol}}\right)$	$\left(\dfrac{\text{kJ}}{\text{mol}}\right)$	$\left(\dfrac{\text{eV}}{\text{bond}}\right)$
C—H	99	414	4.3
O—H	111	464	4.8
C—C	83	347	3.6
C—O	86	360	3.7
H—H	104	435	4.5
C—N	73	305	3.2
N—H	93	389	4.0
O=O	119	498	5.2
C=O	192	803	8.3
C=C	146	611	6.3
N≡N	226	946	9.8

number ($6{\times}10^{23}$) of bonds. The third column gives the energy in the SI units used by most of the world, kilojoules per mole. The final column gives the energy in electron volts (eV) per bond. An electron volt is 1.6×10^{-19} joules. An electron volt is suited for measuring atomic energies, because most bond energies have an easy-to-grasp value of a few electron volts. I wish most of the world used this unit!

Let's tabulate the energies in the combustion of one hydrocarbon unit.

$$
\begin{array}{c}
\text{H} \\
| \\
\text{C}\!-\! \\
| \\
\text{H}
\end{array}
\;+\;
\frac{3}{2}\times
\begin{array}{c}
\text{O} \\
\| \\
\text{O}
\end{array}
\;\longrightarrow\;
\begin{array}{c}
\text{O} \\
\| \\
\text{C} \\
\| \\
\text{O}
\end{array}
\;+\;
\begin{array}{c}
\text{H} \\
\diagup \\
\text{O} \\
\diagdown \\
\text{H}
\end{array}
\qquad (2.4)
$$

		bond energy	
		$\left(\dfrac{\text{kcal}}{\text{mol}}\right)$	$\left(\dfrac{\text{kJ}}{\text{mol}}\right)$
1	$\times$ C—C	1×83	1×347
2	$\times$ C—H	2×99	2×414
1.5	$\times$ O=O	1.5×119	1.5×498
Total		460	1925

The left side of the reaction has two carbon–hydrogen bonds, 1.5 oxygen–oxygen double bonds, and one carbon–carbon bond (connecting the carbon atom in the CH_2 unit to the carbon atom in a neighboring unit). The total, 460 kilocalories or 1925 kilojoules per mole, is the energy required to break the bonds. It is an energy input, so it reduces the net combustion energy.

The right side has two carbon–oxygen double bonds and two oxygen–hydrogen bonds. The total for the right side, 606 kilocalories or 2535 kilojoules per mole, is the energy released in forming these bonds. It is the energy produced, so it increases the net combustion energy.

	bond energy	
	$\left(\dfrac{\text{kcal}}{\text{mol}}\right)$	$\left(\dfrac{\text{kJ}}{\text{mol}}\right)$
$2 \times$ C=O	2×192	2×803
$2 \times$ O—H	2×111	2×464
Total	606	2535

The net result is, per mole of CH_2, an energy release of 606 minus 460 kilocalories, or approximately 145 kilocalories (610 kilojoules). Equivalently, it is also about 6 electron volts per CH_2 unit—about 1.5 chemical bonds' worth of energy. The combustion energy is also useful as an energy per mass rather than per mole. A mole of CH_2 units weighs 14 grams. Therefore, 145 kilocalories per mole is roughly 10 kilocalories or 40 kilojoules per gram. This energy density is worth memorizing because it gives the energy released by burning oil and gasoline or by metabolizing fat (even though fat is not a pure hydrocarbon).

	combustion energy		
	$\left(\dfrac{\text{kcal}}{\text{mol}}\right)$	$\left(\dfrac{\text{kcal}}{\text{g}}\right)$	$\left(\dfrac{\text{kJ}}{\text{g}}\right)$
hydrogen (H_2)	68	34.0	142
methane (CH_4)	213	13.3	56
gasoline (C_8H_{18})	1303	11.5	48
stearic acid ($C_{18}H_{36}O_2$)	2712	9.5	40
glucose ($C_6H_{12}O_6$)	673	3.7	15

The preceding table, adapted from Oxford University's "Virtual Chemistry" site, gives actual combustion energies for plant and animal fuel sources (with pure hydrogen included for fun). The penultimate entry, stearic acid, is a large component of animal fat; animals store energy in a substance with an energy density comparable to the energy density in gasoline—roughly 10 kilocalories or 40 kilojoules per gram. Plants, on the other hand, store energy in starch, which is a chain of glucose units; glucose has an energy density of only roughly 4 kilocalories per gram. This value, the energy density of food carbohydrates (sugars and starches), is also worth memorizing. It is significantly lower than the energy density of fats: Eating fat fills us up much faster than eating starch does.

▷ *How can we explain the different plant and animal energy-storage densities?*

Plants do not need to move, so the extra weight required by using lower-density energy storage is not so important. The benefit of the simpler glucose metabolic pathway outweighs the drawback of the extra weight. For animals, however, the large benefit of lower weight outweighs the metabolic complexity of burning fats.

> **Problem 2.1   Estimating the energy density of common foods**
> In American schools, the traditional lunch is the peanut-butter-and-jelly sandwich. Estimate the energy density in peanut butter and in jelly (or jam).
>
> **Problem 2.2   Peanut butter as fuel**
> If you could convert all the combustion energy in one tablespoon (15 grams) of peanut butter into mechanical work, how many flights of stairs could you climb?
>
> **Problem 2.3   Growth of grass**
> How fast does grass grow? Is the rate limited by rainfall or by sunlight?

2.2 Coin-flip game

The abstractions of atoms, bonds, and bond energies have been made for us by the development of science. But we often have to make new abstractions. To develop this skill, we'll analyze a coin game where two players take turns flipping a (fair) coin; whoever first tosses heads wins.

▷ *What is the probability that the first player wins?*

First get a feel for the game by playing it. Here is one round: TH. The first player missed the chance to win by tossing tails (T); but the second player tossed heads (H) and won.

Playing many games might reveal a pattern to us or suggest how to compute the probability. However, playing many games by flipping a real coin becomes tedious. Instead, a computer can simulate the games, substituting pseudorandom numbers for a real coin. Here are several runs produced by a computer program. Each line begins with 1 or 2 to indicate which player won the game; the rest of the line shows the coin tosses. In these ten iterations, each player won five times. A reasonable conjecture is that each player has an equal chance to win. However, this conjecture, based on only ten games, cannot be believed too strongly.

2	TH
2	TH
1	H
2	TH
1	TTH
2	TTTH
2	TH
1	H
1	H
1	H

Let's try 100 games. Now even counting the wins becomes tedious. My computer counted for me: 68 wins for player 1, and 32 wins for player 2. The probability of player 1's winning now seems closer to 2/3 than to 1/2.

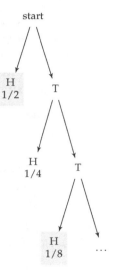

To find the exact value, let's diagram the game as a tree reflecting the alternative endings of the game. Each layer represents one flip. The game ends at a leaf, when one player has tossed heads. The shaded leaves show the first player's wins—for example, after H, TTH, or TTTTH. The probabilities of these winning ways are 1/2 (for H), 1/8 (for TTH), and 1/32 (for TTTTH). The sum of all these winning probabilities is the probability of the first player's winning:

$$\frac{1}{2} + \frac{1}{8} + \frac{1}{32} + \cdots. \tag{2.5}$$

To sum this infinite series without resorting to formulas, make an abstraction: Notice that the tree contains, one level down, a near copy of itself. (In this problem, the abstraction gets reused within the same problem. In computer science, such a structure is called recursive.) For if the first player tosses tails, the second player starts the game in the position of the first player, with the same probability of winning.

To benefit from this equivalence, let's name the reusable idea, namely the probability of the first player's winning, and call it p. The second player wins the game with probability $p/2$: The factor of 1/2 is the probability that the first player tosses tails; the factor of p is the probability that the second player wins, given that the first player blew his chance by tossing tails on the first toss.

Because either the first or the second player wins, the two winning probabilities add to 1:

$$\underbrace{p}_{P(\text{first player wins})} + \underbrace{p/2}_{P(\text{second player wins})} = 1. \tag{2.6}$$

The solution is $p = 2/3$, as suggested by the 100-game simulation. The benefit of the abstraction solution, compared to calculating the infinite probability sum explicitly, is insight. In the abstraction solution, the answer has to be what it is. It leaves almost nothing to remember. An amusing illustration of the same benefit comes from the problem of the fly that zooms back and forth between two approaching trains.

▷ *If the fly starts when the trains are 60 miles apart, each train travels at 20 miles per hour, and the fly travels at 30 miles per hour, how far does the fly travel, in total, before meeting its maker when the trains collide? (Apologies that physics problems are often so violent.)*

Right after hearing the problem, John von Neumann, inventor of game theory and the modern computer, gave the correct distance. "That was quick," said a colleague. "Everyone else tries to sum the infinite series." "What's wrong with that?" said von Neumann. "That's how I did it." In Problem 2.7, you get to work out the infinite-series and the insightful solutions.

Problem 2.4 Summing a geometric series using abstraction
Use abstraction to find the sum of the infinite geometric series
$$1 + r + r^2 + r^3 + \cdots. \tag{2.7}$$

Problem 2.5 Using the geometric-series sum
Use Problem 2.4 to check that the probability of the first player's winning is 2/3:
$$p = \frac{1}{2} + \frac{1}{8} + \frac{1}{32} + \cdots = \frac{2}{3}. \tag{2.8}$$

Problem 2.6 Nested square roots
Evaluate these infinite mixes of arithmetic and square roots:
$$\sqrt{3 \times \sqrt{3 \times \sqrt{3 \times \sqrt{3 \times \cdots}}}}. \tag{2.9}$$
$$\sqrt{2 + \sqrt{2 + \sqrt{2 + \sqrt{2 + \cdots}}}}. \tag{2.10}$$

Problem 2.7 Two trains and a fly
Find the insightful and the infinite-series solution to the problem of the fly and the approaching trains (Section 2.2). Check that they give the same answer for the distance that the fly travels!

Problem 2.8   Resistive ladder
In the following infinite ladder of 1-ohm resistors, what is the resistance between
points A and B? This measurement is indicated by the ohmmeter connected between these points.

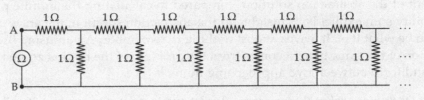

2.3 Purpose of abstraction

The coin game (Section 2.2), like the geometric series (Problem 2.4) or the
resistive ladder (Problem 2.8), contained a copy of itself. Noticing this reuse
greatly simplified the analysis. Abstraction has a second benefit: giving
us a high-level view of a problem or situation. Abstractions then show us
structural similarities between seemingly disparate situations.

As an example, let's revisit the geometric mean, introduced in Section 1.6
to make gut estimates. The geometric mean of two nonnegative quantities
a and b is defined as

$$\text{geometric mean} \equiv \sqrt{ab}. \tag{2.11}$$

This mean is called the geometric mean because it has
a pleasing geometric construction. Divide the diameter
of a circle into two lengths, a and b, and inscribe a right
triangle whose hypotenuse is the diameter. The triangle's
altitude is the geometric mean of a and b.

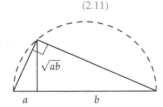

This mean reappears in surprising places, including the
beach. When you stand at the shore and look at the horizon, you are seeing
a geometric mean. The distance to the horizon is the geometric mean of
two important lengths in the problem (Problem 2.9).

For me, its most surprising appearance was in the "Programming and Problem-Solving Seminar" course taught by Donald Knuth [40] (who also created TEX, the typesetting system for this book). The course, taught as a series of two-week problems, helped first-year PhD students transition from
undergraduate homework problems to PhD research problems. A homework problem requires perhaps 1 hour. A research problem requires, say,

1000 hours: roughly a year of work, allowing for other projects. (A few problems stapled together become a PhD.) In the course, each 2-week module required about 30 hours—approximately the geometric mean of the two endpoints. The modules were just the right length to help us cross the bridge from homework to research.

> **Problem 2.9   Horizon distance**
> How far is the horizon when you are standing at the shore? Hint: It's farther for an adult than for a child.
>
> **Problem 2.10   Distance to a ship**
> Standing at the shore, you see a ship (drawn to scale) with a 10-meter mast sail into the distance and disappear from view. How far away was it when it disappeared?

As further evidence that the geometric mean is a useful abstraction, the idea appears even when there is no geometric construction to produce it, such as in making gut estimates. We used this method in Section 1.6 to estimate the population density and then the population of the United States. Let's practice by estimating the oil imports of the United States in barrels per year—without the divide-and-conquer reasoning of Section 1.4.

The method requires that the gut supply a lower and an upper bound. My gut reports back that it would feel fairly surprised if the imports were less than 10 million barrels per year. On the upper end, my gut would be fairly surprised if the imports were higher than 1 trillion barrels per year—a barrel is a lot of oil, and a trillion is a large number!

You might wonder how your gut too can come up with such large numbers and how you can have any confidence in them. Admittedly, I have practiced a lot. But you can practice too. The key is to practice effectively. First, have the courage to guess even when you feel anxious about it (I feel this anxiety still, so I practice this courage often). Second, compare your guess to values in which you can place more confidence—for example, to your own more careful estimates or to official values. The comparison helps calibrate your gut (your right brain) to these large magnitudes. You will find a growing and justified confidence in your judgment of magnitude.

My best guess for the amount is the geometric mean of the lower and upper estimates:

$$\sqrt{10 \text{ million} \times 1 \text{ trillion}} \ \frac{\text{barrels}}{\text{year}}. \tag{2.12}$$

The result is roughly 3 billion barrels per year—which is close to our estimate based on divide-and-conquer reasoning and close to the true value. In contrast, using the arithmetic mean would have produced an estimate of 500 billion barrels per year, which is far too high.

> **Problem 2.11   Arithmetic-mean–geometric-mean inequality**
> Use the geometric construction for the geometric mean to show that the arithmetic mean of a and b (assumed to be nonnegative) is always greater than or equal to their geometric mean. When are the means equal?
>
> **Problem 2.12   Weighted geometric mean**
> A generalization of the arithmetic mean of a and b as $(a + b)/2$ is to give a and b unequal weights. What is the analogous generalization for a geometric mean? (The weighted geometric mean shows up in Problem 6.29 when you estimate the contact time of a ball bouncing from a table.)

2.4 Analogies

Because abstractions are so useful, it is helpful to have methods for making them. One way is to construct an analogy between two systems. Each common feature leads to an abstraction; each abstraction connects our knowledge in one system to our knowledge in the other system. One piece of knowledge does double duty. Like a mental lever, analogy and, more generally, abstraction are intelligence amplifiers.

2.4.1 Electrical–mechanical analogies

An illustration with many abstractions on which we can practice is the analogy between a spring–mass system and an inductor–capacitor (*LC*) circuit.

$$(2.13)$$

In the circuit, the voltage source—the V_{in} on its left side—supplies a current that flows through the inductor (a wire wrapped around an iron rod) and capacitor (two metal plates separated by air). As current flows through the capacitor, it alters the charge on the capacitor. This "charge" is confusingly named, because the net charge on the capacitor remains zero. Instead, "charge" means that the two plates of the capacitor hold opposite charges, Q and $-Q$, with $Q \neq 0$. The current changes Q. The charges on the two plates create an electric field, which produces the output voltage V_{out} equal to Q/C (where C is the capacitance).

For most of us, the circuit is less familiar than the spring–mass system. However, by building an analogy between the systems, we transfer our understanding of the mechanical to the electrical system.

In the mechanical system, the fundamental variable is the mass's displacement x. In the electrical system, it is the charge Q on the capacitor. These variables are analogous so their derivatives should also be analogous: Velocity (v), the derivative of position, should be analogous to current (I), the derivative of charge.

	spring	circuit
variable	x	Q
1st derivative	v	I
2nd derivative	a	dI/dt

Let's build more analogy bridges. The derivative of velocity, which is the second derivative of position, is acceleration (a). Therefore, the derivative of current (dI/dt) is the analog of acceleration. This analogy will be useful shortly when we find the circuit's oscillation frequency.

These variables describe the state of the systems and how that state changes: They are the kinematics. But without the causes of the motion—the dynamics—the systems remain lifeless. In the mechanical system, dynamics results from force, which produces acceleration:

$$a = \frac{F}{m}. \tag{2.14}$$

Acceleration is analogous to change in current dI/dt, which is produced by applying a voltage to the inductor. For an inductor, the governing relation (analogous to Ohm's law for a resistor) is

$$\frac{dI}{dt} = \frac{V}{L}, \tag{2.15}$$

where L is the inductance, and V is the voltage across the inductor. Based on the common structure of the two relations, force F and voltage V must be analogous. Indeed, they both measure effort: Force tries to accelerate the mass, and voltage tries to change the inductor current. Similarly, mass and inductance are analogous: Both measure resistance to the corresponding effort. Large masses are hard to accelerate, and large-L inductors resist changes to their current. (A mass and an inductor, in another similarity, both represent kinetic energy: a mass through its motion, and an inductor through the kinetic energy of the electrons making its magnetic field.)

Turning from the mass–inductor analogy, let's look at the spring–capacitor analogy. These components represent the potential energy in the system: in the spring through the energy in its compression or expansion, and in the capacitor through the electrostatic potential energy due to its charge.

Force tries to stretch the spring but meets a resistance k: The stiffer the spring (the larger its k), the harder it is to stretch.

$$x = \frac{F}{k}. \tag{2.16}$$

Analogously, voltage tries to charge the capacitor but meets a resistance $1/C$: The larger the value of $1/C$, the smaller the resulting charge.

$$Q = \frac{V}{1/C}. \tag{2.17}$$

Based on the common structure of the relations for x and Q, spring constant k must be analogous to inverse capacitance $1/C$. Here are all our analogies.

	mechanical	electrical
kinematics		
fundamental variable	x	Q
first derivative	v	I
second derivative	a	dI/dt
dynamics		
effort	F	V
resistance to effort (kinetic component)	m	L
resistance to effort (potential component)	k	$1/C$

From this table, we can read off our key result. Start with the natural (angular) frequency ω of a spring–mass system: $\omega = \sqrt{k/m}$. Then apply the analogies. Mass m is analogous to inductance L. Spring constant k is analogous to inverse capacitance $1/C$. Therefore, ω for the LC circuit is $1/\sqrt{LC}$:

$$\omega = \sqrt{\frac{1/C}{L}} = \frac{1}{\sqrt{LC}}. \tag{2.18}$$

Because of the analogy bridges, one formula, the natural frequency of a spring–mass system, does double duty. More generally, whatever we learn about one system helps us understand the other system. Because of the analogies, each piece of knowledge does double duty.

2.4.2 Energy density in the gravitational field

With the electrical–mechanical analogy as practice, let's try a less famil-iar analogy: between the electric and the gravitational field. In particular, we'll connect the energy densities (energy per volume) in the correspond-ing fields. An electric field E represents an energy density of $\epsilon_0 E^2/2$, where ϵ_0 is the permittivity of free space appearing in the electrostatic force be-tween two charges q_1 and q_2:

$$F = \frac{q_1 q_2}{4\pi\epsilon_0 r^2}. \tag{2.19}$$

Because electrostatic and gravitational forces are both inverse-square forces (the force is proportional to $1/r^2$), the energy densities should be analogous. Not least, there should be a gravitational energy density. But how is it re-lated to the gravitational field?

To answer that question, our first step is to find the gravitational analog of the electric field. Rather than thinking of the electric field only as something electric, focus on the common idea of a field. In that sense, the electric field is the object that, when multiplied by the charge, gives the force:

$$\text{force} = \text{charge} \times \text{field}. \tag{2.20}$$

We use words rather than the normal symbols, such as E for field or q for charge, because the symbols might bind our thinking to particular cases and prevent us from climbing the abstraction ladder.

This verbal form prompts us to ask: What is gravitational charge? In elec-trostatics, charge is the source of the field. In gravitation, the source of the field is mass. Therefore, gravitational charge is mass. Because field is force per charge, the gravitational field strength is an acceleration:

$$\text{gravitational field} = \frac{\text{force}}{\text{charge}} = \frac{\text{force}}{\text{mass}} = \text{acceleration}. \tag{2.21}$$

Indeed, at the surface of the Earth, the field strength is g, also called the acceleration due to gravity.

The definition of gravitational field is the first half of the puzzle (we are using divide-and-conquer reasoning again). For the second half, we'll use the field to compute the energy density. To do so, let's revisit the route from electric field to electrostatic energy density:

$$E \rightarrow \frac{1}{2}\epsilon_0 E^2. \tag{2.22}$$

With g as the gravitational field, the analogous route is

$$g \to \frac{1}{2} \times \text{something} \times g^2, \tag{2.23}$$

where the "something" represents our ignorance of what to do about ϵ_0.

▷ *What is the gravitational equivalent of ϵ_0?*

To find its equivalent, compare the simplest case in both worlds: the field of a point charge. A point electric charge q produces a field

$$E = \frac{1}{4\pi\epsilon_0} \frac{q}{r^2}. \tag{2.24}$$

A point gravitational charge m (a point mass) produces a gravitational field (an acceleration)

$$g = \frac{Gm}{r^2}, \tag{2.25}$$

where G is Newton's constant.

The gravitational field has a similar structure to the electric field. Both are inverse-square forces, as expected. Both are proportional to the charge. The difference is the constant of proportionality. For the electric field, it is $1/4\pi\epsilon_0$. For the gravitational field, it is simply G. Therefore, G is analogous to $1/4\pi\epsilon_0$; equivalently, ϵ_0 is analogous to $1/4\pi G$.

Then the gravitational energy density becomes

$$\frac{1}{2} \times \frac{1}{4\pi G} \times g^2 = \frac{g^2}{8\pi G}. \tag{2.26}$$

We will use this analogy in Section 9.3.3 when we transfer our hard-won knowledge of electromagnetic radiation to understand the even more subtle physics of gravitational radiation.

Problem 2.13 Gravitational energy of the Sun

What is the energy in the gravitational field of the Sun? (Just consider the field outside the Sun.)

Problem 2.14 Pendulum period including buoyancy

The period of a pendulum in vacuum is (for small amplitudes) $T = 2\pi\sqrt{l/g}$, where l is the bob length and g is the gravitational field strength. Now imagine the pendulum swinging in a fluid (say, air). By replacing g with a modified value, include the effect of buoyancy in the formula for the pendulum period.

Problem 2.15   Comparing field energies
Find the ratio of electrical to gravitational field energies in the fields produced by
a proton.

2.4.3 Parallel combination

Analogies not only reuse work, they help us rewrite expressions in compact,
insightful forms. An example is the idea of parallel combination. It appears
in the analysis of the infinite resistive ladder of Problem 2.8.

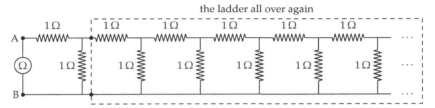

To find the resistance R across the ladder (in other words, what the ohmmeter measures between the nodes A and B), you represent the entire ladder
as a single resistor R. Then the whole ladder is 1 ohm in series with the
parallel combination of 1 ohm and R:

$$\tag{2.27}$$

The next step in finding R usually invokes the parallel-resistance formula:
that the resistance of R_1 and R_2 in parallel is

$$\frac{R_1 R_2}{R_1 + R_2}. \tag{2.28}$$

For our resistive ladder, the parallel combination of 1 ohm with the ladder
is 1 ohm $\times\, R/(1\text{ ohm} + R)$. Placing this combination in series with 1 ohm
gives a resistance

$$1\,\Omega + \frac{1\,\Omega \times R}{1\,\Omega + R}. \tag{2.29}$$

This recursive construction reproduces the ladder, only one unit longer. We
therefore get an equation for R:

$$R = 1\,\Omega + \frac{1\,\Omega \times R}{1\,\Omega + R}. \tag{2.30}$$

The (positive) solution is $R = (1 + \sqrt{5})/2$ ohms. The numerical part is the golden ratio ϕ (approximately 1.618). Thus, the ladder, when built with 1-ohm resistors, offers a resistance of ϕ ohms.

Although the solution is correct, it skips over a reusable idea: the parallel combination. To facilitate its reuse, let's name the idea with a notation:

$$R_1 \parallel R_2. \tag{2.31}$$

This notation is self-documenting, as long as you recognize the symbol $\parallel$ to mean "parallel," a recognition promoted by the parallel bars. A good notation should help thinking, not hinder it by requiring us to remember how the notation works. With this notation, the equation for the ladder resistance R is

$$R = 1\,\Omega \; + \; 1\,\Omega \parallel R \tag{2.32}$$

(the parallel-combination operator has higher priority than—is computed before—the addition). This expression more plainly reflects the structure of the system, and our reasoning about it, than does the version

$$R = 1\,\Omega + \frac{1\,\Omega \times R}{1\,\Omega + R}. \tag{2.33}$$

The $\parallel$ notation organizes the complexity.

Once you name an idea, you find it everywhere. As a child, after my family bought a Volvo, I saw Volvos on every street. Similarly, we'll now look at examples of parallel combination far beyond the original appearance of the idea in circuits. For example, it gives the spring constant of two connected springs (Problem 2.16):

$$\underbrace{\text{WWWWW}}_{k_1}\underbrace{\text{WWWWW}}_{k_2} = \underbrace{\text{WWWWWWWWW}}_{k_1 \parallel k_2} \tag{2.34}$$

Problem 2.16   Springs as capacitors

Using the analogy between springs and capacitors (discussed in Section 2.4.1), explain why springs in series combine using the parallel combination of their spring constants.

Another surprising example is the following spring–mass system with two masses:

The natural frequency ω, expressed without our $\parallel$ abstraction, is

$$\omega = \frac{k(m + M)}{mM}. \tag{2.35}$$

This form looks complicated until we use the $\parallel$ abstraction:

$$\omega = \frac{k}{m \parallel M}. \tag{2.36}$$

Now the frequency makes more sense. The two masses act like their parallel combination $m \parallel M$:

The replacement mass $m \parallel M$ is so useful that it has a special name: the reduced mass. Our abstraction organizes complexity by turning a three-component system (a spring and two masses) into a simpler two-component system.

In the spirit of notation that promotes insight, use lowercase ("small") m for the mass that is probably smaller, and uppercase ("big") M for the mass that is probably larger. Then write $m \parallel M$ rather than $M \parallel m$. These two forms produce the same result, but the $m \parallel M$ order minimizes surprise: The parallel combination of m and M is smaller than either mass (Problem 2.17), so it is closer to m, the smaller mass, than to M. Writing $m \parallel M$, rather than $M \parallel m$, places the most salient information first.

> **Problem 2.17   Using the resistance analogy**
> By using the analogy with parallel resistances, explain why $m \parallel M$ is smaller than m and M.

▷ *Why do the two masses combine like resistors in parallel?*

The answer lies in the analogy between mass and resistance. Resistance appears in Ohm's law:

voltage = resistance × current. $\tag{2.37}$

Voltage is an effort. Current, which results from the effort, is a flow. Therefore, the more general form—one step higher on the abstraction ladder—is

effort = resistance × flow. $\tag{2.38}$

In this form, Newton's second law,

force = mass × acceleration, $\hspace{6cm}$ (2.39)

identifies force as the effort, mass as the resistance, and acceleration as the flow.

Because the spring can wiggle either mass, just as current can flow through either of two parallel resistors, the spring feels a resistance equal to the parallel combination of the resistances—namely, $m \parallel M$.

> **Problem 2.18**   **Three springs connected**
> What is the effective spring constant of three springs connected in a line, with spring constants 2, 3, and 6 newtons per meter, respectively?

2.4.4 Impedance as a higher-level abstraction

Resistance, in the electrical sense, has appeared several times, and it underlies a higher-level abstraction: impedance. Impedance extends the idea of electrical resistance to capacitors and inductors. Capacitors and inductors, along with resistors, are the three linear circuit elements: In these elements, the connection between current and voltage is described by a linear equation: For resistors, it is a linear algebraic relation (Ohm's law); for capacitors or inductors, it is a linear differential equation.

▷ *Why should we extend the idea of resistance?*

Resistors are easy to handle. When a circuit contains only resistors, we can immediately and completely describe how it behaves. In particular, we can write the voltage at any point in the circuit as a linear combination of the voltages at the source nodes. If only we could do the same when the circuit contains capacitors and inductors.

We can! Start with Ohm's law,

$$\text{current} = \frac{\text{voltage}}{\text{resistance}},\hspace{4cm}(2.40)$$

and look at it in the higher-level and expanded form

$$\text{flow} = \frac{1}{\text{resistance}} \times \text{effort}.\hspace{4cm}(2.41)$$

For a capacitor, flow will still be current. But we'll need to find the capacitive analog of effort. This analogy will turn out slightly different from the electrical–mechanical analogy between capacitance and spring constant

(Section 2.4.1), because now we are making an analogy between capacitors and resistors (and, eventually, inductors). For a capacitor,

$$\text{charge} = \text{capacitance} \times \text{voltage}. \tag{2.42}$$

To turn charge into current, we differentiate both sides to get

$$\text{current} = \text{capacitance} \times \frac{d(\text{voltage})}{dt}. \tag{2.43}$$

To make the analogy quantitative, let's apply to the capacitor the simplest voltage whose form is not altered by differentiation:

$$V = V_0\, e^{j\omega t}, \tag{2.44}$$

where V is the input voltage, V_0 is the amplitude, ω is the angular frequency, and j is the imaginary unit $\sqrt{-1}$. The voltage V is a complex number; but the implicit understanding is that the actual voltage is the real part of this complex number. By finding how the current I (the flow) depends on V (the effort), we will extend the idea of resistance to a capacitor.

▷ *With this exponential form, how can we represent the more familiar oscillating voltages $V_1 \cos \omega t$ or $V_1 \sin \omega t$, where V_1 is a real voltage?*

Start with Euler's relation:

$$e^{j\omega t} = \cos \omega t + j \sin \omega t. \tag{2.45}$$

To make $V_1 \cos \omega t$, set $V_0 = V_1$ in $V = V_0\, e^{j\omega t}$. Then

$$V = V_1(\cos \omega t + j \sin \omega t). \tag{2.46}$$

and the real part of V is just $V_1 \cos \omega t$.

Making $V_1 \sin \omega t$ is more tricky. Choosing $V_0 = jV_1$ almost works:

$$V = jV_1(\cos \omega t + j \sin \omega t) = V_1(j \cos \omega t - \sin \omega t). \tag{2.47}$$

The real part is $-V_1 \sin \omega t$, which is correct except for the minus sign. Thus, the correct amplitude is $V_0 = -jV_1$. In summary, our exponential form can compactly represent the more familiar sine and cosine signals.

With this exponential form, differentiation is simpler than with sines or cosines. Differentiating V with respect to time just brings down a factor of $j\omega$, but otherwise leaves the $V_0\, e^{j\omega t}$ alone:

$$\frac{dV}{dt} = j\omega \times \underbrace{V_0\, e^{j\omega t}}_{V} = j\omega V. \tag{2.48}$$

With this changing voltage, the capacitor equation,

$$\text{current} = \text{capacitance} \times \frac{d(\text{voltage})}{dt}, \tag{2.49}$$

becomes

$$\text{current} = \text{capacitance} \times j\omega \times \text{voltage}. \tag{2.50}$$

Let's compare this form to its analog for a resistor (Ohm's law):

$$\text{current} = \frac{1}{\text{resistance}} \times \text{voltage}. \tag{2.51}$$

Matching up the pieces, we find that a capacitor offers a resistance

$$Z_C = \frac{1}{j\omega C}. \tag{2.52}$$

This more general resistance, which depends on the frequency, is called impedance and denoted Z. (In the analogy of Section 2.4.1 between capacitors and springs, we found that capacitor offered a resistance to being charged of $1/C$. Impedance, the result of an analogy between capacitors and resistors, contains $1/C$ as well, but also contains the frequency in the $1/j\omega$ factor.)

Using impedance, we can describe what happens to any sinusoidal signal in a circuit containing capacitors. Our thinking is aided by the compact notation—the capacitive impedance Z_C (or even R_C). The notation hides the details of the capacitor differential equation and allows us to transfer our intuition about resistance and flow to a broader class of circuits.

The simplest circuit with resistors and capacitors is the so-called low-pass RC circuit. Not only is it the simplest interesting circuit, it will also be, thanks to further analogies, a model for heat flow. Let's apply the impedance analogy to this circuit.

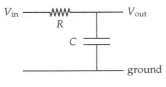

To help us make and use abstractions, let's imagine defocusing our eyes. Under blurry vision, the capacitor looks like a resistor that just happens to have a funny resistance $R_C = 1/j\omega C$. Now the entire circuit looks just like a pure-resistance circuit. Indeed, it is the simplest such circuit, a voltage divider. Its behavior is described by one number: the gain, which is the ratio of output to input voltage $V_{\text{out}}/V_{\text{in}}$.

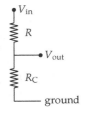

In the RC circuit, thought of as a voltage divider,

$$\text{gain} = \frac{\text{capacitive resistance}}{\text{total resistance from } V_{\text{in}} \text{ to ground}} = \frac{R_C}{R + R_C}. \tag{2.53}$$

Because $R_C = 1/j\omega C$, the gain becomes

$$\text{gain} = \frac{\frac{1}{j\omega C}}{R + \frac{1}{j\omega C}}. \tag{2.54}$$

After clearing the fractions by multiplying by $j\omega C$ in the numerator and denominator, the gain simplifies to

$$\text{gain} = \frac{1}{1 + j\omega RC}. \tag{2.55}$$

▷ *Why is the circuit called a low-pass circuit?*

At high frequencies ($\omega \to \infty$), the $j\omega RC$ term in the denominator makes the gain zero. At low frequencies ($\omega \to 0$), the $j\omega RC$ term disappears and the gain is 1. High-frequency signals are attenuated by the circuit; low-frequency signals pass through mostly unchanged. This abstract, high-level description of the circuit helps us understand the circuit without our getting buried in equations. Soon we will transfer our understanding of this circuit to thermal systems.

The gain contains the circuit parameters as the product RC. In the denominator of the gain, $j\omega RC$ is added to 1; therefore, $j\omega RC$, like 1, must have no dimensions. Because j is dimensionless (is a pure number), ωRC must be itself dimensionless. Therefore, the product RC has dimensions of time. This product is the circuit's time constant—usually denoted τ.

The time constant has two physical interpretations. To construct them, we imagine charging the capacitor using a constant input voltage V_0; eventually (after an infinite time), the capacitor charges up to the input voltage ($V_{\text{out}} = V_0$) and holds a charge $Q = CV_0$. Then, at $t = 0$, we make the input voltage zero by connecting the input to ground.

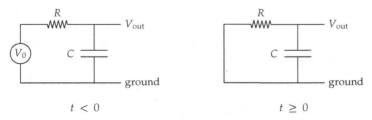

The capacitor discharges through the resistor, and its voltage decays exponentially:

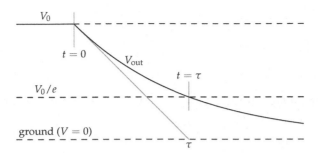

After one time constant τ, the capacitor voltage falls by a factor of e toward
its final value—here, from V_0 to V_0/e. The $1/e$ time is our first interpretation
of the time constant. Furthermore, if the capacitor voltage had decayed at
its initial rate (just after $t = 0$), it would have reached zero voltage after one
time constant τ—the second interpretation of the time constant.

The time-constant abstraction hides—abstracts away—the details that produced it: here, electrical resistance and capacitance. Nonelectrical systems
can also have a time constant but produce it by a different mechanism.
Our high-level understanding of time constants, because it is not limited
to electrical systems, will help us transfer our understanding of the electrical low-pass filter to nonelectrical systems. In particular, we are now ready
to understand heat flow in thermal systems.

> **Problem 2.19   Impedance of an inductor**
> An inductor has the voltage–current relation
>
> $$V = L\frac{dI}{dt},$$ (2.56)
>
> where L is the inductance. Find an inductor's frequency-dependent impedance
> Z_L. After finding this impedance, you can analyze any linear circuit as if it were
> composed only of resistors.

2.4.5 Thermal systems

The *RC* circuit is a model for thermal systems—which are not obviously
connected to circuits. In a thermal system, temperature difference, the analog of voltage difference, produces a current of energy. Energy current, in
less fancy words, is heat flow. Furthermore, the current is proportional to
the temperature difference—just as electric current is proportional to voltage difference. In both systems, flow is proportional to effort. Therefore,
heat flow can be understood by using circuit analogies.

As an example, I often prepare a cup of tea but forget to drink it while it is hot. Like a discharging capacitor, the tea slowly cools toward room temperature and becomes undrinkable. Heat flows out through the mug. Its walls provide a thermal resistance; by analogy to an *RC* circuit, let's denote the thermal resistance R_t. The heat is

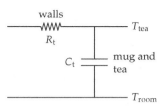

stored in the water and mug, which form a heat reservoir. This reservoir, of heat rather than of charge, provides the thermal capacitance, which we denote C_t. (Thus, the mug participates in the thermal resistance and capacitance.) Resistance and capacitance are transferable ideas.

The product $R_t C_t$ is, by analogy to the *RC* circuit, the thermal time constant τ. To estimate τ with a home experiment (the method we used in Section 1.7), heat up a mug of tea; as it cools, sketch the temperature gap between the tea and room temperature. In my extensive experience of tea neglect, an enjoyably hot cup of tea becomes lukewarm in half an hour. To quantify these temperatures, enjoyably warm may be $130\,°\text{F}$ ($\approx 55\,°\text{C}$), room temperature is $70\,°\text{F}$ ($\approx 20\,°\text{C}$), and lukewarm may be $85\,°\text{F}$ ($\approx 30\,°\text{C}$).

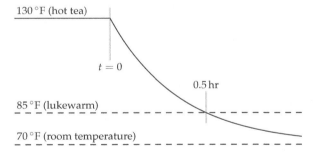

▷ *Based on the preceding data, what is the approximate thermal time constant of the mug of tea?*

In one thermal time constant, the temperature gap falls by a factor of e (just as the voltage gap falls by a factor of e in one electrical time constant). For my mug of tea, the temperature gap between the tea and the room started at $60\,°\text{F}$:

$$\underbrace{\text{enjoyably warm}}_{130\,°\text{F}} - \underbrace{\text{room temperature}}_{70\,°\text{F}} = 60\,°\text{F}. \qquad (2.57)$$

In the half hour while the tea cooled in the microwave, the temperature gap fell to $15\,°\text{F}$:

$$\underbrace{\text{lukewarm}}_{85\,°\text{F}} - \underbrace{\text{room temperature}}_{70\,°\text{F}} = 15\,°\text{F}. \tag{2.58}$$

Therefore, the temperature gap decreased by a factor of 4 in half an hour. Falling by the canonical factor of e (roughly 2.72) would require less time: perhaps 0.3 hours (roughly 20 minutes) instead of 0.5 hours. A more precise calculation would be to divide 0.5 hours by ln 4, which gives 0.36 hours. However, there is little point doing this part of the calculation so precisely when the input data are far less precise. Therefore, let's estimate the thermal time constant τ as roughly 0.3 hours.

Using this estimate, we can understand what happens to the tea mug when, as it often does, it spends a lonely few days in the microwave, subject to the daily variations in room temperature. This analysis will become our model for the daily temperature variations in a house.

▷ *How does a teacup with $\tau \approx 0.3$ hours respond to daily temperature variations?*

First, set up the circuit analogy. The output signal is still the tea's temperature. The input signal is the (sinusoidally) varying room temperature. However, the ground signal, which is our reference temperature, cannot also be the room temperature. Instead, we need a constant reference temperature. The simplest choice is the average room temperature T_{avg}. (After we have transferred this analysis to the temperature variation in houses, we'll see that the conclusion is the same even with a different reference temperature.)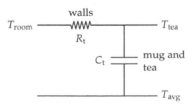

The gain connects the amplitudes of the output and input signals:

$$\text{gain} = \frac{\text{amplitude of the output signal}}{\text{amplitude of the input signal}} = \frac{1}{1 + j\omega\tau}. \tag{2.59}$$

The input signal (room temperature) varies with a frequency f of 1 cycle per day. Then the dimensionless parameter $\omega\tau$ in the gain is roughly 0.1. Here is that calculation:

$$\underbrace{2\pi \times \overbrace{1\,\frac{\text{cycle}}{\text{day}}}^{f}}_{\omega} \times \underbrace{0.3\,\text{hr}}_{\tau} \times \underbrace{\frac{1\,\text{day}}{24\,\text{hr}}}_{1} \approx 0.1. \tag{2.60}$$

The system is driven by a low-frequency signal: ω is not large enough to make $\omega\tau$ comparable to 1. As the gain expression reminds us, the mug of tea is a low-pass filter for temperature variations. It transmits this low-frequency input temperature signal almost unchanged to the output—to the tea temperature. Therefore, the inside (tea) temperature almost exactly follows the outside (room) temperature.

The opposite extreme is a house. Compared to the mug, a house has a much higher mass and therefore thermal capacitance. The resulting time constant $\tau = R_t C_t$ is probably much longer for a house than for the mug. As an example, when I taught in sunny Cape Town, where houses are often unheated even in winter, the mildly insulated house where I stayed had a thermal time constant of approximately 0.5 days.

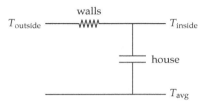

For this house the dimensionless parameter $\omega\tau$ is much larger than it was for the tea mug. Here is the corresponding calculation.

$$2\pi \times \overbrace{1\,\frac{\text{cycle}}{\text{day}}}^{f} \times \underbrace{0.5\,\text{days}}_{\tau} \approx 3. \tag{2.61}$$

where the ω brace spans $2\pi \times 1\,\frac{\text{cycle}}{\text{day}}$.

▷ *What consequence does $\omega\tau \approx 3$ have for the indoor temperature?*

In the Cape Town winter, the outside temperature varied daily between 45 °F and 75 °F; let's also assume that it varied approximately sinusoidally. This 30 °F peak-to-peak variation, after passing through the house low-pass filter, shrinks by a factor of approximately 3. Here is how to find that factor by estimating the magnitude of the gain.

$$|\text{gain}| = \left|\frac{\text{amplitude of } T_{\text{inside}}}{\text{amplitude of } T_{\text{outside}}}\right| = \left|\frac{1}{1 + j\omega\tau}\right|. \tag{2.62}$$

(It is slightly confusing that the outside temperature is the input signal, and the inside temperature is the output signal!) Now plug in $\omega\tau \approx 3$ to get

$$|\text{gain}| \approx \left|\frac{1}{1 + 3j}\right| = \frac{1}{\sqrt{1^2 + 3^2}} \approx \frac{1}{3}. \tag{2.63}$$

In general, when $\omega\tau \gg 1$, the magnitude of the gain is approximately $1/\omega\tau$.

Therefore, the outside peak-to-peak variation of 30 °F becomes a smaller inside peak-to-peak variation of 10 °F. Here is a block diagram showing this effect of the house low-pass filter.

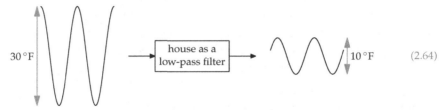

$$(2.64)$$

Our comfort depends not only on the temperature variation (I like a fairly steady temperature), but also on the average temperature.

> *What is the average temperature indoors?*

It turns out that the average temperature indoors is equal to the average temperature outdoors! To see why, let's think carefully about the reference temperature (our thermal analog of ground). Before, in the analysis of the forgotten tea mug, our reference temperature was the average indoor temperature. Because we are now trying to determine this value, let's instead use a known convenient reference temperature—for example, the cool 10 °C, which makes for round numbers in Celsius or Fahrenheit (50 °F).

The input signal (the outside temperature) varied in winter between 45 °F and 75 °F. Therefore, it has two pieces: (1) our usual varying signal with the 30 °F peak-to-peak variation, and (2) a steady signal of 10 °F.

$$(2.65)$$

The steady signal is the difference between the average outside temperature of 60 °F and the reference signal of 50 °F.

Let's handle each piece in turn—we are using divide-and-conquer reasoning again. We just analyzed the varying piece: It passes through the house low-pass filter and, with $\omega\tau \approx 3$, it shrinks significantly in amplitude. In contrast, the nonvarying part, which is the average outside temperature, has zero frequency by definition. Therefore, its dimensionless parameter $\omega\tau$ is exactly 0. This signal passes through the house low-pass filter with a gain of 1. As a result, the average output signal (the inside temperature) is also 60 °F: the same steady 10 °F signal measured relative to the reference temperature of 50 °F.

The 10 °F peak-to-peak inside-temperature amplitude is a variation around 60 °F. Therefore, the inside temperature varies between 55 °F and 65 °F (13 °C to 18 °C). Indoors, when I am not often running or otherwise generating much heat, I feel comfortable at 68 °F (20 °C). So, as this circuit model of heat flow predicts, I wore a sweater day and night in the Cape Town house. (For more on using *RC* circuit analogies for building design, see the "Design masterclass" article by Doug King [30].)

Problem 2.20 When is the house coldest?
Based on the general form for the gain, $1/(1+j\omega\tau)$, when in the day will the Cape Town house be the coldest, assuming that the outside is coldest at midnight?

2.5 Summary and further problems

Geometric means, impedances, low-pass filters—these ideas are all abstractions. An abstraction connects seemingly random details into a higher-level structure that allows us to transfer knowledge and insights. By building abstractions, we amplify our intelligence.

Indeed, each of our reasoning tools is an abstraction or reusable idea. In Chapter 1, for example, we learned how to split hard problems into tractable ones, and we named this process divide-and-conquer reasoning. Don't stop with this one process. Whenever you reuse an idea, identify the transferable process, and name it: Make an abstraction. With a name, you will recognize and reuse it.

Problem 2.21 From circles to spheres
In this problem, you first find the area of a circle from its circumference and then use analogous reasoning to find the volume of a sphere.

a. Divide a circle of radius r into pie wedges. Then snip and unroll the circle:

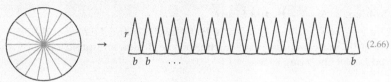

(2.66)

Use the unrolled picture and the knowledge that the circle's circumference is $2\pi r$ to show that its area is πr^2.

b. Now extend the argument to a sphere of radius r: Find its volume given that its surface area is $4\pi r^2$. (This method was invented by the ancient Greeks.)

Problem 2.22   Gain of an *LRC* circuit

Use the impedance of an inductor (Problem 2.19) to find the gain of the classic *LRC* circuit. In this configuration, in which the output voltage measured across the resistor, is the circuit a low-pass filter, a high-pass filter, or a band-pass filter?

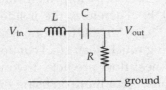

Problem 2.23   Continued fraction

Evaluate the continued fraction

$$1 + \cfrac{1}{1 + \cfrac{1}{1 + \cdots}}. \tag{2.67}$$

Compare this problem with Problem 2.8.

Problem 2.24   Exponent tower

Evaluate

$$\sqrt{2}^{\sqrt{2}^{\sqrt{2}^{\cdot^{\cdot^{\cdot}}}}}. \tag{2.68}$$

Here, a^{b^c} means $a^{(b^c)}$.

Problem 2.25   Coaxial cable termination

In physics and electronics laboratories around the world, the favorite way to connect equipment and transmit signals is with coaxial cable. The standard coaxial cable, RG-58/U, has a capacitance per length of 100 picofarads per meter and an inductance per length of 0.29 microhenries per meter. It can be modeled as a long inductor–capacitor ladder:

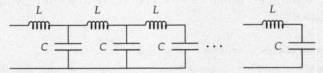

What resistance R placed at the end (in parallel with the last capacitor) makes the cable look like an infinitely long LC cable?

Problem 2.26   UNIX and Linux

Using Mike Gancarz's *The UNIX Philosophy* [17] and *Linux and the Unix Philosophy* [18], find examples of abstraction in the design and philosophy of the UNIX and Linux operating systems.

Part II

Discarding complexity without losing information

You've divided your hard problems into manageable pieces. You've found transferable, reusable ideas. When these tools are not enough and problems are still too complex, you need to discard complexity—the theme of our next three tools. They help us discard complexity without losing information. If a system contains a symmetry (Chapter 3)—or what is closely related, it is subject to a conservation law—using the symmetry greatly simplifies the analysis. Alternatively, we often do not care about a part of an analysis, because it is the same for all the objects in the analysis. To ignore those parts, we use proportional reasoning (Chapter 4). Finally, we can ensure that our equations do not add apples to oranges. This simple idea—dimensional analysis (Chapter 5)—greatly shrinks the space of possible solutions and helps us master complexity.

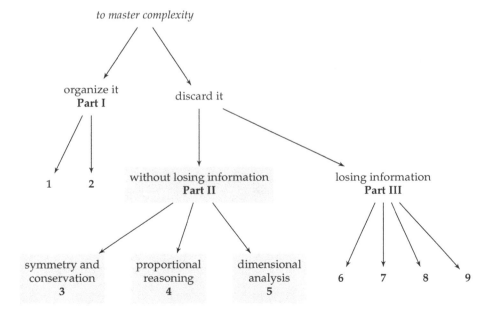

3
Symmetry and conservation

The rain is pouring down and shelter is a few hundred yards away. Do you get less wet by running? On the one hand, running means less time for raindrops to hit you. On the other hand, running means that the raindrops come toward you more directly and therefore more rapidly. The resolution is not obvious—until you apply the new tool of this chapter: symmetry and conservation. (In Section 3.1.1, we'll resolve this run-in-the-rain question.)

3.1 Invariants

We use symmetry and conservation whenever we find a quantity that, despite the surrounding complexity, does not change. This conserved quantity is called an invariant. Finding invariants simplifies many problems.

Our first invariant appeared unannounced in Section 1.2 when we estimated the carrying capacity of a lane of highway. The carrying capacity—the rate at which cars flow down the lane—depends on the separation between the cars and on their speed. We could have tried to estimate each quantity and then the carrying capacity. However, the separation between cars and the cars' speeds vary greatly, so these estimates are hard to make reliably.

Instead, we invoked the 2-second following rule. As long as drivers obey it, the separation between cars equals 2 seconds of driving. Therefore, one car flows by every 2 seconds—which is the lane's carrying capacity (in cars per time). By finding an invariant, we simplified a complex, changing process. *When there is change, look for what does not change!* (This wisdom is from Arthur Engel's *Problem-Solving Strategies* [12].)

3.1.1 To run or walk in the rain?

We'll practice with this tool by deciding whether to run or walk in the rain. It's pouring, your umbrella is sitting at home, and home lies a few hundred meters away.

▷ *To minimize how wet you become, should you run or walk?*

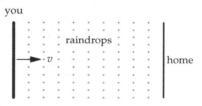

Let's answer this question with three simplifications. First, assume that there is no wind, so the rain is falling vertically. Second, assume that the rain is steady. Third, assume that you are a thin sheet: You have zero thickness along the direction toward your house (this approximation was more valid in my youth). Equivalently, your head is protected by a waterproof cap, so you do not care whether raindrops hit your head. You try to minimize just the amount of water hitting your front.

Your only degree of freedom—the only parameter that you get to choose—is your speed. A high speed leaves you in the rain for less time. However, it also makes the rain come at you more directly (more horizontally). But what remains constant, independent of your speed, is the volume of air that you sweep out. Because the rain is steady, that volume contains a fixed number of raindrops, independent of your speed. Only these raindrops hit your front. Therefore, you get equally wet, no matter your speed.

This surprising conclusion is another application of the principle that when there is change, look for what does not change. Here, we could change our speed by choosing to walk or run. Yet no matter what our speed, we sweep out the same volume of air—our invariant.

Because the conclusion of this invariance analysis, that it makes no difference whether you walk or run, is surprising, you might still harbor a nagging doubt. Surely running in the rain, which we do almost as a reflex, provide some advantage over a leisurely stroll.

▷ *Is it irrational to run to avoid getting wet?*

If you are infinitely thin, and are just a rectangle moving in the rain, then the preceding analysis applies: Whether you run or walk, your front will absorb the same number of raindrops. But most of us have a thickness, and the number of drops landing on our head depends on our speed. If your head is exposed and you care how many drops land on your head, then you should run. But if your head is covered, feel free to save your energy and enjoy the stroll. Running won't keep you any dryer.

3.1.2 Tiling a mouse-eaten chessboard

Often a good way to practice a new tool is on a mathematical problem. Then we do not add the complexity of the physical world to the problem of learning a new tool. Here, therefore, is a mathematical problem: a solitaire game.

A mouse comes and eats two diagonally opposite corners out of a standard, 8 × 8 chessboard. We have a box of rectangular, 2 × 1 dominoes.

▷ *Can these dominoes tile the mouse-eaten chessboard? In other words, can we lay down the dominoes to cover every square exactly once (with no empty squares and no overlapping dominoes)?*

Placing a domino on the board is one move in this solitaire game. For each move, you choose where to place the domino—so you have many choices at each move. The number of possible move sequences grows rapidly. Instead of examining all these sequences, we'll look for an invariant: a quantity unchanged by any move of the game.

Because each domino covers one white square and one black square, the following quantity I is invariant (remains fixed):

$$I = \text{uncovered black squares} - \text{uncovered white squares}. \qquad (3.1)$$

On a regular chess board, with 32 white squares and 32 black squares, the initial position has $I = 0$. The nibbled board has two fewer black squares, so I starts at $30 - 32 = -2$. In the winning position, all squares are covered, so $I = 0$. Because I is invariant, we cannot win: The dominoes cannot tile the nibbled board.

Each move in this game changes the chessboard. By finding what does not change, an invariant, we simplified the analysis.

Invariants are powerful partly because they are abstractions. The details of the empty squares—their exact locations—lie below the abstraction barrier. Above the barrier, we see only the excess of black over white squares. The abstraction contains all the information we need in order to know that we can never tile the chessboard.

Problem 3.1   Cube solitaire

A cube has numbers at each vertex; all vertices start at 0 except for the lower left corner, which starts at 1. The moves are all of the same form: Pick any edge and increment its two vertices by one. The goal of this solitaire game is to make all vertices multiples of 3.

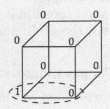

For example, picking the bottom edge of the front face and then the bottom edge of the back face, makes the following sequence of cube configurations:

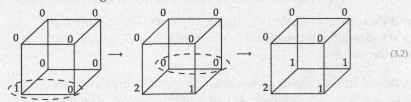

(3.2)

Although no configuration above wins the game, can you win with a different move sequence? If you can win, give a winning sequence. If you cannot win, prove that you cannot.

Hint: Create analogous but simpler versions of this game.

Problem 3.2   Triplet solitaire

Here is another solitaire game. Start with the triplet $(3, 4, 5)$. At each move, choose any two of the three numbers. Call the choices a and b. Then make the following replacements:

$$a \longrightarrow 0.8a - 0.6b;$$
$$b \longrightarrow 0.6a + 0.8b.$$

(3.3)

Can you reach $(4, 4, 4)$? If you can, give a move sequence; otherwise, prove that it is impossible.

Problem 3.3   Triplet-solitaire moves as rotations in space

At each step in triplet solitaire (Problem 3.2), there are three possible moves, depending on which pair of numbers from among a, b, and c you choose to replace. Describe each of the three moves as a rotation in space. That is, for each move, give the rotation axis and the angle of rotation.

Problem 3.4   Conical pendulum

Finding the period of a pendulum, even at small amplitudes, requires calculus because of the pendulum's varying speed. When there is change, look for what does not change. Accordingly, Christiaan Huygens (1629–1695), called "the most ingenious watchmaker of all time" [20, p. 79] by the great physicist Arnold Sommerfeld, analyzed the motion of a pendulum moving in a horizontal circle (a conical pendulum). Projecting its two-dimensional motion onto a vertical screen produces one-dimensional pendulum motion; thus, the period of the two-dimensional motion is the same as the period of one-dimensional pendulum motion! Use that idea to find the period of a pendulum (without calculus!).

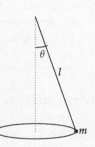

3.1.3 Logarithmic scales

In the solitaire game in Section 3.1.2, a move covered two chessboard squares with a domino. In the game of understanding the world, a frequent move is changing the system of units. As in solitaire, ask, "When such a move is made, what is invariant?" As an exam-

low end of hearing	20 Hz
piano middle C	262 Hz
highest piano C	4186 Hz
high end of hearing	20 kHz

ple to crystallize our thinking, here are frequencies related to human hearing. Let's graph them using kilohertz (kHz) as the unit. The frequencies then arrange themselves as follows:

Now let's change units from kilohertz to hertz (Hz)—and keep the 0, 10, and 20 labels at their current positions on the page. This change magnifies every spacing by a factor of 1000: The new 20 hertz is where 20 kilohertz was (about 4 inches or 10 centimeters to the right of the origin), and 20 kilohertz, the high end of human hearing, sits 100 meters to our right—far beyond the borders of the page. This new scale is not very useful.

However, we missed the chance to use an invariant: the *ratio* between frequencies. For example, the ratio between the upper end of human hearing (20 kilohertz) and middle C (262 hertz) is roughly 80. If we use a representation based on ratio rather than absolute difference, then the spacing between frequencies would not change even when we changed the unit.

We have such a representation: logarithms! On a logarithmic scale, a distance corresponds to a ratio rather than a difference. To see the contrast, let's place the numbers from 1 to 10 on a logarithmic scale.

The physical gap between 1 and 2 represents not their difference but rather their ratio, namely 2. According to my ruler, the gap is approximately 3.16 centimeters. Similarly, the physical gap between 2 and 3—approximately 1.85 centimeters—represents the smaller ratio 1.5. In contrast to their relative positions on a linear scale, 2 and 3 on a logarithmic scale are closer than 1 and 2 are. On a logarithmic scale, 1 and 2 have the same separation as 2 and 4: Both gaps represent a ratio of 2 and therefore have the same physical size (3.16 centimeters).

Problem 3.5   Practice with ratio thinking

On a logarithmic scale, how does the physical gap between 2 and 8 compare to the gap between 1 and 2? Decide based on your understanding of ratios; then check your reasoning by measuring both gaps.

Problem 3.6   More practice with ratio thinking

Is the gap between 1 and 10 less than twice, equal to twice, or more than twice the gap between 1 and 3? Decide based on your understanding of ratios; then check your reasoning by measuring both gaps.

Problem 3.7   Moving along a logarithmic scale

On the logarithmic scale in the text, the gap between 2 and 3 is approximately 1.85 centimeters. Where do you land if you start at 6 and move 1.85 centimeters rightward? Decide based on your understanding of ratios; then check your reasoning by using a ruler to find the new location.

Problem 3.8   Extending the scale to the right

On the logarithmic scale in the text, the gap between 1 and 10 is approximately 10.5 centimeters. If the scale were extended to include numbers up to 1000, how large would the gap between 10 and 1000 be?

Problem 3.9   Extending the scale to the left

If the logarithmic scale were extended to include numbers down to 0.01, how far to the left of 1 would you have to place 0.04?

On a logarithmic scale, the frequencies related to hearing arrange themselves as follows:

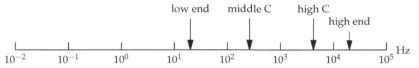

Changing the units to kilohertz just shifts all the frequencies, but leaves their relative positions invariant:

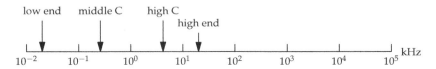

Problem 3.10   Acoustic energy fluxes

In acoustics, sound intensity is measured by energy flux, which is measured in decibels (dB)—a logarithmic representation of watts per square meter. On the decibel scale, 0 decibels corresponds to the reference level of 10^{-12} watts per square meter. Every 10 decibels (or 1 bel) represents an increase in energy flux of a factor of 10 (thus, 20 decibels represents a factor-of-100 increase in energy flux).

a. How many watts per square meter is 60 decibels (the sound level of normal conversation)?

b. Place the following energy fluxes on a decibel scale: 10^{-9} watts per square meter (an empty church), 10^{-2} watts per square meter (front row at an orchestra concert), and 1 watt per square meter (painfully loud).

Logarithmic scales offer two benefits. First, as we saw explicitly, logarithmic scales incorporate invariance. The second benefit was only implicit in the previous discussion: Logarithmic scales, unlike linear scales, allow us to represent a huge range. For example, if we include power-line hum (50 or 60 hertz) on the linear frequency scale on p. 61, we can hardly distinguish its position from 0 kilohertz. The logarithmic scale has no problem.

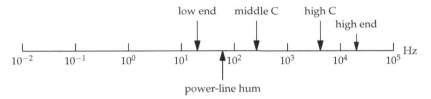

In our fallen world, benefits usually conflict. One usually has to sacrifice one benefit for another (speed for accuracy or justice for mercy). With logarithmic scales, however, we can eat our cake and have it.

Problem 3.11   Labeling a logarithmic scale

Let's make a scale to represent sizes in the universe, from protons (10^{-15} meters) to galaxies (10^{30} meters), with people and bacteria in between. With such a large range, we should use a logarithmic scale for size L. Which one of these two ways of labeling the scale is correct, and which way is nonsense?

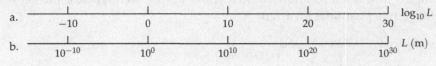

Logarithmic scales can make otherwise obscure symbolic calculations intuitive. An example is the geometric mean, which we used in Section 1.6 to make gut estimates:

$$\text{estimate} = \sqrt{\text{lower bound} \times \text{upper bound}}. \tag{3.4}$$

Geometric means also occur in the physical world. As you found in Problem 2.9, the distance d to the horizon, as seen from a height h above the Earth's surface, is

$$d \approx \sqrt{hD}, \tag{3.5}$$

where $D = 2R_{\text{Earth}}$ is the diameter of the Earth. Imagine a lifeguard sitting with his or her eyes at a height $h = 4$ meters above sea level. Then the distance to the horizon is

$$d \approx \left(\underbrace{4\,\text{m}}_{h} \times \underbrace{12\,000\,\text{km}}_{D} \right)^{1/2}. \tag{3.6}$$

To do the calculation, we convert 12 000 kilometers to 1.2×10^7 meters, calculate $4 \times 1.2 \times 10^7$, and compute the square root:

$$d \approx \sqrt{4 \times 1.2 \times 10^7}\,\text{m} \approx 7000\,\text{m} = 7\,\text{km}. \tag{3.7}$$

In this symbolic form with a square root, the calculation obscures the fundamental structure of the geometric mean. We first calculate hD, which is an area (and often contains, as it did here, a huge number). Then we take the square root to get back a distance. However, the area has nothing to do with the structure of the problem. It is merely a bookkeeping device.

Bookkeeping devices are useful; they are how you tell a computer what to calculate. However, to understand the calculation, we, as humans, should use a logarithmic scale to represent the distances. This scale captures the structure of the problem.

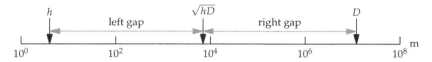

How can we describe the position of the geometric mean $\sqrt{hD}$?

The first clue is that the geometric mean, because it is a mean, lies somewhere between h and D. This property is not obvious from the calculation using the square root. To find where the geometric mean lies, mind the gaps. On a logarithmic scale, a gap represents the ratio of its endpoints. As shown in the table, the left and right gaps represent the

gap	endpoints	ratio
left	$h...\sqrt{hD}$	$\dfrac{\sqrt{hD}}{h} = \sqrt{\dfrac{D}{h}}$
right	$\sqrt{hD}...D$	$\dfrac{\sqrt{D}}{\sqrt{hD}} = \sqrt{\dfrac{D}{h}}$

same ratio, namely $\sqrt{D/h}$! Therefore, on the logarithmic scale, the geometric mean lies exactly halfway between h and D.

Based on this ratio representation, we can rephrase the geometric-mean calculation in a form that we can do mentally.

What distance is as large compared to 4 meters as 12 000 kilometers is compared to it?

For lack of imagination, my first guess is 1 kilometer. It's 12 000 times smaller than the diameter D (which is 12 000 kilometers), but only 250 times larger than the height h (4 meters). My guess of 1 kilometer is therefore somewhat too small.

How is a guess of 10 kilometers?

It's 2500 times larger than h, but only 1200 times smaller than D. I overshot slightly. How about 7 kilometers? It's roughly 1750 times larger than 4 meters, and roughly 1700 times smaller than 12 000 kilometers. Those gaps are close to each other, so 7 kilometers is the approximate geometric mean.

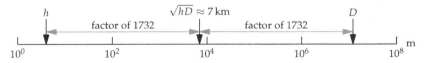

Similarly, when we make gut estimates, we should place our lower and upper estimates on a logarithmic scale. Our best gut estimate is then their midpoint. What a simple picture!

▷ *Should all quantities be placed on a logarithmic scale?*

No. An illustrative contrast is between size and position. Both quantities have the same units. But size ranges from 0 to ∞, whereas position ranges from −∞ to ∞. Position therefore cannot be placed on a logarithmic scale (where would you put −1 meter?). In contrast, size (a magnitude) belongs on a logarithmic scale. In general, location parameters, such as position, should not be placed on a logarithmic scale but magnitudes should.

3.2 From invariant to symmetry operation

In the preceding examples, we knew the moves of the game and sought the invariant. In the mouse-eaten chessboard (Section 3.1.2), the moves are putting down a 2×1 domino on two adjacent empty squares. The invariant was the difference between empty black and white squares. Often, however, the benefit of invariants lies in the other direction: You know the invariant and seek the moves that preserve it. These moves are called the symmetry operations or simply the symmetries.

We'll first examine this idea in a familiar situation: converting units (Section 3.2.1). Then we'll practice it on a sum solved by the three-year-old Carl Friedrich Gauss (Section 3.2.2) and then by finding maxima and minima (Section 3.2.3).

3.2.1 Converting units

We often convert a quantity from one unit system to another—for example, mass from English to metric units or prices from dollars to pounds or euros. A useful physical conversion is writing energy density—energy divided by amount of stuff—in useful units. Let's start with the reasonable energy unit for a chemical bond, namely the electron volt or eV (Section 2.1). Then a useful unit for energy density is

$$\frac{1\,\text{eV}}{\text{molecule}}.$$ (3.8)

This energy density is our invariant. As we convert from one unit system to another, our moves have to preserve the energy density.

▷ *What are the legal moves—the moves that preserve the energy density?*

The legal moves are all ways of multiplying by 1—for example, by

$$\frac{6 \times 10^{23} \text{ molecules}}{1 \text{ mol}} \quad \text{or} \quad \frac{1 \text{ mol}}{6 \times 10^{23} \text{ molecules}}. \tag{3.9}$$

Either quotient is a form of 1, because 1 mole is defined to be Avogadro's number of molecules, and Avogadro's number is 6×10^{23}. I carefully wrote "1 mol" with the number rather than simply as "mol." The more explicit form reminds us that "6×10^{23} molecules per mole" is shorthand for a quotient of two identical quantities: 6×10^{23} molecules and 1 mole.

Multiplying the energy density by the first form of 1 gives

$$\frac{1 \text{ eV}}{\cancel{\text{molecule}}} \times \frac{6 \times 10^{23} \cancel{\text{molecules}}}{1 \text{ mol}} = \frac{6 \times 10^{23} \text{ eV}}{\text{mol}}. \tag{3.10}$$

(If we had multiplied by the second form of 1, the units of molecules would have become molecules squared instead of canceling. The strike-through lines help us check that we got the desired units.) The giant exponent makes this form almost meaningless. To improve it, let's multiply by another form of 1, based on the definition of an electron volt. Two forms of 1 are

$$\frac{1.6 \times 10^{-19} \text{ J}}{1 \text{ eV}} \quad \text{or} \quad \frac{1 \text{ eV}}{1.6 \times 10^{-19} \text{ J}}. \tag{3.11}$$

Multiplying by the first form of 1 gives

$$\frac{1 \cancel{\text{eV}}}{\cancel{\text{molecule}}} \times \frac{6 \times 10^{23} \cancel{\text{molecules}}}{1 \text{ mol}} \times \frac{1.6 \times 10^{-19} \text{ J}}{1 \cancel{\text{eV}}} \approx \frac{10^2 \text{ kJ}}{\text{mol}}. \tag{3.12}$$

(A more exact value is 96 kilojoules per mole.) In the United States, energies related to food are stated in Calories, also known as kilocalories (roughly 4.2 kilojoules). In calorie units, the useful energy-density unit is

$$\frac{96 \cancel{\text{kJ}}}{1 \text{ mol}} \times \frac{1 \text{ kcal}}{4.2 \cancel{\text{kJ}}} \approx \frac{23 \text{ kcal}}{\text{mol}}. \tag{3.13}$$

▶ *Which form is more meaningful:* $23 \frac{kcal}{mol}$ *or* $\frac{23 \, kcal}{mol}$?

The forms are mathematically equivalent: You can multiply by 23 before or after dividing by a mole. However, they are not psychologically equivalent. The first form builds the abstraction of kilocalories per mole, and then says, "Here are 23 of them." In contrast, the second form gives us the energy for 1 mole, a human-sized amount. The second form is more meaningful.

Similarly, the speed of light c is commonly quoted as (approximately)

$$3 \times 10^8 \, \frac{\text{m}}{\text{s}}. \tag{3.14}$$

The psychologically fruitful alternative is

$$c = \frac{3 \times 10^8 \, \text{m}}{1 \, \text{s}}. \tag{3.15}$$

This form suggests that 300 million meters, at least for light, is the same as 1 second. With this idea, you can convert wavelength to frequency (Problem 3.14); with a slight extension, you can convert frequency to energy (Problem 3.15) and energy to temperature (Problem 3.16).

Problem 3.12   Absurd units

By multiplying by suitable forms of 1, convert 1 furlong per fortnight into meters per second.

Problem 3.13   Rainfall units

Rainfall, in nonmetric parts of the world, is sometimes measured in acre feet. By multiplying by suitable forms of 1, convert 1 acre foot to cubic meters. (One square mile is 640 acres.)

Problem 3.14   Converting wavelength to frequency

Convert green-light wavelength, 0.5 micrometers (0.5 μm), to a frequency in cycles per second (hertz or Hz).

Problem 3.15   Converting frequency to energy

Analogously to how you used the speed of light in Problem 3.14, use Planck's constant h to convert the frequency of green light to an energy in joules (J) and in electron volts (eV). This energy is the energy of a green-light photon.

Problem 3.16   Converting energy to temperature

Use Boltzmann's constant k_B to convert the energy of a green-light photon (Problem 3.15) to a temperature (in kelvin). This temperature, except for a factor of 3, is the surface temperature of the Sun!

Conversion factors need not be numerical. Insight often comes from symbolic factors. Here is an example from fluid flow. As we will derive in Section 5.3.2, the drag coefficient c_d is defined as the dimensionless ratio

$$c_d \equiv \frac{F_{\text{drag}}}{\frac{1}{2} \rho v^2 A}, \tag{3.16}$$

where ρ is the fluid density, v is the speed of the object moving in the fluid, and A is the object's cross-sectional area. To give this definition and ratio a physical interpretation, multiply it by d/d, where d is the distance that the object travels:

$$c_{\mathrm{d}} \equiv \frac{F_{\mathrm{drag}}d}{\frac{1}{2}\rho v^2 A d}. \tag{3.17}$$

The numerator, $F_{\mathrm{drag}}d$, is the work done or the energy consumed by drag. In the denominator, the product Ad is the volume of fluid displaced by the object, so ρAd is the corresponding mass of fluid. Therefore, the denominator is also

$$\frac{1}{2} \times \text{mass of fluid displaced} \times v^2. \tag{3.18}$$

The object's speed v is also approximately the speed given to the displaced fluid (which the object shoved out of its way). Therefore, the denominator is roughly

$$\frac{1}{2} \times \text{mass of fluid displaced} \times (\text{speed of displaced fluid})^2. \tag{3.19}$$

This expression is the kinetic energy given to the displaced fluid. The drag coefficient is therefore roughly the ratio

$$c_{\mathrm{d}} \sim \frac{\text{energy consumed by drag}}{\text{energy given to the fluid}}. \tag{3.20}$$

My tenth-grade chemistry teacher, Mr. McCready, told us that unit conversion was the one idea that we should remember from the entire course. Almost every problem in the chemistry textbook could be solved by unit conversion, which says something about the quality of the book but also about the power of the method.

3.2.2 Gauss's childhood sum

A classic example of going from the invariant to the symmetry is the following story of the young Carl Friedrich Gauss. Although maybe just a legend, the story is so instructive that it ought to be true. Once upon a time, when Gauss was 3 years old, his schoolteacher, wanting to occupy the students, assigned them to compute the sum

$$S = 1 + 2 + 3 + \cdots + 100, \tag{3.21}$$

and sat back to enjoy the break. In a few minutes, Gauss returned with an answer of 5050.

▷ *Was Gauss right? If so, how did he compute the sum so quickly?*

Gauss saw that the sum—the invariant—is unchanged when the terms are added backward, from highest to lowest:

$$1 + 2 + 3 + \cdots + 100 = 100 + 99 + 98 + \cdots + 1. \tag{3.22}$$

Then he added the two versions of the sum, the original and the reflected:

$$
\begin{aligned}
1 + \quad 2 + \quad 3 + \cdots + 100 &= \ S \\
+ \quad 100 + 99 + \ 98 + \cdots + \quad 1 &= \ S \\
\hline
101 + 101 + 101 + \cdots + 101 &= 2S.
\end{aligned} \tag{3.23}
$$

In this form, $2S$ is easy to compute: It contains 100 copies of 101. Therefore, $2S = 100 \times 101$, and $S = 50 \times 101$ or 5050—as the young Gauss claimed. He made the problem so simple by finding a symmetry: a transformation that preserved the invariant.

Problem 3.17   Number sum
Use Gauss's method to find the sum of the integers between 200 and 300 (inclusive).

Problem 3.18   Symmetry for algebra
Use symmetry to find the missing coefficients in the expansion of $(a - b)^3$:

$$(a - b)^3 = a^3 - 3a^2b +? \, ab^2 +? \, b^3. \tag{3.24}$$

Problem 3.19   Integrals
Evaluate these definite integrals. Hint: Use symmetry.

(a) $\displaystyle\int_{-10}^{10} x^3 e^{-x^2} \, dx$, (b) $\displaystyle\int_{-\infty}^{\infty} \frac{x^3}{1 + 7x^2 + 18x^8} \, dx$, and (c) $\displaystyle\int_0^{\infty} \frac{\ln x}{1 + x^2} \, dx$.

3.2.3 Finding maxima or minima

To practice finding the symmetry operation, we'll find the maximum of the function $6x - x^2$ without using calculus. Calculus is the elephant gun. It can solve many problems, but only after blasting them into the same form (smithereens). Avoiding calculus forces us to use more particular, but more subtle methods—such as symmetry. As Gauss did in summing $1 + 2 + \cdots + 100$, let's find a symmetry operation that preserves the essential feature of the problem—namely, the location of the maximum.

Symmetry implies moving around an object's pieces. Fortunately, our function $6x - x^2$ factors into pieces:

$$6x - x^2 = x(6 - x). \tag{3.25}$$

This form, along with the idea that multiplication is commutative, suggests the symmetry operation. For if the operation just swaps the two factors, replacing $x(6 - x)$ with $(6 - x)x$, it does not change the location of the maximum. (A parabola has exactly one maximum or minimum.)

The symmetry operation that makes the swap is

$$x \longleftrightarrow 6 - x. \tag{3.26}$$

It turns 2 into 4 (and vice versa) and 1 into 5 (and vice versa). The only value unchanged (left invariant) by the symmetry operation is $x = 3$. Therefore, $6x - x^2$ has its maximum at $x = 3$.

Geometrically, the symmetry operation reflects the graph of $6x - x^2$ through the line $x = 3$. By construction, this symmetry operation preserves the location of the maximum. Therefore, the maximum has to lie on the line $x = 3$.

We could have found this maximum in several other ways, so the use of symmetry might seem superfluous or like overkill. However, it warms us up for the following, more complicated use. The energy required to fly has two pieces: generating lift, which requires an energy A/v^2, and fighting drag, which requires an energy Bv^2. (A and B are constants that we estimate in Sections 3.6.2 and 4.6.1.)

$$E_{\text{flight}} = \frac{A}{v^2} + Bv^2. \tag{3.27}$$

To minimize fuel consumption, planes choose their cruising speed to minimize E_{flight}. More precisely, a cruising speed is selected, and the plane is designed so that this speed minimizes E_{flight}.

▷ *In terms of the constants A and B, what speed minimizes E_{flight}?*

Like the parabola $x(6-x)$, this energy has one extremum. For the parabola, the extremum was a maximum; here, it is a minimum. Also similar to the parabola, this energy has two pieces connected by a commutative operation. For the parabola, the operation was multiplication; here, it is addition. Continuing the analogy, if we find a symmetry operation that transposes the two pieces, then the speed preserved by the operation will be the minimum-energy speed.

Finding this symmetry operation is hard to do in one gulp, because it must transpose $1/v^2$ and v^2 and transpose A and B. These two difficulties suggest that we apply divide-and-conquer reasoning: Find a symmetry operation that transposes $1/v^2$ and v^2, and then modify so that it also transposes A and B.

To transpose $1/v^2$ and v^2, the symmetry operation is the following:

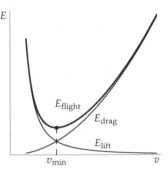

$$v \longleftrightarrow \frac{1}{v}. \tag{3.28}$$

Now let's restore one of the two constants and modify the symmetry operation so that it transposes A/v^2 and v^2:

$$v \longleftrightarrow \frac{\sqrt{A}}{v}. \tag{3.29}$$

Now let's restore the second constant, B, and find the full symmetry operation that transposes A/v^2 and Bv^2:

$$\sqrt{B}\,v \longleftrightarrow \frac{\sqrt{A}}{v}. \tag{3.30}$$

Rewriting it as a replacement for v, the symmetry operation becomes

$$v \longleftrightarrow \frac{\sqrt{A/B}}{v}. \tag{3.31}$$

This symmetry operation transposes the drag energy and lift energy, leaving the total energy E_{flight} unchanged. Solving for the speed preserved by the symmetry operation gives us the minimum-energy speed:

$$v_{\text{min}} = \left(\frac{A}{B}\right)^{1/4}. \tag{3.32}$$

In Section 4.6.1, once we find A and B in terms of the characteristics of the air (its density) and the plane (such as its wingspan), we can estimate the minimum-energy (cruising) speeds of planes and birds.

Problem 3.20   Solving a quadratic equation using symmetry

The equation $6x - x^2 + 7 = 0$ has a solution at $x = -1$. Without using the quadratic formula, find any other solutions.

3.3 **Physical symmetry**

For a physical application of symmetry, imagine a uniform metal sheet, perhaps aluminum foil, cut into the shape of a regular pentagon. Imagine that to the edges are attached heat sources and sinks—big blocks of metal at a fixed temperature—in order to hold the edges at the temperatures marked on the figure. After

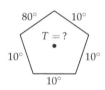

we wait long enough, the temperature distribution in the pentagon stops changing (comes to equilibrium).

▷ *Once the pentagon temperature equilibrates, what is the temperature at its center?*

A brute-force, analytic solution is difficult. Heat flow is described by the heat equation, a linear second-order partial-differential equation:

$$\kappa \nabla^2 T = \frac{\partial T}{\partial t},$$ (3.33)

where T is the temperature as a function of position and time, and κ (kappa) is the thermal diffusivity (which we will study in more detail in Chapter 7). But don't worry: You do not have to understand the equation, only that it is difficult to solve!

Once the temperature settles down, the time derivative becomes zero, and the equation simplifies to $\kappa \nabla^2 T = 0$. However, even this simpler equation has solutions only for simple shapes, and the

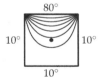

solutions are complicated. For example, the temperature distribution on the simpler square sheet is hardly intuitive (the figure shows contour lines spaced every 10°). For a pentagon, the temperature distribution is worse. However, because the pentagon is regular, symmetry might make the solution flow.

▷ *What is a useful symmetry operation?*

Nature, in the person of the heat equation, does not care about the direction of our coordinate system. Thus, rotating the pentagon about its center does not change the temperature at the center. Therefore, the following five orientations of the pentagon share the same central temperature:

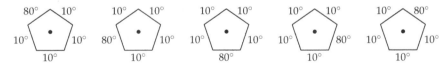

Like Gauss adding the two versions of his sum (Section 3.2.2), stack these sheets mentally and add the temperatures that lie on top of each other to make the temperature profile of a new super sheet (adding the temperatures is valid because the heat equation is linear).

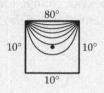

(3.34)

Each super edge contains one 80° edge and four 10° edges, for a temperature of 120°. The super sheet is a regular pentagon where all edges are at 120°. Therefore, the temperature throughout the sheet is 120°—including at the center. Because the symmetry operation has helped us construct a much easier problem, we did not have to solve the heat equation.

One more step tells us the temperature in the center of the original sheet. The symmetry operation rotates the pentagon about its center; when the plates are stacked, the centers align. Each center then contributes one-fifth of the 120° in the center, so the original central temperature is 24°.

To highlight the transferable ideas (abstractions), compare the symmetry solutions to Gauss's sum and to this temperature problem. First, both problems seem complicated. Gauss's sum has many terms, all different; the pentagon problem seems to require solving a difficult differential equation. Second, both problems contain a symmetry operation. In Gauss's sum, the symmetry operation reversed the order of the terms; for the pentagon, the symmetry operation rotates it by 72°. Finally, the symmetry operation leaves an important quantity unchanged. For Gauss's problem, this quantity is the sum; for the pentagon, it is the central temperature.

When there is change, look for what does not change. Look for invariants and the corresponding symmetries: the operations that preserve the invariants.

Problem 3.21   Symmetry solution for a square sheet
Here is the contour plot again of the temperature on a square sheet. The contour lines are separated by 10°. Use that information to label the temperature of each contour line. Based on the symmetry reasoning, what should the temperature at the center of the square be? Is this predicted temperature consistent with what is shown in the contour plot?

Problem 3.22   Simulating the heat equation

Using symmetry, we showed that the temperature at the center of the pentagon is the average of the temperatures of the
sides. Check the solution by simulating the heat equation
with a pentagonal boundary.

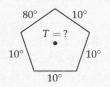

Problem 3.23   Shortest bisecting path

What is the shortest path that bisects an equilateral triangle into two equal areas?
Here are three examples of bisecting paths:

To set your problem-solving gears in motion, first rank these three bisecting paths
according to their lengths.

3.4   Box models and conservation

Invariance underlies a powerful everyday abstraction: box models. We already made a box model in Section 3.1.1, to decide whether to run or walk
in the rain. Now let's examine this method further. The simplest kind of
box contains a fixed amount of stuff—perhaps the volume of fluid or the
number of students at an ideal university (where every student graduates
in a fixed time). Then what goes into the box must come out. This conclusion seems simple, even simplistic, but it has wide application.

3.4.1   Supply and demand

For another example of a box model, return to our
estimate of US oil usage (Section 1.4). The flow
into the box—the push or the supply—is the imported and domestically produced oil. The flow

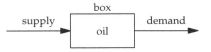

out of the box—the pull or the demand—is the oil usage. The estimate,
literally taken, asks for the supply (how much oil is imported and domestically produced). This estimate is difficult. Fortunately, as long as oil does
not accumulate in the box (for example, as long as oil is not salted away
in underground storage bunkers), then the amount of oil in the box is an
invariant, so the supply equals the demand. To estimate the supply, we accordingly estimated the demand. This conservation reasoning is the basis
of the following estimate of a market size.

▷ *How many taxis are there in Boston, Massachusetts?*

For many car-free years, I lived in an old neighborhood of Boston. I often rode in taxis and wondered about the size of the taxi market—in particular, how many taxis there were. This number seemed hard to estimate, because taxis are scattered throughout the city and hard to count.

The box contains the available taxi driving (mea-sured, for example, as time). It is supplied by taxi drivers. The demand is due to taxi users. As long as the supply and demand match, we can estimate the supply by estimating the demand.

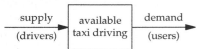

For estimating the demand, the starting point is that Boston has roughly 500 000 residents. As a gut estimate, each resident uses maybe one taxi per month, for a 15-minute ride: Boston taxis are expensive; unless one doesn't own a car, it's hard to imagine using them more often than once a month or for longer than 15 minutes. Then the demand is about 10^5 hours of taxi driving per month:

$$5 \times 10^5 \text{ residents} \times \frac{15 \text{ min}}{\text{resident month}} \times \frac{1 \text{ hr}}{60 \text{ min}} \approx \frac{10^5 \text{ hr}}{\text{month}}. \qquad (3.35)$$

▷ *How many taxi drivers will that many monthly hours support?*

Taxi drivers work long shifts, maybe 60 hours per week. I'd guess that they carry passengers one-half of that time: 30 hours per week or roughly 100 hours per month. At that pace, 10^5 hours of monthly demand could be supplied by 1000 taxi drivers or, assuming each taxi is driven by one driver, by 1000 taxis.

▷ *What about tourists?*

Tourists are very short-term Boston residents, mostly without cars. Tourists, although fewer than residents, use taxis more often and for longer than residents do. To include the tourist contribution to taxi demand, I'll simply double the previous estimate to get 2000 taxis.

This estimate can be checked reliably, because Boston is one of the United States cities where taxis may pick up passengers only with a special permit, the medallion. The number of medallions is strictly controlled, so medal-lions cost a fortune. For about 60 years, their number was restricted to 1525, until a 10-year court battle got the limit raised by 260, to about 1800.

The estimate of 2000 may seem more accurate than it deserves. However, chance favors the prepared mind. We prepared by using good tools: a box model and divide-and-conquer reasoning. In making your own estimates, have confidence in the tools, and expect your estimates to be at least half decent. You will thereby find the courage to start: Optimism oils the rails of estimation.

Problem 3.24 Differential equation for an *RC* circuit

Explain how a box model leads to the differential equation for the low-pass *RC* circuit of Section 2.4.4:

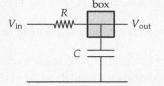

$$RC\frac{dV_{\text{out}}}{dt} + V_{\text{out}} = V_{\text{in}}. \qquad (3.36)$$

(Almost every differential equation arises from a box or conservation argument.)

Problem 3.25 Boston taxicabs tree

Draw a divide-and-conquer tree for estimating the number of Boston taxicabs. First draw it without estimates. Then include your estimates, and propagate the values toward the root.

Problem 3.26 Needles on a Christmas tree

Estimate the number of needles on a Christmas tree.

3.4.2 Flux

Flows, such as the demand for oil or the supply of taxi cabs, are rates—an amount per time. Physical flows are also rates, but they live in a geometry. This embedding allows us to define a related quantity: flux.

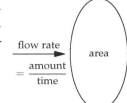

$$\text{flux of stuff} \equiv \frac{\text{rate}}{\text{area}} = \frac{\text{amount of stuff}}{\text{area} \times \text{time}}. \qquad (3.37)$$

For example, particle flux is the rate at which particles (say, molecules) pass through a surface perpendicular to the flow, divided by the area of the surface. Dividing by the surface area, an operation with no counterpart in nonphysical flows (for example, in the demand for taxicabs), makes flux more invariant and useful than rate. For if you double the surface area, you double the rate. This proportionality is not newsworthy, and usually doesn't add insight, only clutter. When there is change, look for what does not change: Even when the area changes, flux does not.

Problem 3.27   Rate versus amount
Explain why rate (amount per time) is more useful than amount.

Problem 3.28   What is current density?
What kind of flux (flux of what?) is current density (current per area)?

The definition of flux leads to a simple and important connection between flux and flow speed. Imagine a tube of stuff (for example, molecules) with cross-sectional area A. The stuff flows through the tube at a speed v.

▷ *In a time t, how much stuff leaves the tube?*

In the time t, the stuff in the shaded chunk, spanning a length vt, leaves the tube. This chunk has volume Avt. The amount of stuff in that volume is

$$\underbrace{\frac{\text{stuff}}{\text{volume}}}_{\text{density of stuff}} \times \underbrace{Avt}_{\text{volume}}. \qquad (3.38)$$

The amount of stuff per volume, the density of stuff, occurs so often that it usually gets a special symbol. When the stuff is particles, the density is labeled n for number density (in contrast to N for the number itself). When the stuff is charge or mass, the density is labeled ρ.

From the amount of stuff, we can find the flux:

$$\text{flux of stuff} = \frac{\text{amount of stuff}}{\text{area} \times \text{time}} = \frac{\text{density of stuff} \times \overbrace{\text{volume}}^{Avt}}{\underbrace{\text{area} \times \text{time}}_{At}}. \qquad (3.39)$$

The product At cancels, leaving the general relation

$$\text{flux of stuff} = \text{density of stuff} \times \text{flow speed}. \qquad (3.40)$$

As a particular example, when the stuff is charge (Problem 3.28), the flux of stuff becomes charge per time per area, which is current per area or current density. With that label for the flux, the general relation becomes

$$\underbrace{\text{current density}}_{J} = \underbrace{\text{charge density}}_{\rho} \times \underbrace{\text{flow speed}}_{v_{\text{drift}}}, \qquad (3.41)$$

where v_{drift} is the flow speed of the charge—which you will estimate in Problem 6.16 for electrons in a wire.

The general relation will be crucial in estimating the power required to fly (Section 3.6) and in understanding heat conduction (Section 7.4.2).

3.4.3 Average solar flux

An important flux is energy flux: the rate at which energy passes through a surface, divided by the area of the surface. Here, rate means energy per time, or power. Therefore, energy flux is power per area. An energy flux essential to life is the solar flux: the solar power per area falling on Earth. This flux drives most of our weather. At the top of the atmosphere, looking directly toward the sun, the flux is roughly $F = 1300$ watts per square meter.

However, this flux is not evenly distributed over the surface of the earth. The simplest reason is night and day. On the night side of the Earth, the solar flux is zero. More subtly, different latitudes have different solar fluxes: The equatorial regions are warmer than the poles because they receive more solar flux than the poles do.

▷ *What is the solar flux averaged over the whole Earth?*

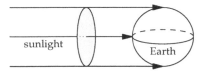

We can find the average flux using a box model (a conservation argument). Here is sunlight coming to the Earth (with parallel rays, because the Sun is so far away). Hold a disk with radius R_{Earth} perpendicular to the sunlight so that it blocks all sunlight that the Earth otherwise would get. The disk absorbs a power that we can find from the energy flux:

$$\text{power} = \text{energy flux} \times \text{area} = F\pi R_{\text{Earth}}^2, \tag{3.42}$$

where F is the solar flux. Now spread this power over the whole Earth, which has surface area $4\pi R_{\text{Earth}}^2$:

$$\text{average flux} = \frac{\text{power}}{\text{surface area}} = \frac{F\pi R_{\text{Earth}}^2}{4\pi R_{\text{Earth}}^2} = \frac{F}{4}. \tag{3.43}$$

Because one-half of the Earth is in night, averaging over the night and daylight parts of the earth accounts for a factor of 2. Therefore, averaging over latitudes must account for another factor of 2 (Problem 3.29).

Problem 3.29   Averaging solar flux over all latitudes
Integrate the solar flux over the whole sunny side of the Earth, accounting for the varying angles between the incident sunlight and the surface. Check that the result agrees with the result of the box model.

The result is roughly 325 watts per square meter. This average flux slightly overestimates what the Earth receives at ground level, because not all of the 1300 watts per square meter hitting the top of the atmosphere reaches the surface. Roughly 30 percent gets reflected near the top of the atmosphere (by clouds). The surviving amount is about 1000 watts per square meter. Averaged over the surface of the Earth, it becomes 250 watts per square meter (which then goes into the surface and the atmosphere), or approximately $F/5$, where F is the flux at the top of the atmosphere.

3.4.4   Rainfall

These 250 watts per square meter determine characteristics of our weather that are essential to life: the average surface temperature and the average rainfall. You get to estimate the surface temperature in Problem 5.43, once you learn the reasoning tool of dimensional analysis. Here, we will estimate the average rainfall.

If the box representing the atmosphere holds a fixed amount of water—and over a long timescale, the amount is constant (it is our invariant)—then what goes into the box must come out of the box.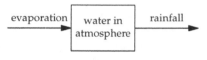
The inflow is evaporation; the outflow is rain. Therefore, to estimate the rainfall, estimate the evaporation—which is produced by the solar flux.

▷ *How much rain falls on Earth?*

Rainfall is measured as a height of water per time—typically, inches or millimeters per year. To estimate global average rainfall, convert the supply of solar energy to the supply of rainwater. In other words, convert power per area to height per time. The structure of the conversion is

$$\frac{\text{power}}{\text{area}} \times \frac{?}{?} = \frac{\text{height}}{\text{time}}, \tag{3.44}$$

where ?/? represents the conversion factor that we need to determine. To find what this conversion factor represents, we multiply both sides by area per power. The result is

$$\frac{?}{?} = \frac{\text{area} \times \text{height}}{\text{power} \times \text{time}} = \frac{\text{volume}}{\text{energy}}. \tag{3.45}$$

▷ *What physical quantity could this volume per energy be?*

We are trying to determine the amount of rain, so the volume in the numerator must be the volume of rain. Evaporating the water requires energy, so the energy in the denominator must be the energy required to evaporate that much water. The conversion factor is then the reciprocal of the heat of vaporization of water L_{vap}, but expressed as an energy per volume. In Section 1.7.3, we estimated L_{vap} as an energy per mass. To make it an energy per volume, just multiply by a mass per volume—namely, by ρ_{water}:

$$\underbrace{\frac{\text{energy}}{\text{mass}}}_{L_{vap}} \times \underbrace{\frac{\text{mass}}{\text{volume}}}_{\rho_{water}} = \underbrace{\frac{\text{energy}}{\text{volume}}}_{\rho_{water}L_{vap}}. \tag{3.46}$$

Our conversion factor, volume per energy, is the reciprocal, $1/\rho_{water}L_{vap}$. Our estimate for the average rainfall then becomes

$$\frac{\text{solar flux going to evaporate water}}{\rho_{water}L_{vap}}. \tag{3.47}$$

For the numerator, we cannot just use F, the full solar flux at the top of the atmosphere. Rather, the numerator incorporates several dimensionless ratios that account for the hoops through which sunlight must jump in order to reach the surface and evaporate water:

0.25	averaging the intercepted flux over the whole surface of the Earth (Section 3.4.3)
0.7	the fraction not reflected at the top of the atmosphere
0.7	of the sunlight not reflected, the fraction reaching the surface (the other 30 percent is absorbed in the atmosphere)
× 0.7	of the sunlight reaching the surface, the fraction reaching the oceans (the other 30 percent mostly warms land)
= 0.09	fraction of full flux F that evaporates water (including averaging the full flux over the whole surface)

The product of these four factors is roughly 9 percent. With $L_{vap} = 2.2 \times 10^6$ joules per kilogram (which we estimated in Section 1.7.3), our rainfall estimate becomes roughly

$$\frac{\overbrace{1300\,\text{W m}^{-2}}^{F} \times \overbrace{0.09}^{\text{fraction}}}{\underbrace{10^3\,\text{kg m}^{-3}}_{\rho_{water}} \times \underbrace{2.2 \times 10^6\,\text{J kg}^{-3}}_{L_{vap}}} \approx \frac{5.3 \times 10^{-8}\,\text{m}}{\text{s}}. \tag{3.48}$$

The length in the numerator is tiny and hard to perceive. Therefore, the common time unit for rainfall is a year rather than a second. To convert the rainfall estimate to meters per year, multiply by 1:

$$\frac{5.3 \times 10^{-8}\,\text{m}}{\cancel{s}} \times \frac{3 \times 10^{7}\,\cancel{s}}{1\,\text{yr}} \approx \frac{1.6\,\text{m}}{\text{yr}} \qquad (3.49)$$

(about 64 inches per year). Not bad: Including all forms of falling water, such as snow, the world average is 0.99 meters per year—slightly higher over the oceans and slightly lower over land (where it is 0.72 meters per year). The moderate discrepancy between our estimate and the actual average arises because some sunlight warms water without evaporating it. To reflect this effect, our table on page 81 needs one more fraction ($\approx 2/3$).

Problem 3.30   Solar luminosity
Estimate the solar luminosity—the power output of the Sun (say, in watts)—based on the solar flux at the top of the Earth's atmosphere.

Problem 3.31   Total solar power falling on Earth
Estimate the total solar power falling on the Earth's surface. How does it compare to the world energy consumption?

Problem 3.32   Explaining the difference between ocean and land rainfall
Why is the average rainfall over land lower than over the ocean?

3.4.5 Residence times

Because of evaporation, the atmosphere contains a lot of water: roughly 1.3×10^{16} kilograms—as vapor, liquid, and solid. This mass tells us the residence time: how long a water molecule remains in the atmosphere before it falls back to the Earth as precipitation (the overall name for rain, snow, or hail). The estimate will illustrate a new way to use box models.

Here is the box representing the water in the atmosphere (assumed to need only one box). The box is filled by evaporation and emptied by rainfall.

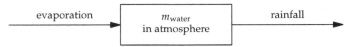

Imagine that the box is a water hose holding a mass m_{water}. How long does a water molecule take to get from one end of the hose to other? This time is the average time taken by a water molecule from evaporation until its return to the Earth as precipitation. In the box model, the time is the time to completely fill the box. This time constant, denoted τ, is

$$\tau = \frac{\text{mass of water in the atmosphere}}{\text{rate of inflow or outflow, as a mass per time}}. \tag{3.50}$$

The numerator is m_{water}. For the denominator, we convert rainfall, which is a speed (for example, in meters per year), to a mass flow rate (mass per time). Let's name the rainfall speed v_{rainfall}. The corresponding mass flux is, using our results from Section 3.4.2, $\rho_{\text{water}}v_{\text{rainfall}}$:

$$\text{mass flux} = \underbrace{\text{density}}_{\rho_{\text{water}}} \times \underbrace{\text{flow speed}}_{v_{\text{rainfall}}} = \rho_{\text{water}}v_{\text{rainfall}}. \tag{3.51}$$

Flux is flow per area, so we multiply mass flux by the Earth's surface area A_{Earth} to get the mass flow:

$$\text{mass flow} = \rho_{\text{water}}v_{\text{rainfall}}A_{\text{Earth}}. \tag{3.52}$$

At this rate, the fill time is

$$\tau = \frac{m_{\text{water}}}{\rho_{\text{water}}v_{\text{rainfall}}A_{\text{Earth}}}. \tag{3.53}$$

There are two ways to evaluate this time: the direct but less insightful method, and the less direct but more insightful method. Let's first do the direct method, so that we at least have an estimate for τ:

$$\tau \sim \frac{1.3 \times 10^{16}\,\text{kg}}{10^3\,\text{kg}\,\text{m}^{-3} \times 1\,\text{m}\,\text{yr}^{-1} \times 4\pi \times (6 \times 10^6\,\text{m})^2} \approx 2.5 \times 10^{-2}\,\text{yr}, \tag{3.54}$$

which is roughly 10 days. Therefore, after evaporating, water remains in the atmosphere for roughly 10 days.

For the less direct but more insightful method, notice which quantities are not reasonably sized—that is, not graspable by our minds—namely, m_{water} and A_{Earth}. But the combination $m_{\text{water}}/\rho_{\text{water}}A_{\text{Earth}}$ is reasonably sized:

$$\frac{m_{\text{water}}}{\rho_{\text{water}}A_{\text{Earth}}} \sim \frac{1.3 \times 10^{16}\,\text{kg}}{10^3\,\text{kg}\,\text{m}^{-3} \times 4\pi \times (6 \times 10^6\,\text{m})^2} \approx 2.5 \times 10^{-2}\,\text{m}. \tag{3.55}$$

This length, 2.5 centimeters, has a physical interpretation. If all water, snow, and vapor fell out of the atmosphere to the surface of the Earth, it would form an additional global ocean 2.5 centimeters deep.

Rainfall takes away 100 centimeters per year. Therefore, draining this ocean, with a 2.5-centimeter depth, requires 2.5×10^{-2} years or about 10 days. This time is, once again, the residence time of water in the atmosphere.

3.5 Drag using conservation of energy

A box model will next help us estimate drag forces. Drag, one of the most difficult subjects in physics, is also one of the most important forces in everyday life. If it weren't for drag, bicycling, flying, and driving would be a breeze. Because of drag, locomotion requires energy. Rigorously calculating a drag force requires solving the Navier–Stokes equations:

$$(\mathbf{v}\cdot\nabla)\mathbf{v} + \frac{\partial \mathbf{v}}{\partial t} = -\frac{1}{\rho}\nabla p + \nu\nabla^2\mathbf{v}. \tag{3.56}$$

They are coupled, nonlinear, partial-differential equations. You could read many volumes describing the mathematics to solve these equations. Even then, solutions are known only in a few circumstances—for example, a sphere moving slowly in a viscous fluid or moving at any speed in a nonviscous fluid. However, a nonviscous fluid—what Feynman [14, Section II-40-2], quoting John von Neumann, rightly disparages as "dry water"—is particularly irrelevant to real life because viscosity is the cause of drag, so a zero-viscosity solution predicts zero drag! Using a box model and conservation of energy is a simple and insightful alternative.

3.5.1 Box model for drag

We will first estimate the energy lost to drag as an object moves through a fluid, as in Section 3.2.1. From the energy, we will find the drag force. To quantify the problem, imagine pushing an object of cross-sectional 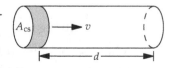 area A_{cs} at speed v for a distance d. The object sweeps out a tube of fluid. (The tube length d is arbitrary, but it will cancel out of the force.)

▷ *How much energy is consumed by drag?*

Energy is consumed because the object gives kinetic energy to the fluid (say, water or air); viscosity, as we will model in Section 6.4.4, then turns this energy into heat. The kinetic energy depends on the mass of the fluid and on the speed it is given. The mass of fluid in the tube is $\rho A_{cs}d$, where ρ is the fluid density. The speed imparted to the fluid is roughly the speed of the object, which is v. Therefore, the kinetic energy given to the fluid is roughly $\rho A_{cs}v^2 d$:

$$E_{\text{kinetic}} \sim \underbrace{\rho A_{cs}d}_{\text{mass}} \times v^2 = \rho A_{cs}v^2 d. \tag{3.57}$$

This calculation ignores the factor of one-half in the definition of kinetic energy. However, the other approximations, such as assuming that only the swept-out fluid is affected or that all the swept-out fluid gets speed v, are at least as inaccurate. For this rough calculation, there is little point in including the factor of one-half.

This kinetic energy is roughly the energy converted into heat. Therefore, the energy lost to drag is roughly $\rho A_{cs} v^2 d$. The drag force is then given by

$$\underbrace{\text{energy lost to drag}}_{\sim \rho A_{cs} v^2 d} = \underbrace{\text{drag force}}_{F_{drag}} \times \underbrace{\text{distance}}_{d} \,. \tag{3.58}$$

Now we can solve for the drag force:

$$F_{drag} \sim \rho A_{cs} v^2. \tag{3.59}$$

As expected, the arbitrary distance d has canceled out.

3.5.2 Testing the analysis with a home experiment

To test this analysis, try the following home experiment. Photocopy or print this page at 200 percent enlargement (a factor of 2 larger in width and height), cut out the template, and tape the two straight edges together to make a cone:

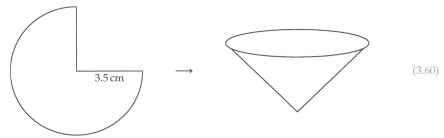

$$\tag{3.60}$$

3.5 cm

We could use many other shapes. However, a cone is easy to construct, and also falls without swishing back and forth (as a sheet of paper would) or flipping over (as long as you drop it point down).

We'll test the analysis by predicting the cone's terminal speed: that is, its steady speed while falling. When the cone is falling at this constant speed, its acceleration is zero, so the net force on it is, by Newton's second law, also zero. Thus, the drag force F_{drag} equals the cone's weight mg (where m is the cone's mass and g is the gravitational acceleration):

$$\rho_{air} v^2 A_{cs} \sim mg. \tag{3.61}$$

The terminal speed thus reveals the drag force. (Even though the drag force equals the weight, the left side is only an approximation to the drag force, so we connect the left and right sides with a single approximation sign ~.) The terminal speed v_{term} is then

$$v_{\text{term}} \sim \sqrt{\frac{mg}{A_{\text{cs}}\rho_{\text{air}}}}. \tag{3.62}$$

The mass of the cone is

$$m = A_{\text{paper}} \times \underbrace{\text{areal density of paper}}_{\sigma_{\text{paper}}}. \tag{3.63}$$

Here, A_{paper} is the area of the cone template; and the areal density σ_{paper}, named in analogy to the regular (volume) density, is the mass per area of paper. Although areal density seems like a strange quantity to define, it is used worldwide to describe the "weight" of different papers.

The quotient m/A_{cs} contains the ratio $A_{\text{paper}}/A_{\text{cs}}$. Rather than estimating both areas and finding their ratio, let's estimate the ratio directly.

▷ *How does the cross-sectional area A_{cs} compare to the area of the paper?*

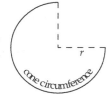

Because the cone's circumference is three-quarters of the circumference of the full circle, its cross-sectional radius is three-quarters of the radius r of the template circle. Therefore,

$$A_{\text{cs}} = \pi\left(\frac{3}{4}r\right)^2. \tag{3.64}$$

Because the template is three-quarters of a full circle,

$$A_{\text{paper}} = \frac{3}{4}\pi r^2. \tag{3.65}$$

The paper area has one factor of three-quarters, whereas the cross-sectional area has two factors of three-quarters, so $A_{\text{paper}}/A_{\text{cs}} = 4/3$. Now v_{term} simplifies as follows:

$$v_{\text{term}} \sim \left(\frac{\overbrace{A_{\text{paper}}\,\sigma_{\text{paper}}}^{m} \times g}{A_{\text{cs}}\,\rho_{\text{air}}}\right)^{1/2} = \left(\frac{\frac{4}{3}\sigma_{\text{paper}}\,g}{\rho_{\text{air}}}\right)^{1/2}. \tag{3.66}$$

The only unfamiliar number is the areal density σ_{paper}, the mass per area of paper. Fortunately, areal density is used commercially, so most reams of printer paper state their areal density: typically, 80 grams per square meter.

▷ *Is this σ_{paper} consistent with the estimates for a dollar bill in Section 1.1?*

There we estimated that the thickness t of a dollar bill, or of paper in general, is approximately 0.01 centimeters. The regular (volume) density ρ would then be 0.8 grams per cubic centimeter:

$$\rho_{paper} = \frac{\sigma_{paper}}{t} \approx \frac{80\,\text{g}\,\text{m}^{-2}}{10^{-2}\,\text{cm}} \times \frac{1\,\text{m}^2}{10^4\,\text{cm}^2} = 0.8\,\frac{\text{g}}{\text{cm}^3}. \tag{3.67}$$

This density, slightly below the density of water, is a good guess for the density of paper, which originates as wood (which barely floats on water). Therefore, our estimate in Section 1.1 is consistent with the proposed areal density of 80 grams per square meter.

After putting in the constants, the cone's terminal speed is predicted to be roughly 0.9 meters per second:

$$v_{term} \sim \left(\frac{4}{3} \times \frac{\overbrace{8 \times 10^{-2}\,\text{kg}\,\text{m}^{-2}}^{\sigma_{paper}} \times \overbrace{10\,\text{m}\,\text{s}^{-2}}^{g}}{\underbrace{1.2\,\text{kg}\,\text{m}^{-3}}_{\rho_{air}}} \right)^{1/2} \approx 0.9\,\text{m}\,\text{s}^{-1}. \tag{3.68}$$

To test the prediction and, with it, the analysis justifying it, I held the cone slightly above my head, from about 2 meters high. After I let the cone go, it fell for almost exactly 2 seconds before it hit the ground—for a speed of roughly 1 meter per second, very close to the prediction. Box models and conservation triumph again!

3.5.3 Cycling

In introducing the analysis of drag, I said that drag is one of the most important physical effects in everyday life. Our analysis of drag will now help us understand the physics of a fantastically efficient form of locomotion—cycling (for its efficiency, see Problem 3.34).

▷ *What is the world-record cycling speed?*

The first task is to define the kind of world record. Let's analyze cycling on level ground using a regular bicycle, even though faster speeds are possible riding downhill or on special bicycles. In bicycling, energy goes into rolling resistance, friction in the chain and gears, and air drag. The importance of drag rises rapidly with speed, due to the factor of v^2 in the drag force, so at high-enough speeds drag is the dominant consumer of energy.

Therefore, let's simplify the analysis by assuming that drag is the only consumer of energy. At the maximum cycling speed, the power consumed by drag equals the maximum power that the rider can supply. The problem therefore divides into two estimates: the power consumed by drag (P_{drag}) and the power that an athlete can supply (P_{athlete}).

Power is force times velocity:

$$\text{power} = \frac{\text{energy}}{\text{time}} = \frac{\text{force} \times \text{distance}}{\text{time}} = \text{force} \times \text{velocity}. \tag{3.69}$$

Therefore,

$$P_{\text{drag}} = F_{\text{drag}} v_{\text{max}} \sim \rho v^3 A_{\text{cs}}. \tag{3.70}$$

Setting $P_{\text{drag}} = P_{\text{athlete}}$ allows us to solve for the maximum speed:

$$v_{\text{max}} \sim \left(\frac{P_{\text{athlete}}}{\rho_{\text{air}} A_{\text{cs}}} \right)^{1/3}, \tag{3.71}$$

where A_{cs} is the cyclist's cross-sectional area. In Section 1.7.2, we estimated P_{athlete} as 300 watts. To estimate the cross-sectional area, divide it into a width and a height. The width is a body width—say, 0.4 meters. A racing cyclist crouches, so the height is roughly 1 meter rather than a full 2 meters. Then A_{cs} is roughly 0.4 square meters.

Plugging in the numbers gives

$$v_{\text{max}} \sim \left(\frac{300\,\text{W}}{1\,\text{kg}\,\text{m}^{-3} \times 0.4\,\text{m}^2} \right)^{1/3}. \tag{3.72}$$

▷ *That formula, with its mix of watts, meters, and seconds, looks suspicious. Are the units correct?*

Let's translate a watt stepwise into meters, kilograms, and seconds, using the definitions of a watt, joule, and newton:

$$W \equiv \frac{J}{s}, \qquad J \equiv N\,m, \qquad N \equiv \frac{\text{kg}\,\text{m}}{s^2}. \tag{3.73}$$

The three definitions are represented in the next divide-and-conquer tree, one definition at each nonleaf node. Propagating the leaves toward the root gives us the following expression for the watt in terms of meters, kilograms, and seconds (the fundamental units in the SI system):

$$W \equiv \frac{\text{kg}\,\text{m}^2}{s^3}. \tag{3.74}$$

The units in v_{max} become

$$\left(\frac{\dfrac{\text{W}}{\cancel{\text{kg}}\ \cancel{\text{m}^2}\ \text{s}^{-3}}}{\cancel{\text{kg}}\ \text{m}^{-3} \times \cancel{\text{m}^2}} \right)^{1/3} = \left(\frac{\text{s}^{-3}}{\text{m}^{-3}} \right)^{1/3}. \qquad (3.75)$$

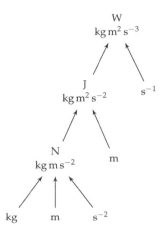

The kilograms cancel, as do the square meters. The cube root then contains only meters cubed over seconds cubed; therefore, the units for v_{max} are meters per second.

Let's estimate how many meters per second. Don't let the cube root frighten you into using a calculator. We can do the arithmetic mentally, if we massage (adjust) the numbers slightly. If only the power were 400 watts (or instead the area were 0.3 square meters)! Instead of wishing, make it so—and don't worry about the loss of accuracy: Because we have neglected the drag coefficient, our speed will be approximate anyway. Then the cube root becomes an easy calculation:

$$v_{\text{max}} \sim \left(\frac{\cancel{300}\ 400\ \text{W}}{1\ \text{kg}\,\text{m}^{-3} \times 0.4\ \text{m}^2} \right)^{1/3} = (1000)^{1/3}\ \text{m}\,\text{s}^{-1} = 10\ \text{m}\,\text{s}^{-1}. \qquad (3.76)$$

In more familiar units, the record speed is 22 miles per hour or 36 kilometers per hour. As a comparison, the world 1-hour record—cycling as far as possible in 1 hour—is 49.7 kilometers or 30.9 miles, set in 2005 by Ondřej Sosenka. Our prediction, based on the conservation analysis of drag, is roughly 70 percent of the actual value.

▷ *How can such an estimate be considered useful?*

High accuracy often requires analyzing and tracking many physical effects. The calculations and bookkeeping can easily obscure the most important effect and its core idea, costing us insight and understanding. Therefore, almost everywhere in this book, the goal is an estimate within a factor of 2 or 3. That level of agreement is usually enough to convince us that our model contains the situation's essential features.

Here, our predicted speed is only 30 percent lower than the actual value, so our model of the energy cost of cycling must be broadly correct. Its main error arises from the factor of one-half that we ignored when estimating the drag force—as you can check by doing Problem 3.33.

3.5.4   **Fuel efficiency of automobiles**

Bicycles, in many places, are overshadowed by cars. From the analysis of drag, we can estimate the fuel consumption of a car (at highway speeds). Most of the world measures fuel consumption in liters of fuel per 100 kilometers of driving. The United States uses the reciprocal quantity, fuel efficiency—distance per volume of fuel—measured in miles per US gallon. To develop unit flexibility, we'll do the calculation using both systems.

For a bicycle, we compared powers: the power consumed by drag with the power supplied by an athlete. For a car, we are interested in the fuel consumption, which is related to the energy contained in the fuel. Therefore, we need to compare energies. For cars traveling at highway speeds, most of the energy is consumed fighting drag. Therefore, the energy consumed by drag equals the energy supplied by the fuel.

Driving a distance d, which will be 100 kilometers, consumes an energy

$$E_{\text{drag}} \sim \rho_{\text{air}} v^2 A_{\text{cs}}\, d. \tag{3.77}$$

The fuel provides an energy

$$E_{\text{fuel}} \sim \underbrace{\text{energy density}}_{\mathcal{E}_{\text{fuel}}} \times \underbrace{\text{fuel mass}}_{\rho_{\text{fuel}} V_{\text{fuel}}} = \mathcal{E}_{\text{fuel}}\, \rho_{\text{fuel}} V_{\text{fuel}}. \tag{3.78}$$

Because $E_{\text{fuel}} \sim E_{\text{drag}}$, the volume of fuel required is given by

$$V_{\text{fuel}} \sim \frac{E_{\text{drag}}}{\rho_{\text{fuel}} \mathcal{E}_{\text{fuel}}} \sim \underbrace{\frac{\rho_{\text{air}}}{\rho_{\text{fuel}}} \frac{v^2 A_{\text{cs}}}{\mathcal{E}_{\text{fuel}}}}_{A_{\text{consumption}}}\, d. \tag{3.79}$$

Because the left-hand side, V_{fuel}, is a volume, the complicated factor in front of the travel distance d must be an area. Let's make an abstraction by naming this area. Because it is proportional to fuel consumption, a self-documenting name is $A_{\text{consumption}}$. Now let's estimate the quantities in it.

1. *Density ratio ρ_{air}/ρ_{fuel}.* The density of gasoline is similar to the density of water, so the density ratio is roughly 10^{-3}.

2. *Speed v.* A highway speed is roughly 100 kilometers per hour (60 miles per hour) or 30 meters per second. (A useful approximation for Americans is that 1 meter per second is roughly 2 miles per hour.)

3. *Energy density $\mathcal{E}_{fuel}$.* We estimated this quantity Section 2.1 as roughly 10 kilocalories per gram or 40 megajoules per kilogram.

4. *Cross-sectional area A_{cs}.* A car's cross section is about 2 meters across by 1.5 meters high, so $A_{cs} \sim 3$ square meters.

car
(cross section)

1.5 m

2 m

With these values,

$$A_{\text{consumption}} \sim 10^{-3} \times \frac{\overbrace{10^3\,\mathrm{m^2\,s^{-2}}}^{v^2} \times \overbrace{3\,\mathrm{m^2}}^{A_{cs}}}{\underbrace{4 \times 10^7\,\mathrm{J\,kg^{-1}}}_{\mathcal{E}_{\text{fuel}}}} \approx 8 \times 10^{-8}\,\mathrm{m^2}. \qquad (3.80)$$

To find the fuel consumption, which is the volume of fuel per 100 kilometers of driving, simply multiply $A_{\text{consumption}}$ by $d = 100$ kilometers or 10^5 meters, and then convert to liters to get 8 liters per 100 kilometers:

$$V_{\text{fuel}} \approx \underbrace{8 \times 10^{-8}\,\mathrm{m^2}}_{A_{\text{consumption}}} \times \underbrace{10^5\,\mathrm{m}}_{d} \times \frac{10^3\,\ell}{1\,\mathrm{m^3}} = 8\,\ell. \qquad (3.81)$$

For the fuel efficiency, we use $A_{\text{consumption}}$ in the form $d = V_{\text{fuel}}/A_{\text{consumption}}$ to find the distance traveled on 1 gallon of fuel, converting the gallon to cubic meters:

$$d \sim \underbrace{\frac{\overbrace{1\,\text{gallon}}^{V_{\text{fuel}}}}{8 \times 10^{-8}\,\mathrm{m^2}}}_{A_{\text{consumption}}} \times \frac{4\,\ell}{1\,\text{gallon}} \times \frac{10^{-3}\,\mathrm{m^{\cancel{3}}}}{1\,\ell} = 5 \times 10^4\,\mathrm{m}. \qquad (3.82)$$

The struck-through exponent of 3 in the $\mathrm{m}^{\cancel{3}}$ indicates that the cubic meters became linear meters, as a result of cancellation with the $\mathrm{m^2}$ in the $A_{\text{consumption}}$. The resulting distance is 50 kilometers or 30 miles. The predicted fuel efficiency is thus roughly 30 miles per gallon.

This prediction is very close to the official values. For example, for new midsize American cars (in 2013), fuel efficiencies of nonelectric vehicles range from 16 to 43 miles per gallon, with a mean and median of 30 miles per gallon (7.8 liters per 100 kilometers).

The fuel-efficiency and fuel-consumption predictions are far more accurate than we deserve, given the many approximations! For example, we ignored all energy losses except for drag. We also used a very rough drag force $\rho_{\text{air}} v^2 A_{cs}$, derived from a reasonable but crude conservation argument. Yet, like Pippi Longstocking, we came out right anyway.

▷ *What went right?*

The analysis neglects two important factors, so such accuracy is possible only if these factors cancel. The first factor is the dimensionless constant hidden in the single approximation sign of the drag force:

$$F_{\text{drag}} \sim \rho_{\text{air}} A_{\text{cs}} v^2. \tag{3.83}$$

Including the dimensionless prefactor (shown in gray), the drag force is

$$F_{\text{drag}} = \frac{1}{2} c_{\text{d}} \, \rho_{\text{air}} A_{\text{cs}} v^2, \tag{3.84}$$

where c_{d} is the drag coefficient (introduced in Section 3.2.1). The factor of one-half comes from the one-half in the definition of kinetic energy. The drag coefficient is the remaining adjustment, and its origin is the subject of Section 5.3.2. For now, we need to know only that, for a typical car, $c_{\text{d}} \approx 1/2$. Therefore, the dimensionless prefactor hidden in the single approximation sign is approximately $1/4$.

▷ *Based on this more accurate drag force, will cars use more or less than 8 liters of fuel per 100 kilometers?*

Including the $c_{\text{d}}/2$ reduces the drag force and the fuel consumption by a factor of 4. Therefore, cars would travel 120 miles on 1 gallon of fuel or would consume only 2 liters per 100 kilometers. This more careful prediction is far too optimistic—and far worse than the original, simpler estimate.

▷ *What other effect did we neglect?*

The engine efficiency—a typical combustion engine, whether gasoline or human, is only about 25 percent efficient: An engine extracts only one-quarter of the combustion energy in the fuel; the remaining three-quarters turns into heat without doing mechanical work. Including this factor increases our estimate of the fuel consumption by a factor of 4.

The engine efficiency and the more accurate drag force together give the following estimate of the fuel consumption, with the new effect in gray:

$$V_{\text{fuel}} \approx \frac{\frac{1}{2} c_{\text{d}}}{0.25} \times \frac{\rho_{\text{air}} v^2 A_{\text{cs}}}{\rho_{\text{fuel}} \mathcal{E}_{\text{fuel}}} d. \tag{3.85}$$

The 0.25 in the denominator, from the engine efficiency, cancels the $\frac{1}{2} c_{\text{d}}$ in the numerator. That is why our carefree estimate, which neglected both factors, was so accurate. The moral, which I intend only half jokingly: Neglect many factors, so that the errors can cancel one another out.

Problem 3.33   Adjusting the cycling record
Our estimate of the world 1-hour record as roughly 35 kilometers (Section 3.5.3) ignored the drag coefficient. For a bicyclist, $c_d \approx 1$. Will including the drag coefficient improve or worsen the prediction in comparison with the actual world record (roughly 50 kilometers)? Answer that question before making the new prediction! What is the revised prediction?

Problem 3.34   Bicyclist fuel efficiency
What is the fuel consumption and efficiency of a bicyclist powered by peanut butter? Express your estimate as an efficiency (miles per gallon of peanut butter) and a consumption (liters of peanut butter per 100 kilometers). How does a bicycle compare with a car?

3.6 Lift using conservation of momentum

If drag is a drag, our next force, which is the companion to drag, should lift our spirits. Using conservation and box models, we will estimate the power required to generate lift. There are two main cases: hovering flight—for example, a hummingbird—and forward flight. Compared to forward flight, hovering flight has one fewer parameter (there is no forward velocity), so let's begin with its analysis, for a bird of mass m.

3.6.1 Hovering: Hummingbirds

▷ *How much power does a hummingbird require to hover?*

Hovering demands power because a hummingbird has weight: The Earth, via the gravitational field, supplies the hummingbird with downward momentum. The Earth therefore loses downward momentum or, equivalently, acquires upward momentum. (Thus, the Earth accelerates upward toward the hummingbird, although very, very slowly.) This flow of momentum can be tracked with a box model. Let's draw the box around the Earth–hummingbird

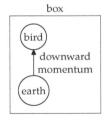

system and imagine the system as the whole universe. The box contains a fixed (constant) amount of downward momentum, so the gravitational field can transfer downward momentum only within the box. In particular, the field transfers downward momentum from the Earth to the hummingbird. This picture is a fancy way of saying that the Earth exerts a downward force on the hummingbird, but the fancy way shows us what the hovering hummingbird must do to stay aloft.

If the hummingbird keeps this downward momentum, it would accumulate downward speed—and crash to the ground. Fortunately, the box has one more constituent: the fluid (air). The hummingbird gives the downward momentum to the air: It flaps its wings and sends air downward. Lift, like drag, requires a fluid. (The air pushes down on the Earth, returning the downward momentum that the Earth lost via the gravitational field. Thus, the Earth does not accelerate.)

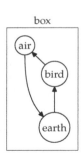

▶ *How much power is required to send air downward?*

Power is force times speed. The force is the gravitational force mg that the hummingbird is unloading onto the air. Estimating the air's downward speed v_z requires careful thought about the flow of momentum. The air carries the downward momentum supplied to the hummingbird. The momentum supply (a momentum rate or momentum per time) is the force mg: Force is simply momentum per time. Because momentum *flux* is momentum per time per area,

$$mg = \text{momentum flux} \times \text{area}. \tag{3.86}$$

When we first studied flux, in Section 3.4.2, we derived that

$$\text{flux of stuff} = \text{density of stuff} \times \text{flow speed}. \tag{3.87}$$

Because our stuff is momentum, this relation takes the particular form

$$\text{momentum flux} = \text{momentum density} \times \text{flow speed}. \tag{3.88}$$

Substituting this momentum flux into mg = momentum flux × area,

$$mg = \text{momentum density} \times \text{flow speed} \times \text{area}. \tag{3.89}$$

Momentum density is momentum ($m_{\text{air}}v_z$) per volume, so it is $\rho_{\text{air}}v_z$. The flow speed is v_z. Thus,

$$mg = \rho_{\text{air}}v_z \times v_z \times \text{area} = \rho_{\text{air}}v_z^2 \times \text{area}. \tag{3.90}$$

To complete this equation, so that it gives us the downward velocity v_z, we need to estimate the area. It is the area over which the hummingbird directs air downward. It is roughly L^2, where L is the wingspan (wingtip to wingtip). Even though the wings do not fill that entire area, the relevant area is still L^2, because the wings disturb air in a region whose size is comparable to their longest dimension. (For this reason, high-efficiency planes, such as gliders, have very long wings.)

Using L^2 as the estimate for the area, we get

$$mg \sim \rho_{\text{air}} v_z^2 L^2, \tag{3.91}$$

so the downward velocity is

$$v_z \sim \sqrt{\frac{mg}{\rho_{\text{air}} L^2}}. \tag{3.92}$$

With this downward velocity and with the downward force mg, the power P (not to be confused with momentum!) is

$$P = F v_z \sim mg \sqrt{\frac{mg}{\rho_{\text{air}} L^2}}. \tag{3.93}$$

Let's estimate this power for an actual hummingbird: the Calliope hummingbird, the smallest bird in North America. Its two relevant characteristics are the following:

$$\text{wingspan } L \approx 11 \, \text{cm,} \tag{3.94}$$
$$\text{mass } m \approx 2.5 \, \text{g.}$$

As the first step in estimating the hovering power, we'll estimate the downward air speed using our formula for v_z. The result is that, to stay aloft, the hummingbird sends air downward at roughly 1.3 meters per second:

$$v_z \sim \left(\frac{\overbrace{2.5 \times 10^{-2} \, \text{N}}^{mg}}{\underbrace{1.2 \, \text{kg} \, \text{m}^{-3}}_{\rho_{\text{air}}} \times \underbrace{1.2 \times 10^{-2} \, \text{m}^2}_{L^2}} \right)^{1/2} \approx 1.3 \, \text{m s}^{-1}. \tag{3.95}$$

The resulting power consumption is roughly 30 milliwatts:

$$P \sim \underbrace{2.5 \times 10^{-2} \, \text{N}}_{mg} \times \underbrace{1.3 \, \text{m s}^{-1}}_{v_z} \approx \underbrace{3 \times 10^{-2} \, \text{W}}_{30 \, \text{mW}}. \tag{3.96}$$

(Because animal metabolism, like a car engine, is only about 25 percent efficient, the hummingbird needs to eat food at a rate corresponding to 120 milliwatts.)

This power seems small: Even an (incandescent) flashlight bulb, for example, requires a few watts. However, as a power per mass, it looks more significant:

$$\frac{P}{m} \sim \frac{3 \times 10^{-2} \, \text{W}}{2.5 \times 10^{-3} \, \text{kg}} \approx 10 \, \frac{\text{W}}{\text{kg}}. \tag{3.97}$$

In comparison, the world-champion cyclist Lance Armstrong, with one of the highest human power outputs, was measured to have a power output of

7 watts per kilogram (Section 1.7.2). However, for a chemically unenhanced world-class athlete, 5 watts per kilogram is a more typical value. According to our estimates, hummingbird muscles should be twice as powerful as this world-class human value! Even for a small bird, hovering is hard work.

Problem 3.35   Fueling hovering
How much nectar must a hummingbird drink, as a fraction of its body mass, in order to hover for its working day (roughly 8 hours)? By mass, nectar is roughly 50 percent sugar.

Problem 3.36   Human hovering
How much power would a person have to put out in order to hover by flapping his or her arms?

3.6.2 Lift in forward flight

Now that we understand the fundamental mechanism of lift—discarding downward momentum by giving it to the air—we are ready to study forward flight: the flight of a migrating bird or of a plane. Forward flight is more complicated than hovering because forward flight has two velocities: the plane's forward velocity v and the downward component v_z of the air's veloc- 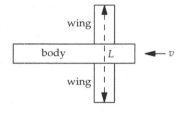 ity after passing around the wing. In forward flight, v_z depends not only on the plane's weight and wingspan, but also on the plane's forward velocity.

To stay aloft, the plane, like the hummingbird, must deflect air downward.

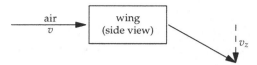

The wing does this magic using complicated fluid mechanics, but we need not investigate it. All the gymnastics are hidden in the box. We need just the downward velocity v_z required to keep the plane aloft, and the power required to give the air that much downward velocity. The power is, as with hovering, mgv_z. However, the downward velocity v_z is not the same as in hovering.

It is determined by a slightly different momentum-flow diagram. It shows the air flow before and after it meets the wing.

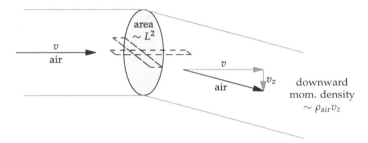

Before the air reaches the wing (the left tube), the air has zero downward momentum. As in the analysis of hovering flight, the Earth supplies downward momentum to the plane, which passes it onto the air. This downward momentum is carried away by the air after the wing (the right tube).

As with any flux, the rate of transfer of downward momentum is

flux of downward momentum × area.   (3.98)

As in the analysis of the hummingbird, this rate must be mg, so that the plane stays aloft. The first factor, the flux of downward momentum, is

density of downward momentum × flow speed.   (3.99)

Therefore,

mg = density of downward momentum × flow speed × area.   (3.100)

As in the analysis of hovering, the density of downward momentum is ρv_z.

In contrast to the analysis of hovering, where the stuff (downward momentum) is carried by the air moving downward, here the stuff is carried by air moving to the right. Thus, where the flow speed in hovering was the downward air speed v_z, in forward flight the flow speed is the forward velocity v.

As in the analysis of hovering, the relevant area is the squared wingspan L^2, because the wings alter the airflow over a distance comparable to their longest dimension, which is their wingspan. You can see this effect in a NASA photograph of an airplane flying through a cloud of smoke. The giant swirl, known as the wake vortex, has a diameter comparable to the plane's wingspan. Large planes can generate vortices that flip over small planes. Thus, when coming in for landing, planes must maintain enough separation to give these vortices time to dissipate.

With these estimates, the equation for v_z becomes

$$\underbrace{mg}_{\text{transfer rate}} \sim \underbrace{\rho_{\text{air}}v_z}_{\text{downward-momentum density}} \times \underbrace{v}_{\text{flow speed}} \times \underbrace{L^2}_{\text{area}}.$$ (3.101)

Now we can solve for the downward air speed:

$$v_z \sim \frac{mg}{\rho_{\text{air}}vL^2}.$$ (3.102)

Now we can estimate the power required to generate lift in forward flight:

$$P = \underbrace{\text{force}}_{mg} \times \underbrace{\text{velocity}}_{v_z} \sim mg \times \frac{mg}{\rho_{\text{air}}vL^2} = \frac{(mg)^2}{\rho_{\text{air}}vL^2}.$$ (3.103)

Here is a comparison of hovering and forward flight.

	hovering	forward flight
deflection area	L^2	L^2
downward-momentum density	$\rho_{\text{air}}v_z$	$\rho_{\text{air}}v_z$
flow speed	v_z	v
downward-momentum flux	$\rho_{\text{air}}v_z^2$	$\rho_{\text{air}}v_z v$
downward-momentum flow mg	$\rho_{\text{air}}v_z^2 L^2$	$\rho_{\text{air}}v_z v L^2$
downward velocity v_z	$\sqrt{mg/\rho_{\text{air}}L^2}$	$mg/\rho_{\text{air}}vL^2$
power to generate lift (mgv_z)	$mg\sqrt{mg/\rho_{\text{air}}L^2}$	$(mg)^2/\rho_{\text{air}}vL^2$

In contrast to hovering, in forward flight the power contains the forward velocity in the denominator—a location that would produce nonsense for hovering, where the forward velocity is zero.

As we did for hovering flight using the Calliope hummingbird, let's apply our knowledge of forward flight to an actual object. The object will be a Boeing 747-400 jumbo jet, and we will estimate the power that it requires in order to take off. A 747 has a wingspan L of approximately 60 meters, and a maximum takeoff mass m of approximately 4×10^5 kilograms (400 tons).

We'll estimate the power in two steps: the weight mg and then the downward air speed v_z. The weight is the easy step: It is just 4×10^6 newtons. The downward air speed v_z is $mg/\rho_{\text{air}}vL^2$. The only unknown quantity is the takeoff speed v. You can estimate it by estimating the plane's acceleration a while taxiing on the runway and by estimating the duration of the acceleration. When I last flew on a 747, I measured the acceleration by suspending my key chain from a string and estimating the angle θ that it made with vertical (perpendicular to the ground). Then $\tan\theta = a/g$. For small θ, the relation simplifies to $a/g \approx \theta$. I found $\theta \approx 0.2$, so the acceleration was about $0.2g$ or 2 meters per second per second. This acceleration lasted for about 40 seconds, giving a takeoff speed of $v \approx 80$ meters per second (180 miles per hour).

The resulting downward speed v_z is roughly 12 meters per second:

$$v_z \sim \frac{\overbrace{4\times10^6\,\text{N}}^{mg}}{\underbrace{1.2\,\text{kg}\,\text{m}^{-3}}_{\rho_{\text{air}}} \times \underbrace{80\,\text{m}\,\text{s}^{-1}}_{v} \times \underbrace{3.6\times10^3\,\text{m}^2}_{L^2}} \approx 12\,\text{m}\,\text{s}^{-1}. \tag{3.104}$$

Then the power required to generate lift is roughly 50 megawatts:

$$P \sim mgv_z \approx 4\times10^6\,\text{N} \times 12\,\text{m}\,\text{s}^{-1} \approx 5\times10^7\,\text{W}. \tag{3.105}$$

Let's see whether these estimates are reasonable. According to the plane's technical documentation, the 747-400's four engines together can provide roughly 1 meganewton of thrust. This thrust can accelerate the plane, with a mass of 4×10^5 kilograms, at 2.5 meters per second per second. This value agrees well with my estimate of 2 meters per second per second, made by suspending a key chain from a string and turning it into a plumb line.

As another check: At takeoff, when v is roughly 80 meters per second, the meganewton of thrust corresponds to a power output Fv of 80 megawatts. This output is comparable to our estimate of 50 megawatts for the power to lift the plane off the ground. After liftoff, the engines use some of their power to lift the plane and some to accelerate the plane, because the plane still needs to reach its cruising speed of 250 meters per second.

Symmetry and conservation make even fluid dynamics tractable.

3.7 Summary and further problems

In the midst of change, find what does not change—the invariant or conserved quantity. Finding these quantities simplifies problems: We focus on the few quantities that do not change rather than on the many ways in which quantities do change. An instance of this idea with wide application is a box model, where what goes in must come out. By choosing suitable boxes, we could estimate rainfalls and drag forces, and understand lift.

Problem 3.37   Raindrop speed

Use the drag force $F_{\text{drag}} \sim \rho A_{\text{cs}} v^2$ to estimate the terminal speed of a typical rain-drop with a diameter of 0.5 centimeters. How could you check the prediction?

Problem 3.38   Average value of sin squared

Use symmetry to find the average value of $\sin^2 t$ over the interval $t = [0, \pi]$.

Problem 3.39   Moment of inertia of a spherical shell

The moment of inertia of an object about an axis of rotation is

$$\sum m_i d_i^2, \qquad\qquad (3.106)$$

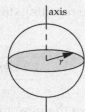

summed over all mass points i, where d_i is the distance of the point from an axis of rotation. Use symmetry to find the moment of inertia of a spherical shell with mass m and radius r about an axis through its center. You shouldn't need to do any integrals!

Problem 3.40   Flying bicyclist

Estimate the wingspan a world-champion bicyclist would require in order to get enough lift for takeoff.

Problem 3.41   Maximum-gain frequency for a second-order system

In this problem, you use symmetry to maximize the gain of an *LRC* circuit or a spring–mass system with damping (using the analogy in Section 2.4.1). The gain G, which is the amplitude ratio $V_{\text{out}}/V_{\text{in}}$, depends on the signal's angular frequency ω:

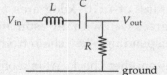

$$G(\omega) = \frac{\dfrac{j\omega}{\omega_0}}{1 + \dfrac{j}{Q}\dfrac{\omega}{\omega_0} - \dfrac{\omega^2}{\omega_0^2}} \qquad\qquad (3.107)$$

where $j = \sqrt{-1}$, ω_0 is the natural frequency of the system, and Q, the quality factor, is a dimensionless measure of the damping. Don't worry about where the gain formula comes from: You can derive it using the impedance method (Problem 2.22), but the purpose of this problem is to maximize its magnitude $|G(\omega)|$. Do so by finding a symmetry operation on ω that leaves $|G(\omega)|$ invariant.

Problem 3.42   Runway length

Estimate the runway length required by a 747 in order to take off.

Problem 3.43   Hovering versus flying

At what forward flight speed does the hummingbird of Section 3.6.1 require as much power to generate lift as it would to hover? How does this speed compare to its typical flight speed?

Problem 3.44   Resistive grid

In an infinite grid of 1-ohm resistors, what is the resistance measured across one resistor?

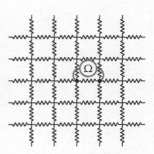

To measure resistance, an ohmmeter injects a current I at one terminal (for simplicity, imagine that $I = 1$ ampere). It removes the same current from the other terminal, and measures the resulting voltage difference V between the terminals. The resistance is $R = V/I$.

Hint: Use symmetry. But it's still a hard problem!

Problem 3.45   Inertia tensor

Here is an inertia tensor (the generalization of moment of inertia) of a particular object, calculated in an ill-chosen (but Cartesian) coordinate system:

$$\begin{pmatrix} 4 & 0 & 0 \\ 0 & 5 & 4 \\ 0 & 4 & 5 \end{pmatrix} \tag{3.108}$$

a. Change the coordinate system to a set of principal axes, where the inertia tensor has the diagonal form

$$\begin{pmatrix} I_{xx} & 0 & 0 \\ 0 & I_{yy} & 0 \\ 0 & 0 & I_{zz} \end{pmatrix} \tag{3.109}$$

and give the principal moments of inertia I_{xx}, I_{yy}, and I_{zz}. Hint: Which properties of a matrix are invariant when changing coordinate systems?

b. Give an example of an object with a similar inertia tensor. Rhetorical question: In which coordinate system is it easier to think of such an object?

This problem was inspired by a problem on the physics written qualifying exam during my days as a PhD student. The problem required diagonalizing an inertia tensor, and there was too little time to rederive or even apply the change-of-basis formulas. Time pressure sometimes pushes one toward better solutions!

Problem 3.46   Temperature distribution on an infinite sheet

On this infinite, uniform sheet, the x axis is held at zero temperature, and the y axis is held at unit temperature ($T = 1$). Find the temperature everywhere (except the origin!). Use Cartesian coordinates $T(x, y)$ or polar coordinates $T(r, \theta)$, whichever choice makes it easier to describe the temperature.

4

Proportional reasoning

When there is change, look for what does not change. That principle, introduced when we studied symmetry and conservation (Chapter 3), is also the basis for our next tool, proportional reasoning.

4.1 Population scaling

An everyday example of proportional reasoning often happens when cooking for a dinner party. When I prepare fish curry, which I normally cook for our family of four, I buy 250 grams of fish. But today another family of four will join us.

▷ *How much fish do I need?*

I need 500 grams. As a general relation,

$$\text{new amount} = \text{old amount} \times \frac{\text{new number of diners}}{\text{usual number of diners}}. \tag{4.1}$$

Another way to state this relation is that the amount of fish is proportional to the number of diners. In symbols,

$$m_{\text{fish}} \propto N_{\text{diners}}, \tag{4.2}$$

where the $\propto$ symbol is read "is proportional to."

▷ *But where in this analysis is the quantity that does* not *change?*

Another way to write the proportionality relation is

$$\frac{\text{new amount of fish}}{\text{new number of diners}} = \frac{\text{old amount of fish}}{\text{old number of diners}}. \tag{4.3}$$

Thus, even when the number of diners changes, the quotient

$$\frac{\text{amount of fish}}{\text{number of diners}} \tag{4.4}$$

does not change.

For an analogous application of proportional reasoning, here's one way to estimate the number of gas stations in the United States. Following the principle of using human-sized numbers, which we discussed in Section 1.4, I did not try to estimate this large number directly. Instead, I started with my small hometown of Summit, New Jersey. It had maybe 20 000 people and maybe five gas stations; the "maybe" indicates that these childhood memories may easily be a factor of 2 too small or too large. If the number of gas stations is proportional to the population ($N_{\text{stations}} \propto N_{\text{people}}$), then

$$N_{\text{stations}}^{\text{US}} = N_{\text{stations}}^{\text{Summit}} \times \frac{\overbrace{3 \times 10^8}^{N_{\text{people}}^{\text{US}}}}{\underbrace{2 \times 10^4}_{N_{\text{people}}^{\text{Summit}}}}. \tag{4.5}$$

The population ratio is roughly 15 000. Therefore, if Summit has five gas stations, the United States should have 75 000. We can check this estimate. The US Census Bureau has an article (from 2008) entitled "A Gas Station for Every 2,500 People"; its title already indicates that an estimate of roughly 10^5 gas stations is reasonably accurate: Summit, in my reckoning, had 4000 people per gas station. Indeed, the article gives the total as 116 855 gas stations—as close to the estimate as we can expect given the uncertainties in childhood memories!

Problem 4.1   Homicide rates

The US homicide rate (in 2011) was roughly 14 000 per year. The UK rate in the same year was roughly 640. Which is the more dangerous country (per person), and by what factor?

4.2 Finding scaling exponents

The dinner example (Section 4.1) used linear proportionality: When the number of dinner guests doubled, so did the amount of food. The relation between the quantities had the form $y \propto x$ or, more explicitly, $y \propto x^1$. The exponent, which here is 1, is called the scaling exponent. For that reason, proportionalities are often called scaling relations. Scaling exponents are a powerful abstraction: Once you know the scaling exponent, you usually do not care about the mechanism underlying it.

4.2.1 Warmup

After linear proportionality, the next simplest and most common type of proportionality is quadratic—a scaling exponent of 2—and its close cousin, a scaling exponent of 1/2. As an example, here is a big circle with diameter $d_{\text{big}} = \sqrt{5}$ cm.

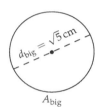

▷ *What is the diameter of the circle with one-half the area of this circle?*

Let's first do the very common brute-force solution, which does not use proportional reasoning, so that you see what not to do. It begins with the area of the big circle:

$$A_{\text{big}} = \frac{\pi}{4} d_{\text{big}}^2 = \frac{5}{4} \pi \ \text{cm}^2. \tag{4.6}$$

The area of the small circle A_{small} is $A_{\text{big}}/2$, so $A_{\text{small}} = 5\pi/8 \ \text{cm}^2$. $A_{\text{small}} = \dfrac{A_{\text{big}}}{2}$

Therefore, the diameter of the small circle is given by

$$d_{\text{small}} = \sqrt{\frac{A_{\text{small}}}{\pi/4}} = \sqrt{\frac{5}{2}} \ \text{cm}. \tag{4.7}$$

Although this result is correct, by including $\pi/4$ and then dividing it out, we run around Robin Hood's barn (all of Sherwood forest) to reach a simple result. There must be a more elegant and insightful approach.

This improved approach also starts with the relation between a circle's area and its diameter: $A = \pi d^2/4$. However, it discards the complexity early—in the next step—rather than carrying it through the analysis and having it vanish only at the end. An everyday analog of this approach is packing for a trip. Rather than dragging around books that you will not read or clothes that you will not wear, prune early and travel light: Pack only what you will use and set aside the rest.

To lighten your problem-solving luggage, observe that all circles, independent of their diameter, have the same prefactor $\pi/4$ connecting d^2 and A. Therefore, when we make a proportionality or scaling relation between A and d, we discard the prefactor. The result is the following quadratic proportionality (one where the scaling exponent is 2):

$$A \propto d^2. \tag{4.8}$$

For finding the new diameter, we need the inverse scaling relation:

$$d \propto A^{1/2}. \tag{4.9}$$

In this form, the scaling exponent is $1/2$. This proportionality is shorthand for the ratio relation

$$\frac{d_{\text{small}}}{d_{\text{big}}} = \left(\frac{A_{\text{small}}}{A_{\text{big}}}\right)^{1/2}. \tag{4.10}$$

The area ratio is $1/2$, so the diameter ratio is $1/\sqrt{2}$. Because the large diameter is $\sqrt{5}$ cm, the small diameter is $\sqrt{5/2}$ cm.

The proportional-reasoning solution is shorter than the brute-force approach, so it offers fewer places to go wrong. It is also more general: It shows that the result does not require that the shape be a circle. As long as the area of the shape is proportional to the square of its size (as a length)—a relation that

$$A_{\text{big}} = \frac{5\pi}{4} \text{ cm}^2 \dashrightarrow A_{\text{small}} = \frac{5\pi}{8} \text{ cm}^2$$
$$\textit{(extra baggage)}$$
$$A = \frac{\pi}{4}d^2 \qquad d = \sqrt{\frac{A}{\pi/4}}$$
$$d_{\text{big}} = \sqrt{5} \text{ cm} \xrightarrow[\text{prop. reasoning}]{d \propto A^{1/2}} d_{\text{small}} = \sqrt{\frac{5}{2}} \text{ cm}$$

holds for all planar shapes—the length ratio is $1/\sqrt{2}$ whenever the area ratio is $1/2$. All that matters is the scaling exponent.

Problem 4.2   Length of the horizontal bisecting path

In Problem 3.23, about the shortest path that bisects an equilateral triangle, one candidate path is a horizontal line. How long is that line relative to a side of the triangle?

Areas are connected to flux, because flux is rate per area. Thus, the scaling exponent for area—namely, 2—appears in flux relationships. For example:

▷ *What is the solar flux at Pluto's orbit?*

The solar flux F at a distance r from the Sun is the solar luminosity L_{Sun}—the radiant power output of the sun—spread over a sphere with radius r:

$$F = \frac{L_{Sun}}{4\pi r^2}.$$ (4.11)

Even as r changes, the solar luminosity remains the same (conservation!), as does the factor of 4π. Therefore, in the spirit of packing light for a trip, simplify the equality $F = L_{Sun}/4\pi r^2$ to the proportionality

$$F \propto r^{-2}$$ (4.12)

by discarding the factors L_{Sun} and 4π. The scaling exponent here is -2: The minus sign indicates the inverse proportionality between flux and area, and the 2 is the scaling exponent connecting r to area.

The scaling relation is shorthand for

$$\frac{F_{\text{Pluto's orbit}}}{F_{\text{Earth's orbit}}} = \left(\frac{r_{\text{Pluto's orbit}}}{r_{\text{Earth's orbit}}}\right)^{-2},$$ (4.13)

or

$$F_{\text{Pluto's orbit}} = F_{\text{Earth's orbit}} \left(\frac{r_{\text{Pluto's orbit}}}{r_{\text{Earth's orbit}}}\right)^{-2}.$$ (4.14)

The ratio of orbital radii is roughly 40. Therefore, the solar flux at Pluto's orbit is roughly 40^{-2} or $1/1600$ of the flux at the Earth's orbit. The resulting flux is roughly 0.8 watts per square meter:

$$F_{\text{Pluto's orbit}} = \frac{1300\,\text{W}}{\text{m}^2} \times \frac{1}{1600} \approx \frac{0.8\,\text{W}}{\text{m}^2}.$$ (4.15)

Receiving such a small amount of sunlight, Pluto must be very cold. We can estimate its surface temperature with a further proportionality.

Surface temperature depends mostly on so-called blackbody radiation. The surface temperature is the temperature at which the radiated flux equals the incoming flux; we are making another box model. 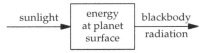 The radiated flux is given by the blackbody formula (which we will derive in Section 5.5.2)

$$F = \sigma T^4,$$ (4.16)

where T is the temperature, and σ is the Stefan–Boltzmann constant:

$$\sigma \approx 5.7 \times 10^{-8} \frac{\text{W}}{\text{m}^2\,\text{K}^4}.$$ (4.17)

▷ *What is the resulting surface temperature on Pluto?*

As with any proportional-reasoning calculation, there is a long-winded, brute-force alternative (try Problem 4.5). The elegant approach directly uses the proportionalities

$$T \propto F^{1/4} \quad \text{and} \quad F \propto r^{-2}, \tag{4.18}$$

where r is the orbital radius. Together, they produce a new proportionality

$$T \propto \left(r^{-2}\right)^{1/4} = r^{-1/2}. \tag{4.19}$$

A compact graphical notation, similar to the divide-and-conquer trees, encapsulates this derivation:

$$r \;-\boxed{-2}\longrightarrow F \;-\;\boxed{\tfrac{1}{4}}\;\longrightarrow T$$

As indicated by the $r \to F$ arrow, changing r changes F. The boxed number along the arrow gives the scaling exponent. Therefore, the $r \to F$ arrow represents $F \propto r^{-2}$. The $F \to T$ arrow indicates that changing F changes T and, in particular, that $T \propto F^{1/4}$.

To find the scaling exponent connecting r to T, multiply the scaling exponents along the path:

$$-2 \times \frac{1}{4} = -\frac{1}{2}. \tag{4.20}$$

Problem 4.3   Explaining the graphical notation
In our graphical representation of scaling relations, why is the final scaling exponent the product, rather than the sum, of the scaling exponents along the way?

This scaling exponent represents the following comparison:

$$\frac{T_{\text{Earth}}}{T_{\text{Pluto}}} = \left(\frac{r_{\text{Earth's orbit}}}{r_{\text{Pluto's orbit}}}\right)^{-1/2} = \left(\frac{r_{\text{Pluto's orbit}}}{r_{\text{Earth's orbit}}}\right)^{1/2}. \tag{4.21}$$

(The rightmost form, with the positive exponent, is more direct than the intermediate form, because it does not first produce a fraction smaller than 1 and then take its reciprocal with a negative exponent.) The ratio of orbital radii is 40, so the ratio of surface temperatures should be $\sqrt{40}$ or roughly 6. Pluto's surface temperature should be roughly 50 K:

$$T_{\text{Pluto}} \approx \frac{T_{\text{Earth}}}{6} \approx \frac{293\,\text{K}}{6} \approx 50\,\text{K}. \tag{4.22}$$

Pluto's actual mean surface temperature is 44 K, very close to our prediction based on proportional reasoning.

Problem 4.4   Explaining the discrepancy
Why is our prediction for Pluto's mean temperature slightly too high?

Problem 4.5   Brute-force calculation of surface temperature
To practice recognizing common but inferior problem-solving methods, use the brute-force method to estimate the surface temperature on Pluto: (a) From the solar flux at Pluto's orbit, calculate the solar flux averaged over the surface; (b) use that flux to estimate a blackbody temperature.

4.2.2 Orbital periods

In the preceding examples, the scaling relations formed chains (trees without branching):

$$\text{food for a dinner party}: \quad N_{\text{guests}} \xrightarrow{\ +1\ } m_{\text{fish}}$$

$$\text{area of a circle}: \quad r \xrightarrow{\ +2\ } A_{\text{circle}} \tag{4.23}$$

$$\text{surface temperature}: \quad r \xrightarrow{\ -2\ } F \xrightarrow{\ \frac{1}{4}\ } T$$

More elaborate relationships also occur—as we will find in rederiving a famous law of planetary motion.

▷ *How does a planet's orbital period depend on its orbital radius r?*

We'll study the special case of circular orbits (many planetary orbits are close to circular). Our exploratory thinking is often aided by making proportionality questions concrete. Therefore, rather than finding the scaling exponent using the abstract notion of "depend on," answer the doubling question: "When I double *this* quantity, what happens to *that* quantity?"

▷ *What is so special about doubling?*

Doubling—multiplying by a factor of 2—is the simplest useful change. A factor of 1 is simpler; however, being no change at all, it is too simple.

▷ *What happens to the period if we double the orbital radius?*

The most direct effect of doubling the orbital radius is that gravity gets weaker. Because the gravitational force is an inverse-square force—that is, $F \propto r^{-2}$—the gravitational force falls by a factor of 4. A compact and intuitive notation for these changes is to mark the change directly under the quantity: A notation of $\times n$ indicates multiplication by a factor of n.

$$\underset{\times\frac{1}{4}}{\underset{\sim}{F}} \propto \underset{\times 2}{\underset{\sim}{r}}^{-2}. \tag{4.24}$$

Because force is proportional to acceleration, the planet's acceleration a falls by the same factor of 4.

In circular motion, acceleration and velocity are related by $a = v^2/r$. (We will derive this relation in Section 5.1.1 and Section 6.3.4, with two different reasoning tools.) Therefore, the orbital velocity v is $\sqrt{ar}$, and doubling the radius increases the orbital speed by a factor of $\sqrt{1/2}$:

$$\underset{\times\sqrt{\frac{1}{2}}}{\underset{\sim}{v}} = \left(\underset{\times\frac{1}{4}}{\underset{\sim}{a}} \times \underset{\times 2}{\underset{\sim}{r}} \right)^{1/2}. \tag{4.25}$$

Although this calculation is correct, when it is stated as a factor of $\sqrt{1/2}$ it confounds our expectations and produces numerical whiplash. As we finish reading "increases by a factor of," we expect a number greater than 1. But we get a number smaller than 1. An increase by a factor smaller than 1 is more simply described as a decrease. Therefore, it is more direct to say that the orbital speed falls by a factor of $\sqrt{2}$.

The orbital period is $T \sim r/v$ (the $\sim$ contains the dimensionless prefactor 2π), so it increases by a factor of $2^{3/2}$:

$$\underset{\times 2^{3/2}}{\underset{\sim}{T}} \sim \underset{\times 2}{\underset{\sim}{r}} \times \underset{\times\sqrt{2}}{\underset{\sim}{v}}^{-1}. \tag{4.26}$$

In summary, doubling the orbital radius multiplies the orbital period by $2^{3/2}$. In general, the connection between r and T is

$$T \propto r^{3/2}. \tag{4.27}$$

This result is Kepler's third law for circular orbits. Our scaling analysis has the following graphical representation:

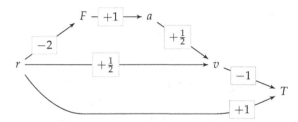

In this structure, the new feature is that two paths reach the orbital velocity v. There we add the incoming exponents.

1. r - $\boxed{+\frac{1}{2}}$ $\longrightarrow$ v represents the $\sqrt{r}$ in $v = \sqrt{ar}$. It carries $+1/2$ powers of r.

2. a - $\boxed{+\frac{1}{2}}$ $\longrightarrow$ v represents the $\sqrt{a}$ in $v = \sqrt{ar}$. To determine how many powers of r flow through this arrow, follow the chain containing it:

$$r - \boxed{-2} \longrightarrow F - \boxed{+1} \longrightarrow a - \boxed{+\tfrac{1}{2}} \longrightarrow v$$

One power of r starts at the left side. It becomes -2 powers after passing through the first scaling exponent and arriving at F. It remains -2 powers on arrival at a. Finally, it becomes -1 power on arrival at v. This path therefore carries -1 power of r.

Its contribution is the result of multiplying the three scaling exponents along the chain:

$$\underbrace{-2}_{r \to F} \times \underbrace{+1}_{F \to a} \times \underbrace{+\frac{1}{2}}_{a \to v} = -1. \tag{4.28}$$

The -1 power carried by the three-link chain, representing r^{-1}, combines with the $+1/2$ from the direct $r \to v$ arrow, which represents $r^{+1/2}$. Adding the exponents on r, the result is that v contains $-1/2$ powers of r:

$$v \propto \underbrace{r^{-1}}_{\text{via } F} \times \underbrace{r^{+1/2}}_{\text{direct}} = r^{-1/2}. \tag{4.29}$$

Let's practice the same reasoning by finding the scaling relation connecting T and r—which is Kepler's third law. The direct $r \to T$ arrow, with a scaling exponent of $+1$, carries $+1$ powers of r. The $v \to T$ arrow carries -1 powers of v. Because v contains $-1/2$ powers of r, the $v \to T$ arrow carries $+1/2$ powers of r:

$$\underbrace{-\frac{1}{2}}_{r \to v} \times \underbrace{-1}_{v \to T} = +\frac{1}{2}. \tag{4.30}$$

Together, the two arrows contribute $+3/2$ powers of r and give us Kepler's third law:

$$T \propto \underbrace{r^{+1}}_{r \to T} \times \underbrace{r^{+1/2}}_{\text{via } v} = r^{+3/2}. \tag{4.31}$$

To summarize the exponent rules, for which we now have several illustrations: (1) Multiply exponents along a path, and (2) add exponents when paths meet.

Now let's apply Kepler's third law to a nearby planet.

▷ *How long is the Martian year?*

The proportionality $T \propto r^{3/2}$ is shorthand for the comparison

$$\frac{T_{\text{Mars}}}{T_{\text{ref}}} = \left(\frac{r_{\text{Mars}}}{r_{\text{ref}}}\right)^{3/2}, \tag{4.32}$$

where r_{Mars} is the orbital radius of Mars, r_{ref} is the orbital radius of a reference planet, and T_{ref} is its orbital period. Because we are most familiar with the Earth, let's choose it as the reference planet. The reference period is 1 (Earth) year; and the reference radius is 1 astronomical unit (AU), which is 1.5×10^{11} meters. The benefit of this choice is that we will obtain the period of Mars's orbit in the familiar unit of Earth years.

The distance of Mars to the Sun varies between 2.07×10^{11} meters (1.38 astronomical units) and 2.49×10^{11} meters (1.67 astronomical units). Thus, the orbit of Mars is not very circular and has no single orbital radius r_{Mars}. (Its significant deviation from circularity allowed Kepler to conclude that planets move in ellipses.) As a proxy for r_{Mars}, let's use the average of the minimum and maximum radii. It is 1.52 astronomical units, making the ratio of orbital periods approximately 1.88:

$$\frac{T_{\text{Mars}}}{T_{\text{Earth}}} = \left(\frac{1.52\,\text{AU}}{1\,\text{AU}}\right)^{3/2} \approx 1.88. \tag{4.33}$$

Therefore, the Martian year is 1.88 Earth years long.

Problem 4.6 Brute-force calculation of the orbital period
To emphasize the contrast between proportional reasoning and the brute-force approach, find the period of Mars's orbit using the brute-force approach by starting with Newton's law of gravitation and then finding the orbital velocity and circumference.

A surprising conclusion about orbits comes from our doubling question introduced on page 109.

▷ *What happens to the period of a planet if you double its mass?*

Using a type of thought experiment due to Galileo, imagine two identical planets, orbiting one just behind the other along the same orbital path. They have the same period. Now tie them together. The rope does not change the period, so the double-mass planet has same period as each individual planet: The scaling exponent is zero ($T \propto m^0$)!

> **Problem 4.7   Pendulum period versus mass**
> How does the period of an ideal pendulum depend on the mass of the bob?

4.2.3 Projectile range

In the previous examples, only one variable was independent; changing it changed all the others. However, many problems contain multiple independent variables. An example is the range R of a rock launched at an angle θ with speed v. The

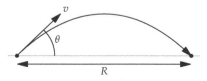

traditional derivation uses calculus. You solve for the position of the rock as a function of time, solve for the time when its height is zero (the ground level), and then insert that time into the horizontal position to find the range. This analysis is not wrong, but its result still seems like magic. I leave unsatisfied, thinking, "The result must be true. But I still do not know *why*."

That "why" insight comes from proportional reasoning, which discards the nonessential complexity. Let's begin with our doubling question.

▷ *How does doubling each of the independent variables affect the range?*

The independent variables include the launch velocity v, the gravitational acceleration g (because gravity returns the rock to Earth), and the launch angle. However, angles do not fit so easily into proportional reasoning, so we won't explore the role of θ here (we will handle it in Section 8.2.2.1 using the tool of easy cases, or you can try Problem 4.9).

Because the only forces are vertical, the rock's horizontal velocity remains constant throughout its flight (an invariant!). Thus, the range is given by

$$R = \text{time aloft} \times \text{initial horizontal velocity.} \qquad (4.34)$$

The time aloft is determined by the initial vertical velocity, because gravity steadily reduces it (at the rate g):

$$\text{time aloft} \sim \frac{\text{initial vertical velocity}}{g}. \qquad (4.35)$$

> **Problem 4.8   Missing dimensionless prefactor**
> What is the missing dimensionless prefactor in the preceding expression for the time that the rock stays aloft?

Now double the launch velocity v. That change doubles the initial horizontal and vertical components of the velocity. Doubling the vertical component doubles the time aloft. Because the range is proportional to the horizontal velocity and to the time aloft, when the launch velocity doubles, the range quadruples. The scaling exponent connecting v to R is 2: $R \propto v^2$.

▷ *What is the effect of doubling g?*

Doubling g doesn't change the horizontal velocity or the initial vertical velocity, but it halves the time aloft and therefore the range as well. The scaling exponent connecting g to R is -1: $R \propto g^{-1}$.

The combined scaling relation, which gives the dependence of R on both g and v, is

$$R \propto \frac{v^2}{g}. \tag{4.36}$$

Using v_x and v_y for the horizontal and vertical components of the launch velocity, the graphical representation of this reasoning is

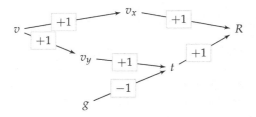

This graph shows a new feature: two independent variables, v and g. We'll need to track the powers of v and g separately.

The range R has two incoming paths. The path via the horizontal velocity v_x contributes $+1 \times +1 = +1$ powers of v but no powers of g. The path via t also contributes $+1$ powers of v, and contributes -1 powers of g. The diagram compactly represents how R became proportional to v^2/g.

The full range formula, including the launch angle θ is

$$R \sim \frac{v^2}{g} \sin \theta \cos \theta. \tag{4.37}$$

The dependence on v and g is just as we predicted.

The moral of this example is that you can derive and understand relations by ignoring constants of proportionality and instead concentrating on the scaling exponents. Furthermore, you can use this ability in order to spot mistakes: Just check each independent variable's scaling exponent. For example, if someone proposes that projectile range R is proportional to v^3/g, think, "The $1/g$ makes sense from the time aloft. But what about the v^3? One power of v comes from the horizontal velocity and one power from the time aloft, which explains two powers of v. But where does the third power comes from? The range should instead contain v^2/g."

> **Problem 4.9   Angular factors in projectile range**
> Explain the $\sin\theta$ and $\cos\theta$ factors by using the relation
>
> range = time aloft × horizontal velocity.   (4.38)

4.2.4 Planetary surface gravity

Scaling, or proportional reasoning, connects independent to dependent variables. Often we have freedom in choosing the independent variables. Here is an example to show you how to use that freedom.

▷ *Assuming that planets are uniform spheres, how does g, the gravitational acceleration at the surface, depend on the planet's radius R?*

We seek the scaling exponent n in $g \propto R^n$. At the planet's surface, the gravitational force F on an object of mass m is GMm/R^2, where G is Newton's constant, and M is the planet's mass. The gravitational acceleration g is F/m or GM/R^2. Because G is the same for all objects, we pack light and eliminate G to make the proportionality

$$g \propto \frac{M}{R^2}.$$ (4.39)

In this form, with mass M and radius R as the independent variables, the scaling exponent n is -2.

However, an alternative relation comes from noticing that the planet's mass M depends on the planet's radius R and density ρ as $M \sim \rho R^3$. Then

$$g \propto \frac{\rho R^3}{R^2} = \rho R.$$ (4.40)

In this form, with density and radius as the independent variables, the scaling exponent on R is now only 1.

▷ *Which scaling relation, with mass or density, is preferable?*

Planets vary widely in their mass: from 3.3×10^{23} kilograms (Mercury) to 1.9×10^{27} kilograms (Jupiter), a range 4 decades wide (a factor of 10^4). They vary greatly in their radius: from 7×10^4 kilometers (Jupiter) down to 2.4×10^3 kilometers (Mercury), a range of a factor of 30. The quotient M/R^2 has huge variations in the numerator and denominator that mostly oppose each other. When there is change, look for what does not change—or, at least, what does not change as much. In contrast to masses, planetary densities vary from only 0.7 grams per cubic centimeter (Saturn) to 5.5 grams per cubic centimeter (Earth)—a range of only a factor of 8. The variations in planetary surface gravity are easier to understand using the planet's radius and density rather than its radius and mass.

This result is general. Mass is an extensive quantity: When two objects combine, their masses add. Density, in contrast, is an intensive quantity. Adding more of a particular substance does not change its density. Using intensive quantities for the independent variables usually leads to more insightful results than using extensive quantities does.

Problem 4.10   Distance to the Moon
The orbital period near the Earth's surface (say, for a low-flying satellite) is roughly 1.5 hours. Use that information to estimate the distance to the Moon.

Problem 4.11   Moon's angular diameter and radius
On a night with a full moon, estimate the Moon's angular diameter—that is, the visual angle subtended by the Moon. Use that angle and the result of Problem 4.10 to estimate the Moon's radius.

Problem 4.12   Surface gravity on the Moon
Assuming that all planets (and moons) have the same density, use the radius of the Moon (Problem 4.11) to estimate its surface gravity. Then compare your estimate with the actual value and suggest an explanation for the discrepancy.

Problem 4.13   Gravitational strength inside a planet
Imagine a uniform, spherical planet with radius R. How does the gravitational acceleration g depend on r, the distance from the center of the planet? Give the scaling exponent when $r < R$ and when $r \geq R$, and sketch $g(r)$.

Problem 4.14   Making toast land butter-side up
As a piece of toast slides off a dining table (starting with almost no horizontal velocity), it picks up angular velocity. Once it leaves the table, its angular velocity remains constant. In everyday experience, a toast usually backflips (rotates 180°) by the time it hits the ground, and lands butter-side down. How high would tables have to be for a piece of toast to land butter-side up?

4.3 Scaling exponents in fluid mechanics

The preceding introductory examples may mislead you into thinking that proportional reasoning is useful only when we could also find the exact solution. As a counterexample, we return to that source of mathematical beauty but also misery, fluid mechanics, where exact solutions exist for hardly any situations of practical interest. To make progress, we need to discard complexity by focusing on the scaling exponents.

4.3.1 Falling cones

In Section 3.5.1, we used conservation reasoning to show that the drag force is given by

$$F_{\text{drag}} \sim \rho A_{\text{cs}} v^2. \tag{4.41}$$

In Section 3.5.2, we tested this result by correctly predicting the terminal speed of a falling paper cone. However, that experiment concerned only one cone of a particular size. A natural generalization of that experiment is to predict a cone's terminal speed as a function of its size.

▷ *How does the terminal speed of a paper cone depend on its size?*

Size is an ambiguous notion. It might refer to an area, a volume, or a length. Here, let's consider size to be the cone's cross-sectional radius r. The quantitative question is to find the scaling exponent n in $v \propto r^n$, where v is the cone's terminal speed.

The doubling question is, "What happens to the terminal speed when we double r?" At the terminal speed, drag and weight mg balance:

$$\underbrace{\rho_{\text{air}} A_{\text{cs}} v^2}_{\text{drag}} \sim \underbrace{mg}_{\text{weight}}. \tag{4.42}$$

Therefore, the terminal speed is just what we found in Section 3.5.2:

$$v \sim \sqrt{\frac{mg}{\rho_{\text{air}} A_{\text{cs}}}}. \tag{4.43}$$

All cones feel the same g and ρ_{air}, so we pack light and eliminate these variables to make the proportionality

$$v \propto \sqrt{\frac{m}{A_{\text{cs}}}}. \tag{4.44}$$

Doubling r quadruples the amount of paper used to make the cone and therefore its mass m. It also quadruples its cross-sectional area A_{cs}. According to the proportionality, the two effects cancel: When r doubles, v should remain constant. All cones of the same shape (and made from the same paper) should fall at the same speed!

This result always surprises me. So I tried the experiment. I printed the cone template in Section 3.5.2 at 400-percent magnification (a factor of 4 increase in length), cut it out, taped the two straight edges together, and raced the small and big cones by dropping them from a height of about 2 meters. After a roughly 2-second fall, they landed almost simultaneously—within 0.1 seconds of each other. Thus, their terminal speeds are the same, give or take 5 percent.

Proportional reasoning triumphs again! Surprisingly, the proportional-reasoning result is much more accurate than the drag-force estimate $\rho_{air}A_{cs}v^2$ on which it is based.

▷ *How can predictions based on proportional reasoning be more accurate than the original relations?*

To see how this happy situation arose, let's redo the calculation but include the dimensionless prefactor in the drag force. With the dimensionless prefactor (shaded in gray), the drag force is

$$F_{drag} = \frac{1}{2}c_d\ \rho_{air}A_{cs}v^2, \tag{4.45}$$

where c_d is the drag coefficient (introduced in Section 3.2.1). The prefactor carries over to the terminal speed:

$$v = \sqrt{\frac{m}{\frac{1}{2}c_d\ A_{cs}}}. \tag{4.46}$$

Ignoring the prefactor decreases v by a factor of $\sqrt{2/c_d}$. (For nonstreamlined objects, $c_d \sim 1$, so the decrease is roughly by a factor of $\sqrt{2}$.)

In contrast, in the ratio of terminal speeds v_{big}/v_{small}, the prefactor drops out. Here is the ratio with the prefactors shaded in gray:

$$\frac{v_{big}}{v_{small}} = \sqrt{\frac{m_{big}}{\frac{1}{2}c_d^{big}\ A_{cs}^{big}}} \Big/ \sqrt{\frac{m_{small}}{\frac{1}{2}c_d^{small}\ A_{cs}^{small}}}. \tag{4.47}$$

As long as the drag coefficients c_d are the same for the big and small cones, they divide out from the ratio. Then the scaling result—that the terminal speed is independent of size—is exact, independent of the drag coefficient.

The moral, once again, is to build on what we know, rather than computing quantities that we do not need and that only clutter the analysis and thereby our thinking. Scaling relations bootstrap our knowledge. Here, if you know the terminal speed of the small cone, use that speed—and the scaling result that speed is independent of size—to find the terminal speed of the large cone. In the next section, we will apply this approach to estimate the fuel consumption of a plane.

Problem 4.15   Taping the cones
The big cone, with twice the radius of the small cone, has four times the weight of paper. But what about the tape? If you tape along the entire radius of the cone template, how does the length of the tape compare between the large and small cones? How should you apply the tape in order to maintain the 4 : 1 weight ratio?

Problem 4.16   Bigger and bigger cones
Further test the scaling relation $v \propto r^0$ (terminal speed is independent of size) by building a huge cone from a template with four times the radius of the template for the small cone. Then race the small, big, and huge cones.

Problem 4.17   Four cones versus one cone
Use the cone template in Section 3.5.2 (at 200-percent enlargement) to make five small cones. Then stack four of the cones to make one heavy small cone. How much faster will the four-cone stack fall in comparison to the single small cone? That is, predict the ratio of terminal speeds

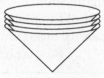

$$\frac{v_{\text{four cones}}}{v_{\text{one cone}}}. \qquad (4.48)$$

Then check your prediction by trying the experiment.

4.3.2 Fuel consumption of a Boeing 747 jumbo jet

In Section 3.5.4, we estimated the fuel consumption of automobiles. For the next example, we'll estimate the fuel consumption of a Boeing 747 jumbo jet. Rather than repeating the structure of the automobile estimate in Section 3.5.4 but with parameters for a plane—which would be the brute-force approach—we will reuse the automobile estimate and supplement it with proportional reasoning.

A plane uses its fuel to generate energy to fight drag and to generate lift. However, for this estimate, forget about lift. At a plane's cruising speed, lift and drag consume comparable energies (as we will show in Section 4.6). Thus, neglecting lift ignores only a factor of 2—in energy, lift plus drag is twice the drag alone—and allows us to estimate fuel consumption before we study lift. Divide and conquer: Don't bite off all the complexity at once!

The energy consumed fighting drag is proportional to the drag force:

$$E \propto \rho_{air} A_{cs} v^2. \tag{4.49}$$

This scaling relation is shorthand for the following comparison between a plane and a car:

$$\frac{E_{plane}}{E_{car}} = \frac{\rho_{air}^{cruising\ altitude}}{\rho_{air}^{sea\ level}} \times \frac{A_{cs}^{plane}}{A_{cs}^{car}} \times \left(\frac{v_{plane}}{v_{car}}\right)^2. \tag{4.50}$$

Thus, the energy-consumption ratio breaks into three ratio estimates.

▶ *What are reasonable estimates for the three ratios?*

1. *Air density.* A plane's cruising altitude is typically 35 000 feet or 10 kilometers, which is slightly above Mount Everest. At that height, mountain climbers require oxygen tanks, so the air, and oxygen, density must be significantly lower than it is at sea level. (Once you learn the reasoning tool of lumping, you can predict the density—see Problem 6.36.) The density ratio turns out to be roughly 3:

$$\frac{\rho_{air}^{cruising\ altitude}}{\rho_{air}^{sea\ level}} \approx \frac{1}{3}. \tag{4.51}$$

The thinner air wins the plane a factor of 3 in fuel efficiency.

2. *Cross-sectional area.* To estimate the cross-sectional area of the plane, we need to estimate the width and height of the plane's body; its wings are very streamlined, so they contribute negligible drag. When estimating lengths, let's make our measuring rod (our unit) a person length. Using this measuring rod has two benefits. First, the measuring rod is easy for our gut to picture, because we feel our own size innately. Second, the lengths that we will measure will be only a small multiple of a person length. The numerical part of the quantity (for example, the 1.5 in "1.5 person lengths") will be comparable to 1, and therefore also easy to picture. As a rule of

thumb, ratios between 1/3 and 3 are easy to picture and feel in our gut because our mental number hardware is exact for the quantities 1, 2, and 3. We choose our measuring rods accordingly.

In applying the person-length measuring rod to the width of the plane's body, I remember the comfortable days of regulated air travel. As a child, I would find three adjacent empty seats in the back of the plane and sleep for the whole trip (which explains why my parents insist that traveling with small children was easy). A jumbo jet is three or four such seat groups across—call it three person lengths—and its cross section is roughly circular. A circle is roughly a square, so the cross-sectional area is roughly 10 square person lengths:

$$(3 \text{ person lengths})^2 \approx 10 \, (\text{person length})^2. \tag{4.52}$$

Although we might worry that this estimate used too many approximations, we should do it anyway: It gives us a rough-and-ready value and allows us to make progress. Our goal is edible jam today, not delicious jam tomorrow!

Now let's estimate the cross-sectional area of a car. In standard units, it's about 3 square meters, as we estimated in Section 3.5.4. But let's apply our person-sized measuring rod. From nocturnal activities in cars, you may have experienced that cars, uncomfortably, are about one person across. A car's cross section is roughly square, so the cross-sectional area is roughly 1 square person length.

car cross section

> **Problem 4.18** **Comparing the cross-sectional areas**
> How well does 1 square person length match 3 square meters?

The ratio of cross-sectional areas is roughly 10; therefore, the plane's larger cross-sectional area costs it a factor of 10 in fuel efficiency.

3. *Speed*. A reason to fly rather than drive is that planes travel faster than cars. A plane travels almost at the speed of sound: 1000 kilometers per hour or 600 miles per hour. A car travels at around 100 kilometers per hour or 60 miles per hour. The speed ratio is roughly a factor of 10, so

$$\left(\frac{v_{\text{plane}}}{v_{\text{car}}} \right)^2 \approx 100. \tag{4.53}$$

The plane's greater speed costs it a factor of 100 in fuel efficiency.

Now we can combine the three ratio estimates to estimate the ratio of energy consumptions:

$$\frac{E_{\text{plane}}}{E_{\text{car}}} \sim \underbrace{\frac{1}{3}}_{\text{density}} \times \underbrace{10}_{\text{area}} \times \underbrace{100}_{(\text{speed})^2} \approx 300. \tag{4.54}$$

A plane should be 300 times less fuel efficient than a car—terrible news for anyone who travels by plane!

▷ *Should you therefore never fly again?*

Our conscience is saved because a plane carries roughly 300 passengers, whereas a typical (Western) car is driven by one person to and from work. Therefore, per person, a plane and a car should have comparable fuel efficiencies: 30 passenger miles per gallon, as we estimated in Section 3.5.4, or 8 liters per 100 passenger kilometers.

According to Boeing's technical data on the 747-400 model, the plane has a range of 13 450 kilometers, its fuel tank contains 216 840 liters, and it carries 416 passengers. These data correspond to a fuel consumption of 4 liters per 100 passenger kilometers:

$$\frac{216\,840\,\ell}{13\,450\,\text{km}} \times \frac{1}{416\,\text{passengers}} \approx \frac{4\,\ell}{100\,\text{passenger km}}. \tag{4.55}$$

Our proportional-reasoning estimate of 8 liters per 100 passenger kilometers is very reasonable, considering the simplicity of the method compared to a full fluid-dynamics analysis.

Problem 4.19   Density of air by proportional reasoning
Make rough estimates of your own top swimming and cycling speeds, or use the data that a record 5-kilometer swim time is 56:16.6 (almost 1 hour). Thereby explain why the density of water must be roughly 1000 times the density of air. Why is this estimate more accurate when based on cycling rather than running speeds?

Problem 4.20   Raindrop speed versus size
How does a raindrop's terminal speed depend on its radius?

Problem 4.21   Estimating an air ticket price from fuel cost
Estimate the fuel cost for a long-distance plane journey—for example, London to Boston, London to Cape Town, or Los Angeles to Sydney. How does the fuel cost compare to the ticket price? (Airlines do not pay many of the fuel taxes paid by ordinary motorists, so their fuel costs are lower than motorists' fuel costs.)

Problem 4.22   Relative fuel efficiency of a cycling
Estimate the fuel efficiency, relative to a car, of a bicyclist powered by peanut butter traveling at a decent speed (say, 10 meters per second or 20 miles per hour). Compare your estimate with your estimate in Problem 3.34.

4.4   Scaling exponents in mathematics

Our scaling relations so far have connected physical quantities. But proportional reasoning can also bring us insight in mathematics. A classic example is the birthday paradox.

▷ *How many people must be in a room before the probability that two people share a birthday (for example, two people are born on July 18) is at least 50 percent?*

Almost everyone, including me, reasons that, with 365 days in the year, we need roughly 50 percent of 365, or 183 people. Let's test this conjecture. The actual probability of having a shared birthday is

$$1 - \left(1 - \frac{1}{365}\right)\left(1 - \frac{2}{365}\right)\left(1 - \frac{3}{365}\right) \cdots \left(1 - \frac{n-1}{365}\right). \qquad (4.56)$$

(A derivation is in *Everyday Probability and Statistics* [50, p. 49] or in the advanced classic *Probability Theory and its Application* [13, vol. 1, p. 33]. But you can explain its structure already: Try Problem 4.23.) For $n = 183$, the probability is $1 - 4.8 \times 10^{-25}$ or almost exactly 1.

Problem 4.23   Explaining the probability of a shared birthday
Explain the pieces of the formula for the probability of a shared birthday: What is the reason for the $1 -$ in front of the product? Why does each factor in the product have a $1 - ?$ And why does the last factor have $(n - 1)/365$ rather than $n/365$?

To make this surprisingly high probability seem plausible, I gave random birthdays to 183 simulated people. Two people shared a birthday on 14 days (and we need only one such day); and three people shared a birthday on three days. According to this simulation and to the exact calculation, the plausible first guess of 183 people is far too high. This surprising result is the birthday paradox.

Even though we could use the exact probability to find the threshold number of people that makes the probability greater than 50 percent, we would still wonder *why*: Why is the plausible argument that $n \approx 0.5 \times 365$ so badly wrong? That insight comes from a scaling analysis.

▷ *At roughly what n does the probability of a shared birthday rise above 50 percent?*

An image often helps me translate a problem into mathematics. Imagine everyone in the room greeting all the others by shaking hands, one handshake at a time, and checking for a shared birthday with each handshake.

▷ *How many handshakes happen?*

Each person shakes hands with $n - 1$ others, making $n(n - 1)$ arrows from one person to another. But a handshake is shared between two people. To avoid counting each handshake twice, we need to divide $n(n-1)$ by 2. Thus, there are $n(n - 1)/2$ or roughly $n^2/2$ handshakes.

With 365 possible birthdays, the probability, per handshake, of a shared birthday is $1/365$. With $n^2/2$ handshakes, the probability that no handshake joins two people with a shared birthday is approximately

$$\left(1 - \frac{1}{365}\right)^{n^2/2}. \tag{4.57}$$

To approximate this probability, take its natural logarithm:

$$\ln\left(1 - \frac{1}{365}\right)^{n^2/2} = \frac{n^2}{2}\ln\left(1 - \frac{1}{365}\right). \tag{4.58}$$

Then approximate the logarithm using $\ln(1 + x) \approx x$ (for $x \ll 1$):

$$\ln\left(1 - \frac{1}{365}\right) \approx -\frac{1}{365}. \tag{4.59}$$

(You can test this useful approximation using a calculator, or see the pictorial explanation in *Street-Fighting Mathematics* [33, Section 4.3].) With that approximation,

$$\ln\left(\text{probability of no shared birthday}\right) \approx -\frac{n^2}{2} \times \frac{1}{365}. \tag{4.60}$$

When the probability of no shared birthday falls below 0.5, the probability of a shared birthday rises above 0.5. Therefore, the condition for having enough people is

$$\ln\left(\text{probability of no shared birthday}\right) < \ln\frac{1}{2}. \tag{4.61}$$

Because $\ln(1/2) = -\ln 2$, the condition simplifies to

$$\frac{n^2}{2 \times 365} > \ln 2. \tag{4.62}$$

Using the approximation $\ln 2 \approx 0.7$, the threshold n is approximately 22.6:

$$n > \sqrt{\ln 2 \times 2 \times 365} \approx 22.6. \tag{4.63}$$

People do not come in fractions, so we need 23 people. Indeed, the exact calculation for $n = 23$ gives the probability of sharing a birthday as 0.507!

From this scaling analysis, we can compactly explain what went wrong with the plausible conjecture. It assumed that the shared-birthday probability p is proportional to the number of people n, reaching $p = 0.5$ when $n = 0.5 \times 365$. The handshake picture, however, shows that the probability is related to the number of handshakes, which is proportional to n^2. What a difference from a simple change in a scaling exponent!

Problem 4.24 Three people sharing a birthday

Extend our scaling analysis to the three-birthday problem: How many people must be in a room before the probability of three people sharing a birthday rises above 0.5? (The results of the exact calculation, along with many approximations, are given by Diaconis and Mosteller [10].)

Problem 4.25 Scaling of bubble sort

The simplest algorithm for sorting—for example, to sort a list of n web pages according to relevance—is bubble sort. You go through the list in passes, comparing neighboring items and swapping any that are out of order.

a. How many passes through the list do you need in order to guarantee that the list is sorted?

b. The running time t (the time to sort the list) is proportional to the number of comparisons. What is the scaling exponent β in $t \propto n^\beta$?

Problem 4.26 Scaling of merge sort

Bubble sort (Problem 4.25) is easy to describe, but there is an alternative that is almost as easy: merge sort. It is a recursive, divide-and-conquer algorithm. You divide the list of n items into two equal parts (assume that n is a power of 2), and sort each half using merge sort. To make the sorted list, just merge the two sorted halves.

a. Here is a list of eight randomly generated numbers: 98, 33, 34, 62, 31, 58, 61, and 15. Draw a tree illustrating how merge sort works on this list. Use two lines for each internal node, showing on one line the original list and on the second line the sorted list.

b. The running time t is proportional to the number of comparison operations. What are the scaling exponents α and β in

$$t \propto n^\alpha (\log n)^\beta? \tag{4.64}$$

c. If you were a revenue agency and had to sort the tax records of all residents in a country, which algorithm would you use, merge sort or bubble sort?

Problem 4.27   Scaling of standard multiplication

In the usual school algorithm for multiplying n digit numbers, you find and add partial products. If the running time t is proportional to the number of single-digit multiplications, what is the scaling exponent β in $t \propto n^{\beta}$?

Problem 4.28   Scaling of Karatsuba multiplication

The Karatsuba algorithm for multiplying two n-digit numbers, discovered by Anatoly Karatsuba in 1960 [28] and published in 1962 [29], was the first development in the theory of multiplication in centuries. Similarly to merge sort, you first break each number into two equal-length halves. For example, you split 2743 into 27 and 43. Using Karatsuba multiplication recursively, you form three products from those halves, and combine the three products to get the original product. The subdividing stops when the numbers are short enough to be multiplied by the computer's hardware (typically, at 32 or 64 bits).

The expensive step, repeated many times, is hardware multiplication, so the running time t is proportional to the number of hardware multiplications. Find the scaling exponent β in $t \propto n^{\beta}$. (You will find a scaling exponent that occurs rarely in physical scalings, namely an irrational number.) How does this β compare with the exponent in the school algorithm for multiplication (Problem 4.27)?

4.5 Logarithmic scales in two dimensions

Scaling relations, which are so helpful in understanding the physical and mathematical worlds, have a natural representation on logarithmic scales, the representation that we introduced in Section 3.1.3 whereby distances correspond to factors or ratios rather than to differences.

As an example, here is the gravitational strength g at a distance r from the center of a planet. Outside the planet, $g \propto r^{-2}$ (the inverse-square law of gravitation). On linear axes, the graph of g versus r looks curved, like a hyperbola. However, its exact shape is hard to identify; the graph does not make the relation between g and r obvious. For example, let's try to represent our favorite scaling analysis: When r doubles, from, say, R_{Earth} to $2R_{\text{Earth}}$, 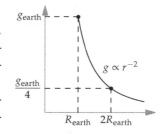 the gravitational acceleration falls from g_{Earth} to $g_{\text{Earth}}/4$. The graph shows that $g(2R)$ is smaller than $g(R)$; unfortunately, the scaling is hard to extract from these points or from the curve.

However, using logarithmic scales for both g and r—called log–log axes—makes the relation clear. Let's try again to represent that when r doubles, the gravitational acceleration falls by a factor of 4. Call 1 unit on either logarithmic scale a factor of 2. The factor of 2 increase in r corresponds to moving 1 unit to the right. The factor of 4 decrease in g corresponds to moving 2 units downward. Therefore, the graph of g versus r is a straight line—and its slope is -2. On logarithmic scales, scaling relations turn into straight lines whose slope is the scaling exponent.

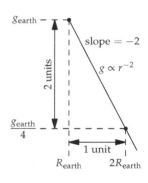

Problem 4.29 Sketching gravitational field strength

Imagine, as in Problem 4.13, a uniform spherical planet with radius R. Sketch, on log–log axes, gravitational field strength g versus r, the distance from the center of the planet. Include the regions $r < R$ and $r \geq R$.

Many natural processes, often for unclear reasons, obey scaling relations. A classic example is Zipf's law. In its simplest form, it states that the frequency of the kth most common word in a language is proportional to $1/k$. The table gives the frequencies of the three most frequent English words. For English, Zipf's law holds reasonably well up to $k \sim 1000$.

word	rank	frequency
the	1	7.0%
of	2	3.5
and	3	2.9

Zipf's law is useful for estimation. For example, suppose that you have to estimate the government budget of Delaware, one of the smallest US states (Problem 4.30). A key step in the estimate is the population. Delaware might be the smallest state, at least in area, and it is probably nearly the smallest in population. To use Zipf's law, we need the data for the most populous state. That information I happen to remember (information about the biggest item is usually easier to remember than information about the smallest item): The most populous state is California, with roughly 40 million people. Because the United States has 50 states, Zipf's law predicts that the smallest state will have a population 1/50th of California's, or roughly 1 million. This estimate is very close to Delaware's actual population: 917 000.

Problem 4.30 Government budget of Delaware

Based on the population of Delaware, estimate its annual government budget. Then look up the value and check your estimate.

4.6　Optimizing flight speed

With our skills in proportional reasoning, let's return to the energy and power consumption in forward flight (Section 3.6.2). In that analysis, we took the flight speed as a given; based on the speed, we estimated the power required to generate lift. Now we can estimate the flight speed itself. We will do so by estimating the speed that minimizes energy consumption.

4.6.1　Finding the optimum speed

Flying requires generating lift and fighting drag. To find the optimum flight speed, we need to estimate the energy required by each process. The lift was the subject of Section 3.6.2, where we estimated the power required as

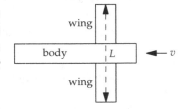

$$P_{\text{lift}} \sim \frac{(mg)^2}{\rho_{\text{air}} v L^2},\tag{4.65}$$

where m is the plane's mass, v is its forward velocity, and L is its wingspan. The lift energy required to fly a distance d is this power multiplied by the travel time d/v:

$$E_{\text{lift}} = P_{\text{lift}} \frac{d}{v} \sim \frac{(mg)^2}{\rho_{\text{air}} v^2 L^2} d.\tag{4.66}$$

In fighting drag, the energy consumed is the drag force times the distance. The drag force is

$$F_{\text{drag}} = \frac{1}{2} c_{\text{d}} \rho_{\text{air}} v^2 A_{\text{cs}},\tag{4.67}$$

where c_{d} is the drag coefficient. To simplify comparing the energies required for lift and drag, let's write the drag force as

$$F_{\text{drag}} = C \rho_{\text{air}} v^2 L^2,\tag{4.68}$$

where C is a modified drag coefficient: It doesn't use the $1/2$ that is part of the usual combination $c_{\text{d}}/2$, and it is measured relative to the squared wingspan L^2 rather than to the cross-sectional area A_{cs}.

▷　*For a 747, which drag coefficient, c_d or C, is smaller?*

The drag force has two equivalent forms:

$$F_{\text{drag}} = \rho_{\text{air}} v^2 \times \begin{cases} CL^2 \\ \frac{1}{2} c_d A_{\text{cs}}. \end{cases} \qquad (4.69)$$

Thus, $CL^2 = c_d A_{\text{cs}}/2$. For a plane, the squared wingspan L^2 is much larger than the cross-sectional area, so C is much smaller than c_d.

With the new form for F_{drag}, the drag energy is

$$E_{\text{drag}} = C \rho_{\text{air}} v^2 L^2 d, \qquad (4.70)$$

and the total energy required for flying is

$$E_{\text{total}} \sim \underbrace{\frac{(mg)^2}{\rho_{\text{air}} v^2 L^2} d}_{E_{\text{lift}}} + \underbrace{C \rho_{\text{air}} v^2 L^2 d}_{E_{\text{drag}}}. \qquad (4.71)$$

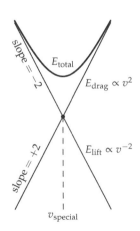

This formula looks intimidating because of the many parameters such as m, g, L, and ρ_{air}. As our interest is the flight speed v (in order to find the total energy), let's use proportional reasoning to reduce the energies to their essentials: $E_{\text{drag}} \propto v^2$ and $E_{\text{lift}} \propto v^{-2}$. On log–log axes, each relation is a straight line with slope $+2$ for drag and slope -2 for lift. Because the relations have different slopes, corresponding to different scaling exponents, their graphs must intersect.

To interpret the intersection, let's incorporate a sketch of the total energy $E_{\text{total}} = E_{\text{drag}} + E_{\text{lift}}$. At low speeds, the dominant consumer of energy is lift because of its v^{-2} dependence, so the sketch of E_{total} follows E_{lift}. At high speeds, the dominant consumer of energy is drag because of its v^2 dependence, so the sketch of E_{total} follows E_{drag}. Between these extremes is an optimum speed that minimizes the energy consumption. (Because the axes are logarithmic, and $\log(a + b) \neq \log a + \log b$, the sum of the two straight lines is not straight!) This optimum speed, marked v_{special}, is the speed at which E_{drag} and E_{lift} cross.

▷ *Why is the optimum right at the crossing speed, rather than faster or slower than the crossing speed?*

Symmetry! Try Problem 4.31 and then look afresh at Section 3.2.3, where we minimized E_{total} without the benefit of log–log axes.

Flying faster or slower than the optimum speed means consuming more energy. That extra consumption cannot always be avoided. A plane is designed so that its cruising speed is its minimum-energy speed. At takeoff and landing, when it flies far below the minimum-energy speed, a plane must work harder to stay aloft, which is one reason that the engines are so loud at takeoff and landing (another reason is that the engine noise reflects off the ground and back to the plane).

The optimization constraint, that a plane flies at the minimum-energy speed, allows us to eliminate v from the total energy. As we saw on the graph, at the minimum-energy speed, the drag and lift energies are equal:

$$\underbrace{\frac{(mg)^2}{\rho_{air}v^2L^2}d}_{E_{lift}} \sim \underbrace{C\rho_{air}v^2L^2d,}_{E_{drag}} \tag{4.72}$$

or, solving for mg (and rejoicing that d canceled out),

$$mg \sim \sqrt{C}\rho_{air}v^2L^2. \tag{4.73}$$

Now we could solve for v explicitly (and you get to find and use that solution in Problems 4.34 and 4.36). However, here we are interested in the total energy, not the flight speed itself. To find the energy without finding the speed, notice a reusable idea, an abstraction, within the right side of the equation for mg—which otherwise looks like a mess.

Namely, the mess contains $\rho_{air}v^2L^2$, which is F_{drag}/C. Therefore, when the plane is flying at the minimum-energy speed,

$$mg \sim \sqrt{C}\,\frac{F_{drag}}{C}, \tag{4.74}$$

so

$$F_{drag} \sim \sqrt{C}\,mg. \tag{4.75}$$

Thanks to our abstraction and all the surrounding approximations, we learn a surprisingly simple relation between the drag force and the plane's weight, connected through the square root of our modified drag coefficient C.

The energy consumed by drag—which, within a factor of 2, is also the total energy—is the drag force times the distance d. Therefore, the total energy is given by

$$E_{total} \sim \underbrace{E_{drag}}_{F_{drag}d} \sim \sqrt{C}\,mgd. \tag{4.76}$$

By using an optimum flight speed, we have eliminated the flight speed v from the total energy.

▷ *Does the total energy depend in reasonable ways upon m, g, C, and d?*

Yes! First, lift overcomes the weight mg; therefore, the energy should, and does, increase with mg. Second, a streamlined plane (low C) should use less energy than a bluff, blocky plane (high C); the energy should, and does, increase with the modified drag coefficient C. Finally, because the plane is flying at a constant speed, the energy should be, and is, proportional to the travel distance d.

4.6.2 Flight range

From the total energy, we can estimate the range of the 747—the distance that it can fly on a full tank of fuel. The energy is $\sqrt{C}\,mgd$, so the range d is

$$d \sim \frac{E_{\text{fuel}}}{\sqrt{C}\,mg}, \tag{4.77}$$

where E_{fuel} is the energy in the full tank of fuel. To estimate d, we need to estimate E_{fuel}, the modified drag coefficient C, and maybe also the plane's mass m.

▷ *How much energy is in a full tank of fuel?*

The fuel energy is the fuel mass times its energy density (as an energy per mass). Let's describe the fuel mass relative to the plane's mass m, as a fraction β of the plane's mass: $m_{\text{fuel}} = \beta m$. Using $\mathcal{E}_{\text{fuel}}$ for the energy density of fuel,

$$E_{\text{fuel}} = \beta m \mathcal{E}_{\text{fuel}}. \tag{4.78}$$

When we put this expression into the range d, the plane's mass m in E_{fuel}, which is the numerator of d, cancels the m in the denominator $\sqrt{C}\,mg$. The range then simplifies to a relation independent of mass:

$$d \sim \frac{\beta \mathcal{E}_{\text{fuel}}}{\sqrt{C}\,g}. \tag{4.79}$$

For the fuel fraction β, a reasonable guess for long-range flight is $\beta \approx 0.4$: a large portion of the payload is fuel. For the energy density $\mathcal{E}_{\text{fuel}}$, we learned in Section 2.1 that $\mathcal{E}_{\text{fuel}}$ is roughly 4×10^7 joules per kilogram (9 kilocalories per gram). This energy density is what a perfect engine would extract. At the usual engine efficiency of one-fourth, $\mathcal{E}_{\text{fuel}}$ becomes roughly 10^7 joules per kilogram.

▷ *How do we find the modified drag coefficient C?*

This estimate is the trickiest part of the range estimate, because C needs to be converted from more readily available data. According to Boeing, the manufacturer of the 747, a 747 has a drag coefficient of $C' \approx 0.022$. As the prime symbol indicates, C' is yet another drag coefficient. It is measured using the wing area A_{wing} and uses the traditional $1/2$:

$$F_{\text{drag}} = \frac{1}{2}C'A_{\text{wing}}\rho_{\text{air}}v^2. \tag{4.80}$$

So we have three drag coefficients, depending on whether the coefficient is referenced to the cross-sectional area A_{cs} (the usual definition of c_{d}), to the wing area A_{wing} (Boeing's definition of C'), or to the squared wingspan L^2 (our definition of C, which also lacks the factor of $1/2$ that the others have).

To convert between these definitions, compare the three ways to express the drag force:

$$F_{\text{drag}} = \rho_{\text{air}}v^2 \times \begin{cases} \dfrac{1}{2}C'A_{\text{wing}} & \text{(Boeing's definition),} \\[2mm] CL^2 & \text{(our definition),} \\[2mm] \dfrac{1}{2}c_{\text{d}}A_{\text{cs}} & \text{(the usual definition).} \end{cases} \tag{4.81}$$

From this comparison,

$$\frac{1}{2}C'A_{\text{wing}} = CL^2 = \frac{1}{2}c_{\text{d}}A_{\text{cs}}. \tag{4.82}$$

The conversion from C' to C is therefore

$$C = \frac{1}{2}C'\frac{A_{\text{wing}}}{L^2}. \tag{4.83}$$

Using l for the wing's front-to-back length, the wing area becomes $A_{\text{wing}} = Ll$. Then the area ratio A_{wing}/L^2 simplifies to l/L (the reciprocal of the wing aspect ratio), and the drag coefficient C simplifies to

$$C = \frac{1}{2}C'\frac{l}{L}. \tag{4.84}$$

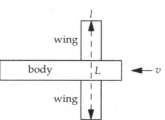

For a 747, $l \approx 10$ meters, $L \approx 60$ meters, and $C' \approx 0.022$, so $C \approx 1/600$:

$$C \sim \frac{1}{2} \times 0.022 \times \frac{10\,\text{m}}{60\,\text{m}} \approx \frac{1}{600}. \tag{4.85}$$

The resulting range is roughly $10\,000$ kilometers:

$$d \sim \frac{\overbrace{0.4 \times 10^7\,\text{J}\,\text{kg}^{-1}}^{\mathcal{E}_{\text{fuel}}}}{\underbrace{\sqrt{1/600}}_{\sqrt{C}} \times \underbrace{10\,\text{m}\,\text{s}^{-2}}_{g}} \approx 10^7\,\text{m} = 10^4\,\text{km}. \tag{4.86}$$

with $\overbrace{}^{\beta}$ over the 0.4.

The actual maximum range of a 747-400 is $13\,450$ kilometers (a figure we used in Section 4.3.2): Our approximate analysis is amazingly accurate.

Problem 4.31   Graphical interpretation of minimization using symmetry

In Section 3.2.3, we started with the general form for the total energy

$$E \sim \underbrace{\frac{A}{v^2}}_{E_{\text{lift}}} + \underbrace{Bv^2}_{E_{\text{drag}}} \tag{4.87}$$

and found the optimum (minimum-energy) speed by constructing the following symmetry operation:

$$v \longleftrightarrow \frac{\sqrt{A/B}}{v}. \tag{4.88}$$

On the log–log plot of the energies versus v, what is the geometric interpretation of this symmetry operation?

Problem 4.32   Minimum-power speed

We estimated the flight speed v_E that minimizes energy consumption. We could also have estimated v_P, the speed that minimizes power consumption. What is the ratio v_P/v_E? Before doing an exact calculation, sketch P_{lift}, P_{drag}, and $P_{\text{lift}} + P_{\text{drag}}$ (on log–log axes) and place v_P on the correct side of v_E. Then check your placement against the result of the exact calculation.

Problem 4.33   Coefficient of lift

Just as we can define a dimensionless drag coefficient c_d, where

$$F_{\text{drag}} = \frac{1}{2}c_d\rho_{\text{air}}Av^2, \tag{4.89}$$

we can define a dimensionless lift coefficient c_L, where

$$F_{\text{lift}} = \frac{1}{2}c_L\rho_{\text{air}}Av^2, \tag{4.90}$$

where the area A here is usually the wing area. (The same formula structure shows up in drag and lift—abstraction!) Estimate c_L for a 747 in cruising flight.

4.6.3 Flight range versus size

In proportional reasoning, we ask, "How does one quantity change when we change an independent variable (for example, if we double an independent variable)?" Because flying objects come in such a wide range of sizes, a natural independent variable is size.

▷ *How does the flight range depend on the plane's size?*

Let's assume that all planes are geometrically similar—that is, that they have the same shape and differ only in their size. Now let's see how changing the plane's size changes the quantities in the range

$$d \sim \frac{\beta \mathcal{E}_{\text{fuel}}}{\sqrt{C} g}. \tag{4.91}$$

The gravitational acceleration g remains fixed. The fuel energy density $\mathcal{E}_{\text{fuel}}$ and the fuel fraction β also remain fixed—which shows again the benefit of using intensive quantities, as we discussed in Section 4.2.4 when we made a scaling relation for planetary surface gravity. Finally, the drag coefficient C depends only on the plane's shape (on how streamlined it is), not on its size, so it too remains constant. Therefore, the flight range is independent of the plane's size!

A further surprise comes from comparing the range of planes with the range of migrating birds. The proportionality $d \propto \beta \mathcal{E}_{\text{fuel}} / \sqrt{C}$ is shorthand for the comparison

$$\frac{d_{\text{plane}}}{d_{\text{bird}}} = \frac{\beta_{\text{plane}}}{\beta_{\text{bird}}} \frac{\mathcal{E}_{\text{plane}}^{\text{fuel}}}{\mathcal{E}_{\text{bird}}^{\text{fuel}}} \left(\frac{C_{\text{plane}}}{C_{\text{bird}}} \right)^{-1/2}. \tag{4.92}$$

The ratio of fuel fractions is roughly 1: For the plane, $\beta \approx 0.4$; and a bird, having eaten all summer, is perhaps 40 percent fat (the bird's fuel). The energy density in jet fuel and fat is similar, as is the efficiency of engines and animal metabolism (at about 25 percent). Therefore, the ratio of energy densities is also roughly 1. Finally, a bird has a similar shape to a plane, so the ratio of drag coefficients is also roughly 1. Therefore, planes and well-fattened migrating birds should have a similar maximum range, about 10 000 kilometers.

Let's check. The longest known nonstop flight by an animal is 11 680 kilometers: made by a bar-tailed godwit tracked by satellite by Robert Gill and his colleagues [19] as the bird flew nonstop from Alaska to New Zealand!

4.7 Summary and further problems

Proportional reasoning focuses our attention on how one quantity determines another. (A wonderful collection of pointers to further reading is the *American Journal of Physics*'s "Resource Letter" on scaling laws [49].) By guiding us toward what is often the most important characteristic of a problem, the scaling exponent, it helps us discard spurious complexity.

Problem 4.34 Cruising speed versus mass
For geometrically similar animals (the same shape and composition but different sizes) in forward flight, how does the animal's minimum-energy flight speed v depend on its mass m? In other words, what is the scaling exponent β in $v \propto m^\beta$?

Problem 4.35 Hovering power versus size
In Section 3.6.1, we derived the power required to hover. For geometrically similar birds, how does the power per mass depend on the animal's size L? In other words, what is the scaling exponent γ in $P_{\text{hover}}/m \propto L^\gamma$? Why are there no large hummingbirds?

Problem 4.36 Cruising speed versus air density
How does a plane's (or a bird's) minimum-energy speed v depend on ρ_{air}? In other words, what is the scaling exponent γ in $v \propto \rho_{\text{air}}^\gamma$?

Problem 4.37 Speed of a bar-tailed godwit
Use the results of Problem 4.34 and Problem 4.36 to write the ratio $v_{747}/v_{\text{godwit}}$ as a product of dimensionless factors, where v_{747} is the minimum-energy (cruising) speed of a 747, and v_{godwit} is the minimum-energy (cruising) speed of a bar-tailed godwit. Using $m_{\text{godwit}} \approx 400$ grams, estimate the cruising speed of a bar-tailed godwit. Compare your result to the average speed of the record-setting bar-tailed godwit that was studied by Robert Gill and his colleagues [19], which made its 11 680-kilometer journey in 8.1 days.

Problem 4.38 Thermal resistance of a house versus a tea mug
When we developed the analogy between low-pass electrical and thermal filters (Section 2.4.5)—whether RC circuits, tea mugs, or houses—we introduced the abstraction of thermal resistance R_{thermal}. In this problem, you estimate the ratio of thermal resistances $R_{\text{thermal}}^{\text{house}}/R_{\text{thermal}}^{\text{tea mug}}$.

House walls are thicker than teacup walls. Because thermal resistance, like electrical resistance, is proportional to the length of the resistor, the house's thicker walls raise its thermal resistance. However, the house's larger surface area, like having many resistors in parallel, lowers the house's thermal resistance. Estimate the size of these two effects and thus the ratio of the two thermal resistances.

Problem 4.39   General birthday problem

Extend the analysis of Problem 4.24 to k people sharing a birthday. Then compare your predictions to the exact results given by Diaconis and Mosteller in [10].

Problem 4.40   Flight of the housefly

Estimate the mechanical power required for a common housefly *Musca domestica* ($m \approx 12$ milligrams) to hover. From everyday experience, estimate its typical flight speed. At this flight speed, compare the power requirements for forward flight and for hovering.

5
Dimensions

In 1906, Los Angeles received 540 millimeters of precipitation (rain, snow, sleet, and hail).

▷ *Is this rainfall large or small?*

On the one hand, 540 is a large number, so the rainfall is large. On the other hand, the rainfall is also 0.00054 kilometers, and 0.00054 is a tiny number, so the rainfall is small. These arguments contradict each other, so at least one must be wrong. Here, both are nonsense.

A valid argument comes from a meaningful comparison—for example, comparing 540 millimeters per year with worldwide average rainfall—which we estimated in Section 3.4.3 as 1 meter per year. In comparison to this rainfall, Los Angeles in 1906 was dry. Another meaningful comparison is with the average rainfall in Los Angeles, which is roughly 350 millimeters per year. In comparison, 1906 was a wet year in Los Angeles.

In the nonsense arguments, changing the units of length changed the result of the comparison. In contrast, the meaningful comparisons are independent of the system of units: No matter what units we select for length and time, the ratio of rainfalls does not change. In the language of symmetry, which we met in Chapter 3, changing units is the symmetry operation, and meaningful comparisons are the invariants. They are invariant because they have no dimensions. When there is change, look for what does not change: Make only dimensionless comparisons.

This criterion is necessary for avoiding nonsense; however, it is not sufficient. To illustrate the difficulty, let's compare rainfall with the orbital speed of the Earth. Both quantities have dimensions of speed, so their ratio is invariant under a change of units. However, judging the wetness or dryness of Los Angeles by comparing its rainfall to the Earth's orbital speed produces nonsense.

Here is the moral of the preceding comparisons. A quantity with dimensions is, by itself, meaningless. It acquires meaning only when compared with a relevant quantity that has the same dimensions. This principle underlies our next tool: dimensional analysis.

Problem 5.1   Book boxes are heavy
In Problem 1.1, you estimated the mass of a small moving-box packed with books. Modify your calculation to use a medium moving-box, with a volume of roughly 0.1 cubic meters. Can you think of a (meaningful!) comparison to convince someone that the resulting mass is large?

Problem 5.2   Making energy consumption meaningful
The United States' annual energy consumption is roughly 10^{20} joules. Suggest two comparisons to make this quantity meaningful. (Look up any quantities that you need to make the estimate, except the energy consumption itself!)

Problem 5.3   Making solar power meaningful
In Problem 3.31, you should have found that the solar power falling on Earth is roughly 10^{17} watts. Suggest a comparison to make this quantity meaningful.

Problem 5.4   Energy consumption by the brain
The human brain consumes about 20 watts, and has a mass of 1–2 kilograms.

a. Make the power more meaningful by estimating the brain's fraction of the body's power consumption:

$$\frac{\text{brain power}}{\text{basal metabolism}}. \tag{5.1}$$

b. Make this fraction even more meaningful by estimating the ratio

$$\frac{\text{the brain's fraction of the body's power consumption}}{\text{the brain's fraction of the body's mass}}. \tag{5.2}$$

> **Problem 5.5   Making oil imports meaningful**
>
> In Section 1.4, we estimated that the United States imports roughly 3×10^9 barrels of oil per year. This quantity needs a comparison to make it meaningful. As one possibility, estimate the ratio
>
> $$\frac{\text{cost of the imported oil}}{\text{US military spending to "defend" oil-rich regions}}.$$ (5.3)
>
> If this ratio is less than 1, suggest why the US government does not cancel that part of the military budget and use the savings to provide US consumers with free imported oil.
>
> **Problem 5.6   Making the energy in a 9-volt battery meaningful**
>
> Using your estimate in Problem 1.11 for the energy in a 9–volt battery, estimate
>
> $$\frac{\text{energy content of the battery}}{\text{cost of the battery}}.$$ (5.4)
>
> Compare that quotient to the same quotient for electricity from the wall socket.

5.1 Dimensionless groups

Because dimensionless quantities are the only meaningful quantities, we can understand the world better by describing it in terms of dimensionless quantities. But we need to find them.

5.1.1 Finding dimensionless groups

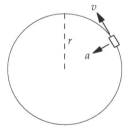

To illustrate finding these dimensionless quantities, let's try an example that uses familiar physics: When learning a new idea, it is helpful to try it on a familiar example. We'll find a train's inward acceleration as it moves on a curved track. The larger the acceleration, the more the track or the train needs to tilt so that the passengers do not feel uncomfortable and (if the track is not tilted enough) the train does not tip over.

Our goal is the relation between the train's acceleration a, its speed v, and the track's radius of curvature r. In our state of knowledge now, the relation could be almost anything. Here are a few possibilities:

$$\frac{a + v^2}{r} = \frac{v^3}{a}; \qquad \frac{r + a^2}{v} = \frac{a^2}{v + r}; \qquad \frac{v}{ra + v^3} = \frac{a + v}{r^2}.$$ (5.5)

Although those possibilities are bogus, among the vast sea of possible relations, one relation is correct.

To find the constraint that will shrink the sea to a few drops, first defocus our eyes and see all the choices as examples of the general form

$$\boxed{\text{blob A}} = \left\langle \text{blob B} \right\rangle \tag{5.6}$$

some function of   another function of
$a, v,$ and r $a, v,$ and r

Even though the blobs might be complicated functions, they must have identical dimensions. By dividing both sides by blob B, we get a simpler form:

$$\frac{\boxed{\text{blob A}}}{\left\langle \text{blob B} \right\rangle} = 1. \tag{5.7}$$

Now each side is dimensionless. Therefore, whatever the relation between $a, v,$ and r, we can write it in dimensionless form.

The preceding process is not limited to this problem. In any valid equation, all the terms have identical dimensions. For example, here is the total energy in a spring–mass system:

$$E_{\text{total}} = \frac{1}{2}mv^2 + \frac{1}{2}kx^2. \tag{5.8}$$

The kinetic-energy term ($mv^2/2$) and the potential-energy term ($kx^2/2$) have the same dimensions (of energy). The same process of dividing by one of the terms turns any equation into a dimensionless equation. As a result, any equation can be written in dimensionless form. Because the dimensionless forms are a small fraction of all possible forms, this restriction offers us a huge reduction in complexity.

To benefit from this reduction, we must compactly describe all such forms: Any dimensionless form can be built from *dimensionless groups*. A dimensionless group is the product of the quantities, each raised to a power, such that the product has no dimensions. In our example, where the quantities are $a, v,$ and r, any dimensionless group G has the form

$$G \equiv a^x v^y r^z, \tag{5.9}$$

where G has no dimensions and where the exponents x, y, and z are real numbers (possibly negative or zero).

Because any equation describing the world can be written in dimensionless form, and because any dimensionless form can be written using dimensionless groups, any equation describing the world can be written using dimensionless groups.

▷ *That news is welcome, but how do we find these groups?*

The first step is to tabulate the quantities with their descriptions. By convention, capital letters are used to represent dimensions. For now, the possible dimensions are length (L), mass (M), and time (T). Then, for example, the dimensions of v are length per time or, more compactly, LT^{-1}.

a	LT^{-2}	acceleration
v	LT^{-1}	speed
r	L	radius

Then, by staring at the table, we find all possible dimensionless groups. A dimensionless group contains no length, mass, or time dimensions. For our example, let's start by getting rid of the time dimension. It occurs in a as T^{-2} and in v as T^{-1}. Therefore, any dimensionless group must contain a/v^2. This quotient has dimensions of L^{-1}. To make it dimensionless, multiply it by the only quantity that is purely a length, which is the radius r. The result, ar/v^2, is a dimensionless group. In the $a^x v^y r^z$ form, it is $a^1 v^{-2} r^1$.

▷ *Are there other dimensionless groups?*

To get rid of time, we started with a/v^2 and then ended, inevitably, with the group ar/v^2. To make another dimensionless group, we would have to choose another starting point. However, the only starting points that get rid of time are powers of a/v^2—for example, v^2/a or a^2/v^4—and those choices lead to the corresponding power of ar/v^2. Therefore, any dimensionless group can be formed from ar/v^2. Our three quantities a, r, and v produce exactly one *independent* dimensionless group.

As a result, any statement about the circular acceleration can be written using only ar/v^2. All dimensionless statements using only ar/v^2 have the general form

$$\frac{ar}{v^2} = \text{dimensionless constant},\tag{5.10}$$

because there are no other independent dimensionless groups to use on the right side.

▷ *Why can't we use ar/v^2 on the right side?*

We can, but it doesn't create new possibilities. As an example, let's try the following dimensionless form:

$$\frac{ar}{v^2} = 3\left(\frac{ar}{v^2}\right) - 1. \tag{5.11}$$

Its solution is $ar/v^2 = 1/2$, which is another example of our general form

$$\frac{ar}{v^2} = \text{dimensionless constant}. \tag{5.12}$$

▷ *But what if we use a more complicated function?*

Let's try one:

$$\frac{ar}{v^2} = \left(\frac{ar}{v^2}\right)^2 - 1. \tag{5.13}$$

Its solutions are

$$\frac{ar}{v^2} = \begin{cases} \phi, \\ -1/\phi, \end{cases} \tag{5.14}$$

where ϕ is the golden ratio (1.618...). Deciding between these solutions would require additional information or requirements, such as the sign convention for the acceleration. Even so, each solution is another example of the general form

$$\frac{ar}{v^2} = \text{dimensionless constant}. \tag{5.15}$$

Using that form, which we cannot escape, the acceleration of the train is

$$a \sim \frac{v^2}{r}, \tag{5.16}$$

where the $\sim$ contains the (unknown) dimensionless constant. In this case, the dimensionless constant is 1. However, dimensional analysis, as this procedure is called, does not tell us its value—which would come from a calculus analysis or, approximately, from a lumping analysis (Section 6.3.4).

Using $a \sim v^2/r$, we can now estimate the inward acceleration of the train going around a curve. Imagine a moderately high-speed train traveling at $v \approx 60$ meters per second (approximately 220 kilometers or 135 miles per hour). At such speeds, railway engineers specify that the track's radius of curvature be at least 2 or 3 kilometers. Using the smaller radius of curvature, the inward acceleration becomes

$$a \sim \frac{(60 \, \text{m s}^{-1})^2}{2 \times 10^3 \, \text{m}} = 1.8 \, \text{m s}^{-2}. \qquad (5.17)$$

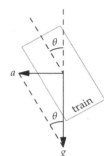

Because no quantity with dimensions is large or small on its own, this acceleration by itself is meaningless. It acquires meaning in comparison with a relevant acceleration: the gravitational acceleration g. For this train, the dimensionless ratio a/g is approximately 0.18. This ratio is also $\tan \theta$, where θ is the train's tilt that would make the passengers feel a net force perpendicular to the floor. With $a/g \approx 0.18$, the comfortable tilt angle is approximately 10°. Indeed, tilting trains can tilt up to 8°. (This range is usually sufficient, as a full tilt is disconcerting: One would see a tilted ground but would still feel gravity acting as it normally does, along one's body axis.)

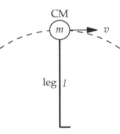

Using our formula for circular acceleration, we can also estimate the maximum walking speed. In walking, one foot is always in contact with the ground. As a rough model of walking, the whole body, represented as a point mass at its center of mass (CM), pivots around the foot in contact with the ground—as if the body were an inverted pendulum. If you walk at speed v and have leg length l, then the resulting circular acceleration (the acceleration toward the foot) is v^2/l. When you walk fast enough such that this acceleration is more than g, gravity cannot supply enough acceleration.

This change happens when $v \sim \sqrt{gl}$. Then your foot leaves the ground, and the walk turns into a run. Therefore, gait is determined by the dimensionless ratio v^2/gl. This ratio, which also determines the speed of water waves (Problem 5.15) and of ships (Problem 5.64), is called the Froude number and abbreviated Fr.

Here are three ways to check whether our pendulum model of walking is reasonable. First, the resulting formula,

$$v_{\text{max}} \sim \sqrt{gl}, \qquad (5.18)$$

explains why tall people, with a longer leg length, generally walk faster than short people.

Second, it predicts a reasonable maximum walking speed. With a leg length of $l \sim 1$ meter, the limit is 3 meters per second or about 7 miles per hour:

$$v_{\text{max}} \sim \sqrt{10 \, \text{m s}^{-2} \times 1 \, \text{m}} \approx 3 \, \text{m s}^{-1}. \qquad (5.19)$$

This prediction is consistent with world-record racewalking speeds, where the back toe may not leave the ground until the front heel has touched the ground. The world records for 20-kilometer racewalking are 1h:24m:50s for women and 1h:17m:16s for men. The corresponding speeds are 3.9 and 4.3 meters per second.

The third test is based on gait. In a fascinating experiment, Rodger Kram and colleagues [31] reduced the effective gravity g. This reduction changed the speed of the walking–running transition, but the speed still satisfied $v^2/gl \sim 0.5$. The universe cares only about dimensionless quantities!

Another moral of this introduction to dimensional analysis is that every dimensionless group is an abstraction. Here, the group ar/v^2 tells us that the universe cares about a, r, or v through the combination ar/v^2. This relation is described by the tree diagram (the edge label of -2 is the exponent of v in v^{-2}). Because there is only one independent dimensionless group, the universe cares only about ar/v^2. Therefore, the universe cares about a, r, and v *only* through the combination ar/v^2. This freedom not to worry about individual quantities simplifies our picture of the world.

5.1.2 Counting dimensionless groups

Finding the circular acceleration required finding all possible dimensionless groups and showing that all groups could be constructed from one group—say, ar/v^2. That reasoning followed a chain of constraints: To get rid of the time, the dimensionless group had to contain the quotient a/v^2; to get rid of the length, the quotient needed to be multiplied by the correct power of r.

▷ *For each problem, do we have to construct a similar chain of reasoning in order to count the dimensionless groups?*

Following the constraints is useful for finding dimensionless groups, but there is a shortcut for counting independent dimensionless groups. The number of independent groups is, roughly, the number of quantities minus the number of dimensions. A more precise statement will come later, but this version is enough to get us started.

Let's test it using the acceleration example. There are three quantities: a, v, and r. There are two dimensions: length (L) and time (T). There should be, and there is, one independent dimensionless group.

Let's also test the shortcut on a familiar physics formula:
$W = mg$, where W is an object's weight, m is its mass, and g
is the gravitational acceleration. There are three quantities
(W, m, and g) made from three dimensions (M, L, and T).

W	MLT^{-2}	weight
m	M	mass
g	LT^{-2}	gravity

Our shortcut predicts no dimensionless groups at all. However, W/mg is
dimensionless, which refutes the prediction!

▷ *What went wrong?*

Although the three quantities seem to contain three dimensions, the three
dimensional combinations actually used—force (MLT^{-2}), mass (M), and
acceleration (LT^{-2})—can be constructed from just two dimensions. For example, we can construct them from mass and acceleration.

▷ *But acceleration is not a fundamental dimension, so how can we use it?*

The notion of fundamental dimensions is a human convention, part of our
system of measurement. Dimensional analysis, however, is a mathematical
process. It cares neither about the universe nor about our conventions for
describing the universe. That lack of care may appear heartless and seem
like a disadvantage. However, it means that dimensional analysis is independent of our arbitrary choices, giving it power and generality.

As far as dimensional analysis is concerned, we can choose
any set of dimensions as our fundamental dimensions. The
only requirement is that they suffice to describe the dimensions of our quantities. Here, mass and acceleration suffice,

W	$M \times [a]$	weight
m	M	mass
g	$[a]$	gravity

as the rewritten dimensions table shows: Using the notation $[a]$ for "the
dimensions of a," the dimensions of W are $M \times [a]$ and the dimensions of
g are just $[a]$.

In summary, the three quantities contain only two *independent* dimensions.
Three quantities minus two independent dimensions produce one independent dimensionless group. Accordingly, the revised counting shortcut is

$$\begin{array}{l} \text{number of quantities} \\ - \ \text{number of independent dimensions} \\ \hline \text{number of independent dimensionless groups} \end{array} \qquad (5.20)$$

This shortcut, known as the Buckingham Pi theorem [4], is named after
Edgar Buckingham and his uppercase Pi (Π) notation for dimensionless
groups. (It is also the rank–nullity theorem of linear algebra [42].)

Problem 5.7   Bounding the number of independent dimensionless groups

Why is the number of independent dimensionless groups never more than the number of quantities?

Problem 5.8   Counting dimensionless groups

How many independent dimensionless groups do the following sets of quantities produce?

a. period T of a spring–mass system in a gravitational field: T, k (spring constant), m, x_0 (amplitude), and g.

b. impact speed v of a free-falling object: v, g, and h (initial drop height).

c. impact speed v of a downward-thrown free-falling object: v, g, h, and v_0 (launch velocity).

Problem 5.9   Using angular frequency instead of speed

Redo the dimensional-analysis derivation of circular acceleration using the radius r and the angular frequency ω as independent variables. Using $a = v^2/r$, find the dimensionless constant in the general dimensionless form

   dimensionless group proportional to a = dimensionless constant.   (5.21)

Problem 5.10   Impact speed of a dropped object

Use dimensional analysis to estimate the impact speed of a freely falling rock that is dropped from a height h.

Problem 5.11   Speed of gravity waves on deep water

In this problem, you use dimensional analysis to find the speed of waves on the open ocean. These waves are the ones you would see from an airplane and are driven by gravity. Their speed could depend on gravity g, their angular frequency ω, and the density of water ρ. Do the analysis as follows.

v	LT^{-1}	wave speed
g	LT^{-2}	gravity
ω	T^{-1}	angular freq.
ρ	ML^{-3}	water density

a. Explain why these quantities produce one independent dimensionless group.

b. What is the group proportional to v?

c. With the further information that the dimensionless constant is 1, predict the speed of waves with a period of 17 seconds (you can measure the period by timing the interval between each wave's arrival at the shore). This speed is also the speed of the winds that produced the waves. Is it a reasonable wind speed?

d. What would the dimensionless constant be if, in the table of quantities, angular frequency ω is replaced by the period T?

> **Problem 5.12   Using period instead of speed**
> In finding a formula for circular acceleration (Section 5.1.1), our independent vari-
> ables were the radius r and the speed v. Redo the dimensional-analysis derivation
> using the radius r and the period T as the independent variables.
>
> a. Explain why there is still only one independent dimensionless group.
>
> b. What is the independent dimensionless group proportional to a?
>
> c. What dimensionless constant is this group equal to?

5.2 One dimensionless group

The most frequent use of dimensional analysis involves one independent
dimensionless group—for example, ar/v^2 in our analysis of circular acceleration (Section 5.1.1). Let's look at this kind of case more closely, which
has lessons for more complicated problems.

5.2.1 Universal constants

What we will find is that dimensional analysis reduces complexity by
generating results with wide application. That is, in the general form

$$\text{independent dimensionless group} = \text{dimensionless constant}, \quad (5.22)$$

the dimensionless constant is universal. Let's see what universal means
through an example—the analysis of a small-amplitude pendulum. So,
imagine releasing a pendulum from a small angle θ_0.

▷ *What is its period of oscillation?*

The first step in dimensional analysis is to list the relevant
quantities. The list begins with the goal quantity—here,
the period T. It depends on gravity g, the string length
l, perhaps the mass m of the bob, and perhaps the amplitude θ_0. Fortunately, when the amplitude is small, as it is
here, then the amplitude turns out not to matter, so we won't list it.

T	T	period
g	LT^{-2}	gravity
l	L	string length
m	M	mass of bob

▷ *What are the dimensionless groups?*

These four quantities contain three independent dimensions (M, L, and T).
By the Buckingham Pi theorem (Section 5.1.2), they produce one independent dimensionless group. This group cannot contain m, because m is the

only quantity with dimensions of mass. A simple dimensionless group is gT^2/l. However, because our eventual goal is the period T, let's make a dimensionless group proportional to T itself rather than to T^2. This group is $T\sqrt{g/l}$. Then the most general dimensionless statement is

$$T\sqrt{\frac{g}{l}} = C, \tag{5.23}$$

where C is a dimensionless constant. Even though its value is, for the moment, unknown, it is *universal*. The same constant applies to a pendulum with a shorter string or, perhaps more surprisingly, to a pendulum on Mars, with its different gravitational strength. If we find the constant for one pendulum on one planet, we know it for all pendulums on all planets.

There are at least three ways to find C. The first is to solve the pendulum differential equation—which is hard work. The second approach is to solve a cleverly designed simpler problem (Problem 3.4). Although the approach is clever, it is not as general as the third method.

The third method is to measure C with a home experiment! For that purpose, I turned my key ring into a pendulum by hanging it from a string. The string was roughly twice the length of American letter paper (thus, 2 × 11 inches) and the period was roughly 1.5 seconds. Therefore,

$$C = T\sqrt{\frac{g}{l}} \approx 1.5\,\text{s} \times \sqrt{\frac{32\,\text{ft}\,\text{s}^{-2}}{2\,\text{ft}}} = 6. \tag{5.24}$$

key ring

▷ *Would the constant be different in a modern system of units?*

In metric units, $g \approx 9.8$ meters per second per second, $l \approx 0.6$ meters, and C is, within calculation inaccuracies, still 6:

$$C \approx 1.5\,\text{s} \times \sqrt{\frac{9.8\,\text{m/s}^2}{0.6\,\text{m}}} \approx 6. \tag{5.25}$$

The system of units does not matter—which is the reason for using dimensionless groups: They are invariant under a change of units. Even so, an explicit 6 probably would not appear if we solved the pendulum differential equation honestly. But a more precise measurement of C might suggest a closed form for this dimensionless constant.

The pendulum length, from the knot where I hold it to the center of the key ring is 0.65 meters (slightly longer than the rough estimate of 0.6 meters). Meanwhile, 10 periods took 15.97 seconds. Then C is close to 6.20:

$$C \approx 1.597\,\text{s} \times \sqrt{\frac{9.8\,\text{m s}^{-2}}{0.65\,\text{m}}} \approx 6.20. \tag{5.26}$$

▷ *What dimensionless numbers could produce 6.20?*

This value is close to 2π, which is approximately 6.28. That guess feels like a leap, but have courage. The guess is plausible once we remember that pendulums oscillate and oscillations often involve 2π, or that we are asking for the period rather than the angular frequency, a choice that often introduces a 2π (as in Problem 5.11(d)). The resulting period T is

$$T = C\sqrt{\frac{l}{g}} = 2\pi\sqrt{\frac{l}{g}}. \tag{5.27}$$

(For a physical explanation of the 2π, try Problem 3.4.)

This example shows how dimensional analysis, a mathematical approach, sits between two physical approaches. First we used our physical knowledge to list the relevant quantities. Then we reduced the space of possible relations by using dimensional analysis. Finally, to find the universal constant C, which dimensional analysis could not tell us, we again used physical knowledge (a home experiment).

Problem 5.13   Your own measurement
Make your own pendulum and measure the universal constant $T\sqrt{g/l}$.

Problem 5.14   Period of a spring–mass system
Use dimensional analysis to find, except for a dimensionless constant, the period T of a spring–mass system with spring constant k, mass m, and amplitude x_0. Find the dimensionless constant in the most general dimensionless statement

group proportional to T = dimensionless constant. (5.28)

Problem 5.15   Speed of waves in shallow water
In shallow water, where the wavelength is much larger than the depth, waves driven by gravity travel at a speed that could depend on gravity g, the water depth h, and the density of water ρ.

a. Find the independent dimensionless group proportional to v^2, where v is the wave speed. Compare this group to the Froude number, the dimensionless ratio that we introduced to study walking (Section 5.1.1).

b. What therefore is the scaling exponent β in the scaling relation $v \propto h^\beta$? Test your prediction by measuring $v(1\,\text{cm})$ and $v(4\,\text{cm})$. To make the measurement, fill

a baking dish with water and excite a sloshing wave by slightly lifting and
quickly setting down one end of the dish.

c. Using your data, estimate the universal (dimensionless) constant in the relation

dimensionless group from part (a) = dimensionless constant.   (5.29)

d. Predict the speed of tidal waves, which are (shallow-water!) waves created by
underwater earthquakes. How long does a tidal wave take to cross an ocean?

5.2.2 Atomic blast energy

Problems with only one dimensionless group need not be simple or
easy to solve by other methods. A classic example is finding the energy released by the first atomic bomb, detonated in the New Mexico
desert in 1945. The blast energy, or yield, was top secret. However,
declassified photographs of the blast, because they had a scale bar,
provided data on the radius of the fireball at several times after the
explosion. The data must have seemed innocuous enough to release.

t (ms)	R (m)
3.26	59.0
4.61	67.3
15.0	106.5
62.0	185.0

But from these data, G. I. Taylor of Cambridge University predicted the yield [44]. The analysis, like many calculations in
fluid mechanics, is long and complicated. We'll instead use
dimensional analysis to find the relation between the blast ra-
dius R, the time t since the blast, the blast energy E, and the air density ρ.
Then we'll use the $R(t)$ data to predict E.

These four quantities contain three independent dimensions. Thus, they produce one independent dimensionless group. To find it, let's eliminate one
dimension at a time. Here, a bit of luck simplifies
the process, because we have dimensions that are
easy to eliminate.

E	ML^2T^{-2}	blast energy
R	L	blast radius
t	T	time since blast
ρ_{air}	ML^{-3}	air density

▷ *Do any dimensions occur in only one or two quantities?*

Yes: Mass occurs in E as M^1 and in ρ_{air} also as M^1; time appears in E as T^{-2}
and in t as T^1. Therefore, to eliminate mass, the dimensionless group must
contain E/ρ_{air}. To eliminate time, the dimensionless group must contain Et^2.
Thus, Et^2/ρ_{air} eliminates mass and time. Because it has dimensions of L^5,
the only way to remove these dimensions without introducing any powers
of time or mass is to divide by R^5. The result, $Et^2/\rho_{air}R^5$, is an independent

dimensionless group. It is also the only dimensionless group proportional to our goal, the blast energy E.

With only one dimensionless group, the most general dimensionless statement about the blast energy is

$$\frac{Et^2}{\rho_{air}R^5} \sim 1. \tag{5.30}$$

For a particular explosion, E is fixed (although unknown), as is ρ_{air}. Therefore, these quantities drop out of the corresponding scaling relation, which becomes $t^2 \propto R^5$ or $R \propto t^{2/5}$. On log–log axes, the data on the blast radius should fall along a line of slope 2/5—as they almost exactly do:

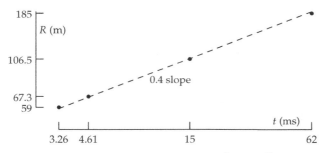

With the dimensional analysis result $Et^2/\rho_{air}R^5 \sim 1$, each point predicts a blast energy according to $E \sim \rho_{air}R^5/t^2$. Using 1 kilogram per cubic meter for ρ_{air}, the E estimates lie between 5.6 and 6.7×10^{13} joules. Unfortunately for judging the accuracy of the estimate, joules are an unfamiliar unit for the energy of a bomb blast. The familiar unit is tons of TNT.

▷ *What is the predicted blast energy as a mass of TNT?*

One gram of TNT releases 1 kilocalorie or roughly 4 kilojoules. It contains only one-fourth the energy density of sugar, but the energy is released much more rapidly! One kiloton is 10^9 grams, which release 4×10^{12} joules. Our predicted yield of 6×10^{13} joules is thus roughly 15 kilotons of TNT.

Just for fun, if we reestimate the yield using a more accurate air density of 1.2 kilograms per cubic meter, the blast energy would be 18 kilotons. This result is shockingly accurate considering the number of approximations it contains: The classified yield was 20 kilotons. (The missing universal constant for shock-wave explosions must be very close to 1.)

Dimensional analysis is powerful!

5.3 More dimensionless groups

The biggest change in dimensional analysis comes with a second independent dimensionless group. With only one dimensionless group, the most general statement that the universe could make was

$$\text{dimensionless group} \sim 1. \tag{5.31}$$

With a second group, the universe gains freedom on the right side:

$$\text{group 1} = f(\text{group 2}), \tag{5.32}$$

where f is a dimensionless function: It takes a dimensionless number as its input, and it produces a dimensionless number as its output.

As an example of two independent dimensionless groups, let's extend the analysis of Problem 5.10, where you predicted the impact speed of a rock dropped from a height h. The impact speed depends on g and h, so a reasonable choice for the independent dimensionless group is $v/\sqrt{gh}$. Therefore, its value is a universal, dimensionless constant (which turns out to be $\sqrt{2}$). To make a second independent dimensionless group, we add a degree of freedom to the problem: that the rock is thrown downward with speed v_0 (the previous problem is the case $v_0 = 0$).

▷ *With this complication, what are independent dimensionless groups?*

Adding a quantity but no new independent dimension creates one more independent dimensionless group. Therefore, there are two independent groups—for example, $v/\sqrt{gh}$ and $v_0/\sqrt{gh}$. The most general dimensionless statement, which has the form group 1 = f(group 2), is

$$\frac{v}{\sqrt{gh}} = f\left(\frac{v_0}{\sqrt{gh}}\right). \tag{5.33}$$

This point is as far as dimensional analysis can take us. To go further requires adding physics knowledge (Problem 5.16). But dimensional analysis already tells us that this function f is a universal *function*: It describes the impact speed of every thrown object in the universe, no matter the launch velocity, the drop height, or the gravitational field strength.

Problem 5.16   Impact speed of a thrown rock

For a rock thrown downward with speed v_0, use conservation of energy to find the form of the dimensionless function f in $v/\sqrt{gh} = f(v_0/\sqrt{gh})$.

Problem 5.17   Nonideal spring

Imagine a mass connected to a spring with force law $F \propto x^3$ (instead of the ideal-spring force $F \propto x$) and therefore with potential energy $V \sim Cx^4$ (where C is a constant). Which curve shows how the system's oscillation period T depends on the amplitude x_0?

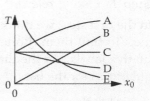

Problem 5.18   Rolling down the plane

In this problem, you use dimensional analysis to simplify finding the acceleration of a ring rolling (without slipping) down an inclined plane.

a. List the quantities on which the ring's acceleration depends. Hint: Include the ring's moment of inertia and its radius. Do you also need to include its mass?

b. Form independent dimensionless groups to write the dimensionless statement

$$\text{group proportional to } a = f(\text{groups not containing } a). \qquad (5.34)$$

c. Does a bigger ring roll faster than a smaller ring?

d. Does a denser ring roll faster than a less dense ring (of the same radius)?

5.3.1 Bending of starlight by the Sun

Our next example of two independent dimensionless groups—the deflection of starlight by the Sun—will illustrate how to incorporate physical knowledge into the mathematical results from dimensional analysis.

Rocks, birds, and people feel the effect of gravity. So why not light? The analysis of its deflection is a triumph of Einstein's theory of general relativity. However, the theory is based on ten coupled, nonlinear partial-differential equations. Rather than solving these difficult equations, let's use dimensional analysis.

Because this problem is more complicated than the previous examples, let's organize the complexity by making the steps explicit, including the step of incorporating physical knowledge. Divide and conquer!

1. List the relevant quantities.

2. Form independent dimensionless groups.

3. Use the groups to make the most general statement about deflection.

4. Narrow the possibilities by incorporating physical knowledge.

Step 1: Listing relevant quantities

In the first step, we think of and list the quantities that determine the bending. To find them, I often draw a diagram with verbal but without quantitative labels. As the diagram cries out for quantitative labels, it suggests quantities for the list.

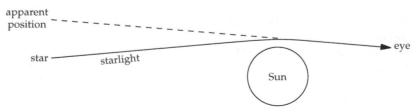

The bent path means that the star shifts its apparent position. The shift is most naturally measured not as an absolute distance but rather as an angle θ. For example, if $\theta = 180°$ (π radians), much larger than the likely deflection, the star would be shifted halfway around the sky.

The labels and the list must include our goal, which is the deflection angle θ. Because deflection is produced by gravity, we should also include Newton's gravitational constant G and the Sun's mass m (we'll

θ	1	angle
Gm	$L^3 T^{-2}$	Sun's gravity
r	L	closest approach

use the more general symbol m rather than M_{Sun}, because we may apply the formula to light paths around other stellar objects). These quantities could join the list as two separate quantities. However, the physical consequences of gravity—for example, the gravitational force—depend on G and m only through the product Gm. Therefore, let's include just the Gm abstraction on the list. (Problem 5.19 shows you how to find its dimensions.)

The final quantity on the list is based on our knowledge that gravity gets weaker with distance. Therefore, we include the distance from the Sun to the light beam. The phrase "distance from the Sun" is ambiguous, because the beam is at various distances. Our quantity r will be the shortest distance from the center of the Sun to the beam (the distance of closest approach).

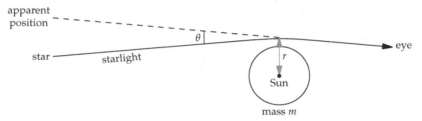

> **Problem 5.19   Dimensions of *Gm* and *G***
> Use Newton's law of universal gravitation, $F = Gm_1m_2/r^2$, to find the dimensions
> of *Gm* and *G*.

Step 2: Form independent dimensionless groups

The second step is to form independent dimensionless groups. One group
is easy: The angle θ is already dimensionless. Unfortunately, three quantities and two independent dimensions (L and T) produce only one independent dimensionless group. With only one dimensionless group, the most
general statement is merely θ = constant.

This prediction is absurd. The bending, if it is produced by gravity, has to
depend on *Gm* and r. These quantities have to appear in a second dimensionless group. Creating a second group, we know from the Buckingham
Pi theorem, requires at least one more quantity. Its absence indicates that
our analysis is missing essential physics.

▷ *What physics is missing?*

No quantity so far distinguishes between the path
of light and of, say, a planet. A crucial difference
is that light travels far more rapidly than a planet.
Let's represent this important characteristic of light
by including the c.

θ	1	angle
Gm	L^3T^{-2}	Sun's gravity
r	L	closest approach
c	LT^{-1}	speed of light

This quantity, a speed, does not introduce a new independent dimension (it
uses length and time). Therefore, it increases the number of independent
dimensionless groups by one, from one to two. To find the new group, first
check whether any dimension appears in only two quantities. If it does, the
search simplifies. Time appears in only two quantities: in *Gm* as T^{-2} and
in c as T^{-1}. To cancel out time, the new dimensionless group must contain
Gm/c^2. This quotient contains one power of length, so our dimensionless
group is Gm/rc^2.

As we hoped, the new group contains *Gm* and r. Its form illustrates again
that quantities with dimensions are not meaningful alone. For knowing just
Gm is not enough to decide whether or not gravity is strong. As a quantity
with dimensions, *Gm* must be compared to another relevant quantity with
the same dimensions. Here, that quantity is rc^2, and the comparison leads
to the dimensionless ratio Gm/rc^2.

▷ *Can we choose other pairs of independent dimensionless groups?*

Yes. For example, θ and $Gm\theta/rc^2$ also make a set of independent dimensionless groups. Mathematically, all pairs of independent dimensionless groups are equivalent, in that any pair can represent any quantitative statement about light bending.

However, look ahead to the goal: We hope to solve for θ. If θ appears in both groups, we will end up with an implicit equation for θ, where θ appears on both sides of the equals sign. Although such an equation is mathematically legitimate, it is much harder to think about and to solve than is an explicit equation, where θ is on the left side only.

Therefore, when choosing independent dimensionless groups, place the goal quantity in only one group. This rule of thumb does not remove all our freedom in choosing the groups, but it greatly limits the choices.

> **Problem 5.20   Physical interpretation of the new group**
> Interpret the dimensionless group Gm/rc^2 by multiplying by m_{light}/m_{light} and regrouping the quantities until you find physical interpretations for the numerator and denominator.

Step 3: Make the most general dimensionless statement

The third step is to use the independent dimensionless groups to write the most general statement about the bending angle. It has the form group 1 $=$ f(group 2). Here,

$$\theta = f\left(\frac{Gm}{rc^2}\right),$$ (5.35)

where f is a universal, dimensionless function. Dimensional analysis cannot determine f. However, it has told us that f is a function only of Gm/rc^2 and not of the four quantities $G, m, r,$ and c separately. That information is the great simplification.

Step 4: Use physical knowledge to narrow the possibilities

The space of possible functions—here, all nonpathological functions of one variable—is vast. Therefore, the fourth and final step is to narrow the possibilities for f by incorporating physical knowledge. First, imagine increasing the effect of gravity by increasing the mass m—which increases Gm/rc^2.

This change should, based on our physical intuition about gravity, also increase the bending angle. Therefore, f should be a monotonically increasing function of Gm/rc^2. Second, imagine an antigravity world, where the gravitational constant G is negative. Then the Sun would deflect light away from it, making the bending angle negative. In terms of $x \equiv Gm/rc^2$, this constraint eliminates even functions of x, such as $f(x) \sim x^2$, which produce the same sign for the bending angle independent of the sign of x.

The simplest function meeting both the monotonicity and sign constraints is $f(x) \sim x$. In terms of the second group GM/rc^2, the form $f(x) \sim x$ is the dimensionless statement about bending

$$\theta \sim \frac{Gm}{rc^2}.$$
(5.36)

All reasonable theories of gravity will predict this relation, because it is almost entirely a mathematical requirement. The theories differ only in the dimensionless factor hidden in the single approximation sign $\sim$:

$$\theta = \frac{Gm}{rc^2} \times \begin{cases} 1 & \text{(simplest guess);} \\ 2 & \text{(Newtonian gravity);} \\ 4 & \text{(general relativity).} \end{cases}$$
(5.37)

The factor of 2 for Newtonian gravity results from solving for the trajectory of a rock roving past the Sun with speed c. The factor of 4 for general relativity, double the Newtonian value, results from solving the ten partial-differential equations in the limit that the gravitational field is weak.

▷ *How large are these angles?*

Let's first estimate the bending angles closer to home—produced by the Earth's gravity. For a light ray just grazing the surface of the Earth, the bending angle (in radians!) is roughly 10^{-9}:

$$\theta_{\text{Earth}} \sim \frac{\overbrace{6.7 \times 10^{-11} \text{ kg}^{-1} \text{ m}^3 \text{ s}^{-2}}^{G} \times \overbrace{6 \times 10^{24} \text{ kg}}^{m_{\text{Earth}}}}{\underbrace{6.4 \times 10^6 \text{ m}}_{R_{\text{Earth}}} \times \underbrace{10^{17} \text{ m}^2 \text{ s}^{-2}}_{c^2}} \approx 0.7 \times 10^{-9}.$$
(5.38)

▷ *Can we observe this angle?*

The bending angle is the angular shift in the position of the star making the starlight. To observe these shifts, astronomers compare a telescope picture of the star and surrounding sky, with and without the deflection. A telescope with lens of diameter D can resolve angles roughly as small as λ/D, where λ is the wavelength of light. As a result, a lens that can resolve 0.7×10^{-9} radians has a diameter of at least 700 meters:

$$D \sim \frac{\lambda}{\theta_{\text{Earth}}} \sim \frac{0.5 \times 10^{-6}\,\text{m}}{0.7 \times 10^{-9}} \approx 700\,\text{m}. \tag{5.39}$$

On its own, this length does not mean much. However, the largest telescope lens has a diameter of roughly 1 meter; the largest telescope mirror, in a different telescope design, is still only 10 meters. No practical lens or mirror can be 700 meters in diameter. Thus, there is no way to see the deflection produced by the Earth's gravitational field.

Physicists therefore searched for a stronger source of light bending. The bending angle is proportional to m/r. The largest mass in the solar system is the Sun. For a light ray grazing the surface of the Sun, $r = R_{\text{Sun}}$ and $m = M_{\text{Sun}}$. So the ratio of Sun-to-Earth-produced bending angles is

$$\frac{\theta_{\text{Sun}}}{\theta_{\text{Earth}}} = \frac{M_{\text{Sun}}}{m_{\text{Earth}}} \times \left(\frac{R_{\text{Sun}}}{R_{\text{Earth}}} \right)^{-1}. \tag{5.40}$$

The mass ratio is roughly 3×10^5; the radius ratio is roughly 100. The ratio of deflection angles is therefore roughly 3000. The required lens diameter, which is inversely proportional to θ, is correspondingly smaller than 700 meters by a factor of 3000: roughly 25 centimeters or 10 inches. That lens size is plausible, and the deflection might just be measurable.

Between 1909 and 1916, Einstein believed that a correct theory of gravity would predict the Newtonian value of 4.2×10^{-6} radians or 0.87 arcseconds:

$$\underbrace{0.7 \times 10^{-9}\,\text{rad}}_{\sim \theta_{\text{Earth}}} \times \underbrace{2}_{\text{Newtonian gravity}} \times \underbrace{3000}_{\theta_{\text{Sun}}/\theta_{\text{Earth}}} \sim 4.2 \times 10^{-6}\,\text{rad}. \tag{5.41}$$

The German astronomer Soldner had derived the same result in 1803. The eclipse expeditions to test his (and Soldner's) prediction got rained out or clouded out. By the time an expedition got lucky with the weather, in 1919, Einstein had invented a new theory of gravity—general relativity—and it predicted a deflection twice as large, or 1.75 arcseconds.

The eclipse expedition of 1919, led by Arthur Eddington of Cambridge University and using a 13-inch lens, measured the deflection. The measurements are difficult, and the results were not accurate enough to decide clearly which theory was right. But 1919 was the first year after World War One, in which Germany and Britain had fought each other almost to oblivion. A theory invented by a German, confirmed by an Englishman (from

Newton's university, no less)—such a picture was welcome after the war. The world press and scientific community declared Einstein vindicated.

A proper confirmation of Einstein's prediction came only with the advent of radio astronomy, which allowed small deflections to be measured accurately (Problem 5.21). The results, described as the dimensionless factor multiplying Gm/rc^2, were around 4 ± 0.2—definitely different from the Newtonian prediction of 2 and consistent with general relativity.

Problem 5.21   Accuracy of radio-telescope measurements

The angular resolution of a telescope is λ/D, where λ is the wavelength and D is the telescope's diameter. But radio waves have a much longer wavelength than light. How can measurements of the bending angle that are made with radio telescopes be so much more accurate than measurements made with optical telescopes?

Problem 5.22   Another theory of gravity

An alternative to Newtonian gravity and to general relativity is the Brans–Dicke theory of gravitation [3]. Look up what it predicts for the angle that the Sun would deflect starlight. Express your answer as the dimensionless prefactor in $\theta \sim Gm/rc^2$.

5.3.2 Drag

Even more difficult than the equations of general relativity, which at least have been solved for realistic problems, are the Navier–Stokes equations of fluid mechanics. We first met them in Section 3.5, where their complexity pushed us to find an alternative to solving them directly. We used a conservation argument, which we tested using a falling cone, and concluded that the drag force on an object traveling through a fluid is

$$F_{\mathrm{drag}} \sim \rho v^2 A_{\mathrm{cs}}, \tag{5.42}$$

where ρ is the density of the fluid, v is the speed of the object, and A_{cs} is its cross-sectional area. This result is the starting point for our dimensional analysis.

With F_{drag} dependent on ρ, v, and A_{cs}, there are four quantities and three independent dimensions. Therefore, by the Buckingham Pi theorem, there is one independent dimensionless group. Our expression for the drag force already gives one such group:

$$\frac{F_{\mathrm{drag}}}{\rho v^2 A_{\mathrm{cs}}}. \tag{5.43}$$

Because the ρv^2 in the denominator looks like mv^2 in kinetic energy, a factor of one-half is traditionally included in the denominator:

$$\text{group 1} \equiv \frac{F_{\text{drag}}}{\frac{1}{2}\rho v^2 A_{\text{cs}}}. \tag{5.44}$$

The resulting dimensionless group is called the drag coefficient c_{d}.

$$c_{\text{d}} \equiv \frac{F_{\text{drag}}}{\frac{1}{2}\rho v^2 A_{\text{cs}}} \tag{5.45}$$

As the only dimensionless group, it has to be a constant: There is no other group on which it could depend. Therefore,

$$\frac{F_{\text{drag}}}{\frac{1}{2}\rho v^2 A_{\text{cs}}} = \text{dimensionless constant.} \tag{5.46}$$

This result is equivalent to our prediction based on conservation of energy, that $F_{\text{drag}} \sim \rho v^2 A_{\text{cs}}$. However, trouble happens when we wonder about the dimensionless constant's value.

Dimensional analysis, as a mathematical technique, cannot predict the constant. Doing so requires physical reasoning, which starts with the observation that drag consumes energy. Here, no quantity on which F_{drag} depends—namely, ρ, v, or A_{cs}—represents a mechanism of energy loss. Therefore, the drag coefficient should be zero. Although this value is consistent with the prediction that the drag coefficient is constant, it contradicts all our experience of fluids!

Our analysis needs to include a mechanism of energy loss. In fluids, loss is due to viscosity (for an object moving at a constant speed and generating no waves). Its physics is the topic of Section 7.3.2, as an example of a diffusion constant. For our purposes here, we need just its dimensions in order to incorporate viscosity into a dimensionless group: The dimensions of kinematic viscosity ν are L^2T^{-1}. (Unfortunately, in almost every font, the standard symbols for velocity and kinematic viscosity, v and ν, look similar; however, even more confusion would result from selecting new symbols.)

Thus, the kinematic viscosity joins our list. Adding one quantity without adding a new independent dimension creates a new independent dimensionless group. Similarly, in our analysis of the deflection of starlight (Section), adding the speed of light c created a second independent dimensionless group.

▷ *What is this new group containing viscosity?*

Before finding the group, look ahead to how it will be used. With a second group, the most general statement has the form

$$\underbrace{\text{group 1}}_{c_{\mathrm{d}}} = f(\text{group 2}). \tag{5.47}$$

Our goal is F_{drag}, which is already part of group 1. Including F_{drag} also in group 2 would be a bad tactic, because it would produce an equation with F_{drag} on both sides and, additionally, wrapped on the right side in the unknown function f. Keep F_{drag} out of group 2.

To make this group from the other quantities v, A_{cs}, ρ, and ν, look for dimensions that appear in only one or two quantities. Mass appears only in the density; thus, the density cannot appear in group 2: If it were part of group 2, there would be no way to cancel the dimensions of mass, and group 2 could not be dimensionless.

v	LT^{-1}	speed
A_{cs}	L^2	area
ρ	ML^{-3}	fluid density
ν	$\mathrm{L}^2\mathrm{T}^{-1}$	viscosity

Time appears in two quantities: speed v and viscosity ν. Because each quantity contains the same power of time (T^{-1}), group 2 has to contain the quotient v/ν—otherwise the dimensions of time would not cancel.

Here's what we know so far about this dimensionless group: So that mass disappears, it cannot contain the density ρ; so that time disappears, it must contain v and ν as their quotient v/ν. The remaining task is to make length disappear—for which purpose we use the object's cross-sectional area A_{cs}. The quotient v/ν has dimensions of L^{-1}, so $\sqrt{A_{\mathrm{cs}}}\,v/\nu$ is a new, independent dimensionless group.

Instead of $\sqrt{A_{\mathrm{cs}}}$, usually one uses the diameter D. With that choice, our group 2 is called the Reynolds number Re:

$$\mathsf{Re} \equiv \frac{vD}{\nu}. \tag{5.48}$$

The conclusion from dimensional analysis is then

$$\underbrace{\text{drag coefficient}}_{c_{\mathrm{d}}} = f(\underbrace{\text{Reynolds number}}_{\mathsf{Re}}). \tag{5.49}$$

For each shape, the dimensionless function f is a universal function; it depends on the shape of the object, but not on its size. For example, a sphere, a cylinder, and a cone have different functions f_{sphere}, f_{cylinder}, and f_{cone}. Similarly, a narrow-angle cone and a wide-angle cone have different functions.

But a small and a large sphere are described by the same function, as are a large and a small cone of the same opening angle. Any difference in the drag coefficients for different-sized objects of the same shape results only from the difference in Reynolds numbers (different because the sizes are different).

Here we see the power of dimensional analysis. It shows us that the universe doesn't care about the size, speed, or viscosity individually. The universe cares about them only through the abstraction known as the Reynolds number. It is the only information needed to determine the drag coefficient (for a given shape).

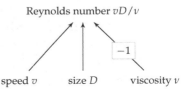

Now let's use this dimensionless framework to analyze the cone experiment: The experimental data showed that the small and large cones fell at the same speed—roughly 1 meter per second.

▷ *What are the corresponding Reynolds numbers?*

The small cone has a diameter of 0.75×7 centimeters (0.75 because one-quarter of the original circumference was removed), which is roughly 5.3 centimeters. The viscosity of air is roughly 1.5×10^{-5} square meters per second (an estimate that we will make in Section 7.3.2). The resulting Reynolds number is roughly 3500:

$$\mathrm{Re} = \frac{\overbrace{1\,\mathrm{m\,s^{-1}}}^{v} \times \overbrace{0.053\,\mathrm{m}}^{D}}{\underbrace{1.5 \times 10^{-5}\,\mathrm{m^2\,s^{-1}}}_{\nu}} \approx 3500. \tag{5.50}$$

▷ *What is the Reynolds number for the large cone?*

Never make a calculation from scratch when you can use proportional reasoning. Both cones experience the same viscosity and share the same fall speed. Therefore, we can consider ν and v to be constants, leaving the Reynolds number vD/ν proportional just to the diameter D.

Because the large cone has twice the diameter of the small cone, it has twice the Reynolds number:

$$\mathrm{Re}_{\mathrm{large}} = 2 \times \mathrm{Re}_{\mathrm{small}} \approx 7000. \tag{5.51}$$

▷ *At that Reynolds number, what is the cone's drag coefficient?*

The drag coefficient is

$$c_{\text{d}} \equiv \frac{F_{\text{drag}}}{\frac{1}{2}\rho_{\text{air}}v^2 A_{\text{cs}}} \tag{5.52}$$

The cone falls at its terminal speed, so the drag force is also its weight W:

$$F_{\text{drag}} = W = A_{\text{paper}}\sigma_{\text{paper}}g, \tag{5.53}$$

where σ_{paper} is the areal density (mass per area) of paper and A_{paper} is the area of the cone template. The drag coefficient is then

$$c_{\text{d}} = \frac{\overbrace{A_{\text{paper}}\sigma_{\text{paper}}g}^{F_{\text{drag}}}}{\frac{1}{2}\rho_{\text{air}}A_{\text{cs}}v^2}. \tag{5.54}$$

As we showed in Section 3.5.2,

$$A_{\text{cs}} = \frac{3}{4}A_{\text{paper}}. \tag{5.55}$$

This proportionality means that the areas cancel out of the drag coefficient:

$$c_{\text{d}} = \frac{\sigma_{\text{paper}}g}{\frac{1}{2}\rho_{\text{air}} \times \frac{3}{4}v^2}. \tag{5.56}$$

To compute c_{d}, plug in the areal density $\sigma_{\text{paper}} \approx 80$ grams per square meter and the measured speed $v \approx 1$ meter per second:

$$c_{\text{d}} \approx \frac{\overbrace{8 \times 10^{-2}\,\text{kg}\,\text{m}^{-2}}^{\sigma_{\text{paper}}} \times \overbrace{10\,\text{m}\,\text{s}^{-2}}^{g}}{\underbrace{\frac{1}{2} \times 1.2\,\text{kg}\,\text{m}^{-3}}_{\rho_{\text{air}}} \times \frac{3}{4} \times \underbrace{1\,\text{m}^2\,\text{s}^{-2}}_{v^2}} \approx 1.8. \tag{5.57}$$

Because no quantity in this calculation depends on the cone's size, both cones have the same drag coefficient. (Our estimated drag coefficient is significantly larger than the canonical drag coefficient for a solid cone, roughly 0.7, and is approximately the drag coefficient for a wedge.)

Thus, the drag coefficient is independent of Reynolds number—at least, for Reynolds numbers between 3500 and 7000. The giant-cone experiment of Problem 4.16 shows that the independence holds even to Re $\sim 14\,000$. Within this range, the dimensionless function f in

$$\text{drag coefficient} = f_{\text{cone}}(\text{Reynolds number}) \tag{5.58}$$

is a constant. What a simple description of the complexity of fluid flow!

This conclusion for f_{cone} is valid for most shapes. The most extensive drag data is for a sphere—plotted below on log–log axes (data adapted from *Fluid-Dynamic Drag: Practical Information on Aerodynamic Drag and Hydrodynamic Resistance* [23]):

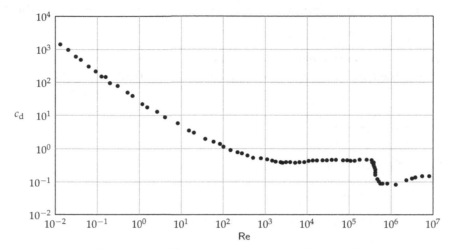

Like the drag coefficient for the cone, the drag coefficient for a sphere is almost constant as the Reynolds number varies from 3500 to 7000. The drag coefficient holds constant even throughout the wider range 2000...3×10^5. Around $\mathtt{Re} \approx 3 \times 10^5$, the drag coefficient falls by a factor of 5, from 0.5 to roughly 0.1. This drop is the reason that golf balls have dimples (you will explain the connection in Problem 7.24).

At low Reynolds numbers, the drag coefficient becomes large. This behavior, which represents a small object oozing through honey, will be explained in Section 8.3.1.2 using the tool of easy cases. The main point here is that, for most everyday flows, where the Reynolds number is in the "few thousand to few hundred thousand" range, the drag coefficient is a constant that depends only on the shape of the object.

Problem 5.23   Giant cone
How large, as measured by the diameter of its cross section, would a paper cone need to be in order for its fall to have a Reynolds number of roughly 3×10^5 (the Reynolds number at which the drag coefficient of a sphere drops significantly)?

Problem 5.24   Reynolds numbers
Estimate the Reynolds number for (a) a falling raindrop, (b) a flying mosquito, (c) a person walking, and (d) a jumbo jet flying at its cruising speed.

Problem 5.25 Compound pendulum .

Use dimensional analysis to deduce as much as you can about the
period T of a compound pendulum—that is, a pendulum where the
bob is not a point mass but is an extended object of mass m. The
light rod (no longer a string) of length l is fixed to the center of mass
of the object, which has a moment of inertia I_{CM} about the point of
attachment. (Assume that the oscillation amplitude is small and
therefore doesn't affect the period.)

Problem 5.26 Terminal velocity of a raindrop

In Problem 3.37, you estimated the terminal speed of a raindrop using

$$F_{\text{drag}} \sim \rho_{\text{air}} A_{\text{cs}} v^2. \qquad (5.59)$$

Redo the calculation using the more precise information that $c_{\text{d}} \approx 0.5$ and retaining the factor of $4\pi/3$ in the volume of a spherical raindrop.

5.4 Temperature and charge

The preceding examples of dimensional analysis have been mechanical, using the dimensions of length, mass, and time. Temperature and charge are
also essential to describing the world and are equally amenable to dimensional analysis. Let's start with temperature.

5.4.1 Temperature

▷ *To handle temperature, do we need to add another fundamental dimension?*

Representing temperature as a new fundamental dimension is one method,
and the dimension is symbolized by Θ. However, there's a simpler method.
It uses another fundamental constant of nature: Boltzmann's constant k_{B}.
It has dimensions of energy per temperature; thus, it connects temperature
to energy. In SI units, k_{B} is roughly 1.6×10^{-23} joules per kelvin. When a
temperature T appears, convert it to the corresponding energy $k_{\text{B}}T$ (just as
we convert G, in gravity problems, to Gm).

For our first temperature example, let's estimate the speed of air molecules.
We will then use that knowledge to estimate the speed of sound. Because
the speed v of air molecules is a result of thermal energy, it depends on
$k_{\text{B}}T$. But v and $k_{\text{B}}T$—two quantities made from two independent dimensions—cannot make a dimensionless group. We need one more quantity:
the mass m of an air molecule.

Although these three quantities contain three dimensions, only two of the dimensions are independent—for example, M and LT^{-1}. Therefore, the three quantities produce one independent dimensionless group. A reasonable choice for it is $mv^2/k_B T$. Then

v	LT^{-1}	thermal speed
$k_B T$	ML^2T^{-2}	thermal energy
m	M	molecular mass

$$v \sim \sqrt{\frac{k_B T}{m}}. \tag{5.60}$$

This result is quite accurate. The missing dimensionless prefactor is $\sqrt{3}$ for the root-mean-square speed and $\sqrt{8/\pi}$ for the mean speed.

The dimensional-analysis result helps predict the speed of sound. In a gas, sound travels because of pressure, which is a result of thermal motion. Thus, it is plausible that the speed of sound c_s should be comparable to the thermal speed. Then,

$$c_s \sim \sqrt{\frac{k_B T}{m}}. \tag{5.61}$$

To evaluate the speed numerically, multiply by a form of 1 based on Avogadro's number:

$$c_s \sim \sqrt{\frac{k_B T}{m} \times \frac{N_A}{N_A}}. \tag{5.62}$$

The numerator now contains $k_B N_A$, which is the universal gas constant R:

$$R \approx \frac{8\,\text{J}}{\text{mol}\,\text{K}}. \tag{5.63}$$

The denominator in the square root is $m N_A$: the mass of one molecule multiplied by Avogadro's number. It is therefore the mass of one mole of air molecules. Air is mostly N_2, with an atomic mass of 28. Including the oxygen for us animals, the molar mass of air is roughly 30 grams per mole. Plugging in 300 K for room temperature gives a thermal speed of roughly 300 meters per second:

$$c_s \sim \sqrt{\frac{8\,\text{J}\,\text{mol}^{-1}\,\text{K}^{-1} \times 300\,\text{K}}{3 \times 10^{-2}\,\text{kg}\,\text{mol}^{-1}}} \approx 300\,\text{m}\,\text{s}^{-1}. \tag{5.64}$$

The actual sound speed is 340 meters per second, not far from our estimate based on dimensional analysis and a bit of physical reasoning. (The discrepancy remaining is due to the difference between isothermal and adiabatic compression and expansion. For a related effect, later try Problem 8.27.)

With this understanding of how to handle temperature, let's use dimensional analysis to estimate the height of the atmosphere. This height, called the scale height H, is the length over which the atmosphere's properties, such as pressure and density, change significantly. At the height H, the atmospheric density and pressure will be significantly lower than at sea level.

The first step in dimensional analysis is to list the quantities that determine the height. That list requires a physical model: The atmosphere is a competition between gravity and thermal motion. Gravity drags molecules toward Earth;

H	L	atmosphere height
g	LT^{-2}	gravity
k_BT	ML^2T^{-2}	thermal energy
m	M	molecular mass

thermal motion spreads them all over the universe. Our list must include quantities representing both sides of that competition: m and g for gravity, and k_BT for thermal energy.

Four quantities (including the goal H) built from three independent dimensions produce one independent dimensionless group. A reasonable choice for the group—reasonable because it is proportional to the goal H—is the ratio mgH/k_BT. Therefore, the atmosphere's scale height is given by

$$H \sim \frac{k_BT}{mg}. \tag{5.65}$$

To evaluate this height numerically, again convert Boltzmann's constant k_B to the universal gas constant R by multiplying by N_A/N_A:

$$H \sim \frac{\overbrace{k_B N_A}^{R} T}{\underbrace{m N_A}_{m_{molar}} g} \approx \frac{\overbrace{8\,\mathrm{J\,mol^{-1}\,K^{-1}}}^{R} \times 300\,\mathrm{K}}{\underbrace{3\times 10^{-2}\,\mathrm{kg\,mol^{-1}}}_{m_{molar}} \times 10\,\mathrm{m\,s^{-2}}} \approx 8\,\mathrm{km}. \tag{5.66}$$

Thus, by 8 kilometers above sea level, atmospheric pressure and density should be significantly lower than at sea level. This conclusion is reasonable: Mount Everest is 9 kilometers high, where the significantly thinner air requires climbers to carry oxygen tanks.

As a rule of thumb for a decaying function such as atmospheric pressure or density, a "significant change" can be estimated as a rise or, here, a fall by a factor of e. (For more on this rule of thumb, try Problem 6.36 after you have studied lumping; see also Section 3.2.1 of *Street-Fighting Mathematics* [33].) Indeed, at 8 kilometers, the standardized atmospheric parameters relative to sea level are $\rho/\rho_0 \approx 0.43$ and $p/p_0 \approx 0.35$. For comparison, $1/e$ is approximately 0.37.

5.4.2 **Charge**

Our second new dimension is charge, symbolized by Q. By introducing charge, we can apply dimensional analysis to electrical phenomena. Let's begin by finding the dimensions of electrical quantities.

▷ *What are the dimensions of current, voltage, and resistance?*

Current is the flow of charge; it has dimensions of charge per time. Using I to denote current and $[I]$ to denote its dimensions,

$$[I] = QT^{-1}. \tag{5.67}$$

The dimensions of voltage can be determined from the relation

$$\text{voltage} \times \text{current} = \text{power}. \tag{5.68}$$

The corresponding dimensional equation is

$$[\text{voltage}] = \frac{[\text{power}]}{[\text{current}]}. \tag{5.69}$$

Because the dimensions of power (energy per time) are ML^2T^{-3}, and the dimensions of current are QT^{-1}, the dimensions of voltage are $ML^2T^{-2}Q^{-1}$:

$$[\text{voltage}] = \frac{ML^2T^{-3}}{QT^{-1}} = ML^2T^{-2}Q^{-1}. \tag{5.70}$$

In this expression, the dimensional combination ML^2T^{-2} is energy. Therefore, voltage is also energy per charge. As a result, an electron volt (1 electron charge times 1 volt) is an energy. A few electron volts is the energy in chemical bonds. Because chemical bonds reflect a change in one or two electrons, atomic and molecular voltages must be a few volts. (An everyday consequence is that typical batteries supply a few volts.)

To find the dimensions of resistance, we'll write Ohm's law as a dimensional equation:

$$[\text{resistance}] = \frac{[\text{voltage}]}{[\text{current}]} = \frac{ML^2T^{-2}Q^{-1}}{QT^{-1}} = ML^2T^{-1}Q^{-2}. \tag{5.71}$$

To simplify these dimensions, use $[V]$—the dimensions of voltage—as an abstraction for the dimensional mess $ML^2T^{-2}Q^{-1}$. Then,

$$[\text{resistance}] = [V] \times TQ^{-1}. \tag{5.72}$$

This alternative is useful when the input to and output of a circuit is voltage. Then treating the dimensions of voltage as a fundamental dimension simplifies the task of finding dimensionless groups. (For practice in using the dimensions of voltage, try Problem 5.27.)

Problem 5.27   Capacitance and inductance
Express the dimensions of capacitance and inductance using L, M, T, and Q. Then write them using $[V]$ (the dimensions of voltage) along with any fundamental dimensions that you need.

Problem 5.28   *RC* circuit

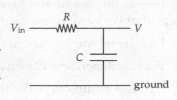

In this problem, you apply dimensional analysis to the low-pass *RC* circuit that we introduced in Section 2.4.4. In particular, make the input voltage V_{in} zero for time $t < 0$ and a fixed voltage V_0 for $t \geq 0$. The goal is the most general dimensionless statement about the output voltage V, which depends on V_0, t, R, and C.

a. Using $[V]$ to represent the dimensions of voltage, fill in a dimensional-analysis table for the quantities V, V_0, t, R, and C.

b. How many independent dimensions are contained in these five quantities?

c. Form independent dimensionless groups and write the most general statement in the form

   group containing V but not $t = f($group containing t but not $V, ...)$,   (5.73)

   where the ... stands for the third dimensionless group (if it exists). Compare your expression to the analysis of the *RC* circuit in Section 2.4.4.

Problem 5.29   Dimensional analysis of an *LRC* circuit

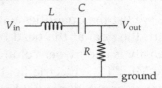

In the *LRC* circuit, the input voltage V_{in} is the real part of a complex-exponential signal $V_0 e^{j\omega t}$, where V_0 is the input amplitude and ω is the angular oscillation frequency. The output voltage V_{out} is then the real part of $V_1 e^{j\omega t}$, where V_1 is the (possibly complex) output amplitude. What quantities determine the gain G, defined as the ratio V_1/V_0? What is a good set of independent dimensionless groups built from G and these quantities?

Along with the potential (or voltage) V, a leading actor in electromagnetics is the electric field. (Once you understand how to handle electric fields, you can practice by analyzing magnetic fields—try Problems 5.31, 5.32, and 5.33.) Electric fields not only transmit force; they also contain energy. This energy is important, not least because its transport is how the Sun warms the Earth. We will estimate the energy contained in electric fields by using dimensional analysis. This analysis will give us the result that we used in Section 2.4.2 to estimate, by analogy, the energy in a gravitational field. Like the gravitational field, the electric field extends over space. Therefore, we usually want not the energy itself, but rather the energy per volume—the energy density.

▷ *What is the energy density in an electric field?*

To answer this question with dimensional analysis, let's follow most of the steps that we used in estimating the bending of starlight (Section 5.3.1).

1. List the relevant quantities.

2. Form independent dimensionless groups.

3. Use the groups to make the most general statement about the energy density.

4. Narrow the possibilities by incorporating physical knowledge. For this problem, we will be able to skip this step.

Thus, the first step is to tabulate the quantities on which our goal, the energy density $\mathcal{E}$, depends. It definitely depends on the electric field E. It also probably depends on ϵ_0, the permittivity of free space appearing in Coulomb's law, because most results in electrostatics contain ϵ_0. But $\mathcal{E}$ shouldn't depend on the speed of light: The speed of light suggests radiation, which requires a changing electric field; however, even a constant electric field contains energy. Thus, our list might be complete. Let's see, by trying to make a dimensionless group.

So we list the quantities with their dimensions, with the goal at the head of the table. Energy density is energy per volume, so its dimensions are $ML^{-1}T^{-2}$. Electric field is force per charge

$\mathcal{E}$	$ML^{-1}T^{-2}$	energy density
E	$MLT^{-2}Q^{-1}$	electric field
ϵ_0	$M^{-1}L^{-3}T^2Q^2$	SI constant

(just as gravitational field is force per mass—analogy!), so its dimensions are $MLT^{-2}Q^{-1}$.

▷ *What are the dimensions of ϵ_0?*

This quantity is the trickiest. It shows up in Coulomb's law:

$$\text{electrostatic force} = \frac{q^2}{4\pi\epsilon_0}\frac{1}{r^2}. \tag{5.74}$$

As a dimensional equation,

$$[\epsilon_0] = \frac{Q^2}{[F]\,L^2}, \tag{5.75}$$

where $[F] = \mathrm{MLT}^{-2}$ represents the dimensions of force. Therefore, ϵ_0 has dimensions of $\mathrm{M}^{-1}\mathrm{L}^{-3}\mathrm{T}^2\mathrm{Q}^2$.

The second step is to find independent dimensionless groups. Any search is aided by knowing how many items to find. This count is provided by the Buckingham Pi theorem. To apply the theorem, we first need to count the independent dimensions.

▷ *How many independent dimensions do the three quantities contain?*

At first glance, the quantities contain four independent dimensions: M, L, T, and Q. However, the Buckingham Pi theorem would then predict −1 dimensionless groups, which is nonsense. Indeed, the number of independent dimensions cannot exceed the number of quantities (a restriction that you explained in Problem 5.7). Here, as you verify in Problem 5.30, there are only two independent dimensions—for example, $\mathrm{MLT}^{-2}\mathrm{Q}^{-1}$ (the dimensions of electric field) and $\mathrm{L}^2\mathrm{Q}^{-1}$.

Three quantities constructed from two independent dimensions produce one independent dimensionless group. A useful choice for this group, because it is proportional to the goal $\mathcal{E}$, is $\mathcal{E}/\epsilon_0 E^2$.

The third step is to use the independent dimensionless group to make the most general statement about the energy density. With only one group, the most general statement is

$$\mathcal{E} \sim \epsilon_0 E^2. \tag{5.76}$$

The fourth step is to narrow the possibilities by using physical knowledge. With only one independent dimensionless group, the space of possibilities is already narrow. The only freedom is the dimensionless prefactor hidden in the single approximation sign $\sim$; it turns out to be $1/2$. In Section 5.4.3, the scaling $\mathcal{E} \propto E^2$ will help us explain the surprising behavior of the electric field produced by an accelerating charge—and thereby explain why stars are visible and radios work.

Problem 5.30 **Rewriting the dimensions**
Express the dimensions of $\mathcal{E}$, E, and ϵ_0 in terms of $[E]$ (the dimensions of electric field) and $\mathrm{L}^2\mathrm{Q}^{-1}$. Thus, show that the three quantities contain only two independent dimensions.

Problem 5.31   Dimensions of magnetic field
A magnetic field **B** produces a force on a moving charge q given by

$$\mathbf{F} = q(\mathbf{v} \times \mathbf{B}), \qquad (5.77)$$

where **v** is the charge's velocity. Use this relation to find the dimensions of magnetic field in terms of M, L, T, and Q. Therefore, give the definition of a tesla, the SI unit of magnetic field.

Problem 5.32   Magnetic energy density
Just as electric fields depend on ϵ_0, the permittivity of free space, magnetic fields depend on the constant μ_0 called the permeability of free space. It is defined as

$$\mu_0 \equiv 4\pi \times 10^{-7}\,\mathrm{N\,A^{-2}}, \qquad (5.78)$$

where N is a newton and A is an ampere (a coulomb per second). Express the dimensions of μ_0 in terms of M, L, T, and Q.

Then use the dimensions of magnetic field (Problem 5.31) to find the energy density in a magnetic field B. (As with the electric field, the dimensionless prefactor will be 1/2, but dimensional analysis does not give us that information.) What analogies can you make between electrostatics and magnetism?

Problem 5.33   Magnetic field due to a wire
Use the dimensions of B (Problem 5.31) and of the permeability of free space μ_0 (Problem 5.32) to find the magnetic field B at a distance r from an infinitely long wire carrying a current I. The missing dimensionless prefactor, which dimensional analysis cannot tell us, turns out to be $1/2\pi$.

Magnetic-resonance imaging (MRI) machines for medical diagnosis use fields of the order of 1 tesla. If this field were produced by a current-carrying wire 0.5 meters away, what current would be required? Therefore, explain why these magnetic fields are produced by superconducting magnets.

Problem 5.34   Fields due to a uniform sheet of charge
Imagine an infinite, uniform sheet of charge containing a charge per area σ. Use dimensional analysis to find the electric field E at a distance z from the sheet. (The missing dimensionless prefactor turns out to be 1/2.) Therefore, give the scaling exponent n in $E \propto z^n$. (In Problem 6.21, you'll investigate a physical explanation for this surprising result.)

Use the analogy between electric and gravitational fields (Section 2.4.2) to find the gravitational field above a uniform sheet with mass per area σ.

5.4.3 Power radiated by a moving charge

In our next example, we'll estimate the power radiated by a moving charge, which is how a radio broadcasting antenna works. This power, when combined with our long-standing understanding of flux (Section 3.4.2) and our new understanding of the energy density in an electric field (Section 5.4.2), will predict a surprising scaling relation for the strength of the radiation field, one responsible for our ability to see the world. The example also illustrates a simple way to manage the dimensions of charge.

▷ *How does the power radiated by a moving charge depend on the acceleration of the charge?*

For the moment, let's ignore the charge's velocity. Most likely, the dependence on acceleration is a scaling relation

$$\text{power} \propto (\text{acceleration})^n. \tag{5.79}$$

We will find the scaling exponent n using dimensional analysis.

First, we list the quantities on which the goal, the radiated power P, depends. It certainly depends on the charge's motion, which is represented by its acceleration a. It also depends on the amount of charge q, because more charge probably means more radiation. The list should also include the speed of light c, because the radiation travels at the speed of light.

It probably also needs the permittivity of free space ϵ_0. However, rather than including ϵ_0 directly, let's reuse the shortcut for gravitation, where we combined Newton's constant G with a mass (in Section 5.3.1). Similarly, ϵ_0 always appears as $q^2/4\pi\epsilon_0$, whether in electrostatic energy or force. Therefore, rather than including q and ϵ_0 separately, we can include only $q^2/4\pi\epsilon_0$.

▷ *What happened to the charge's velocity?*

If the radiated power depended on the velocity, then we could use the principle of relativity to make a perpetual-motion machine: We generate energy simply by using a different (inertial) reference frame, one in which the charge moves faster. Rather than believing in perpetual motion, we should conclude that the velocity does not affect the power. Equivalently, we can eliminate the velocity by switching to a reference frame where the charge has zero velocity.

▷ *Why doesn't the reference-frame argument allow us to eliminate the acceleration?*

It depends on changing to another *inertial* reference frame—that is, a frame that moves at a constant velocity relative to the original frame. This relative motion does not affect the charge's acceleration, only its velocity.

However, if we change to a noninertial—an accelerating—reference frame, we must modify the equations of motion, adding terms for the Coriolis, centrifugal, and Euler forces. (For more on reference frames, a wonderful exposition is John Taylor's *Classical Mechanics* [45].) If we switch to a noninertial frame, all bets are off about the radiated power. In summary, acceleration is different from velocity.

These four quantities, built from three independent dimensions, produce just one independent dimensionless group. Because mass appears in only P and $q^2/4\pi\epsilon_0$ (as M^1), the group must contain $P/(q^2/4\pi\epsilon_0)$. This quotient

P	ML^2T^{-3}	radiated power
$q^2/4\pi\epsilon_0$	ML^3T^{-2}	electrostatic mess
c	LT^{-1}	speed of light
a	LT^{-2}	acceleration

has dimensions of $L^{-1}T^{-1}$. Multiplying it by c^3/a^2, which has dimensions of LT, makes the result dimensionless. Therefore, the independent dimensionless group proportional to P is

$$\frac{P}{q^2/4\pi\epsilon_0}\frac{c^3}{a^2}. \tag{5.80}$$

As the only group, it must be a dimensionless constant, so

$$P \sim \frac{q^2}{4\pi\epsilon_0}\frac{a^2}{c^3}. \tag{5.81}$$

The exact result is almost identical:

$$P = \frac{q^2}{6\pi\epsilon_0}\frac{a^2}{c^3}. \tag{5.82}$$

This result is even correct at relativistic speeds, as long as we use the relativistic acceleration (the four-acceleration) as the generalization of a.

As a scaling relation between P and a, it is simple: $P \propto a^2$. Doubling the acceleration quadruples the radiated power. The steep dependence on the acceleration is an important part of the reason that the sky is blue (we'll do the analysis in Section 9.4.1).

This power is carried by a changing electric field. By using the energy density in the electric field, we can estimate and explain the surprising strength of this field. We start by estimating the energy flux (the power per area) at a distance r, by spreading the radiated power over a sphere of radius r. The sphere's surface area is comparable to r^2, so

$$\text{energy flux} \sim \frac{P}{r^2} \sim \frac{q^2}{4\pi\epsilon_0}\frac{a^2}{c^3}\frac{1}{r^2}. \tag{5.83}$$

Energy flux, based on what we learned in Section 3.4.2, is connected to energy density by

$$\text{energy flux} = \text{energy density} \times \text{transport speed}. \tag{5.84}$$

The transport speed is the speed of light c, so

$$\underbrace{\frac{q^2}{4\pi\epsilon_0}\frac{a^2}{c^3}\frac{1}{r^2}}_{\text{energy flux}} \sim \underbrace{\epsilon_0 E^2}_{\text{energy density}} \times \underbrace{c}_{\text{speed}}. \tag{5.85}$$

Now we can solve for the electric field:

$$E \sim \frac{qa}{\epsilon_0 c^2}\frac{1}{r}. \tag{5.86}$$

As a scaling relation, $E \propto r^{-1}$. Compare it to the -2 scaling exponent for an electrostatic field, where $E \propto r^{-2}$ (Coulomb's law). The radiation field therefore falls off more slowly with distance than the electrostatic field. This important difference explains why we can receive radio signals and see stars. If radiation fields were proportional to $1/r^2$, stars, and most of the world, would be invisible. What difference a scaling exponent can make!

5.5 Atoms, molecules, and materials

With our growing repertoire of dimensional analyses, we can explore ever more of the world. Perhaps the most fundamental property of the world is that it is composed of atoms. Feynman argued for the importance of the atomic theory in his famous lectures on physics [14, Vol. 1, p. 1-2]:

> If, in some cataclysm, all of scientific knowledge were to be destroyed, and only one sentence passed on to the next generation of creatures, what statement would contain the most information in the fewest words? I believe it is the *atomic hypothesis* (or the atomic *fact*, or whatever you wish to call it) that *all things are made of atoms—little particles that move around in perpetual motion, attracting each other when they are a little distance apart, but repelling upon being squeezed into one another*. In that one sentence, you will see, there is an *enormous* amount of information about the world.... [emphasis in original]

The atomic theory was first stated over 2000 years ago by the ancient Greek philosopher Democritus. Using quantum mechanics, we can predict the properties of atoms in great detail—but the analysis involves complicated mathematics that buries the core ideas. By using dimensional analysis, we can keep the core ideas in sight.

Dimensional analysis of hydrogen

We'll study the simplest atom: hydrogen. Dimensional analysis will explain its size, and its size will in turn explain the size of more complex atoms and molecules. Dimensional analysis will also help us estimate the energy needed to disassemble a hydrogen atom. This energy will in turn explain the bond energies in molecules. These energies will explain the stiffness of materials, the speed of sound, and the energy content of fat and sugar—all starting from hydrogen!

In the dimensional analysis of the size of hydrogen, the zeroth step is to give the size a name. The usual name and symbol for the radius of hydrogen is the Bohr radius a_0.

The first step is to find the quantities that determine the size. That determination requires a physical model of hydrogen. A simple model is an electron orbiting a proton at a distance a_0. Their electrostatic attraction provides the force F holding the electron in orbit:

$$F = \frac{e^2}{4\pi\epsilon_0}\frac{1}{a_0^2},$$ (5.87)

where e is the electron charge. Our list of quantities should include the quantities in this equation; in that way, we teach dimensional analysis about what kind of force holds the atom together. Thus, we should somehow include e and ϵ_0. As when we estimated the power radiated by an accelerating charge (Section 5.4.3), we'll include e and ϵ_0 as one quantity, $e^2/4\pi\epsilon_0$.

The force on the electron does not by itself determine the electron's motion. Rather, the motion is determined by its acceleration. Computing the acceleration from the force requires the electron mass m_e, so our list should include m_e.

a_0	L	size
$e^2/4\pi\epsilon_0$	ML^3T^{-2}	electrostatics
m_e	M	electron mass

These three quantities made from three independent dimensions produce zero dimensionless groups (another application of the Buckingham Pi theorem introduced in Section 5.1.2). Without any dimensionless groups, we cannot say anything about the size of hydrogen. The lack of a dimensionless group tells us that our simple model of hydrogen is too simple.

There are two possible resolutions, each involving new physics that adds one new quantity to the list. The first possibility is to add special relativity by adding the speed of light c. That choice produces a dimensionless group and therefore a size. However, this size has nothing to do with hydrogen. Rather, it is the size of an electron, considering it as a cloud of charge (Problem 5.37).

The other problem with this approach is that electromagnetic radiation travels at the speed of light, so once the list includes the speed of light, the electron might radiate. As we found in Section 5.4.3, the power radiated by an accelerating charge is

$$P \sim \frac{q^2}{4\pi\epsilon_0} \frac{a^2}{c^3}. \tag{5.88}$$

An orbiting electron is an accelerating electron (accelerating inward with the circular acceleration $a = v^2/a_0$), so the electron would radiate. Radiation would carry away energy from the electron, the electron would spiral into the proton, and hydrogen would not exist—nor would any other atom. So adding the speed of light only compounds our problem.

The second resolution is instead to add quantum mechanics. Its fundamental equation is the Schrödinger equation:

$$\left(-\frac{\hbar^2}{2m}\nabla^2 + V\right)\psi = E\psi. \tag{5.89}$$

Most of the symbols in this partial-differential equation are not important for dimensional analysis—this disregard is how dimensional analysis simplifies problems. For dimensional analysis, the crucial point is that Schrödinger's equation contains a new constant of nature: $\hbar$, which is Planck's constant h divided by 2π. We can include quantum mechanics in our model of hydrogen simply by including $\hbar$ on our list of relevant quantities. In that way, we teach dimensional analysis about quantum mechanics.

The new quantity $\hbar$ is an angular momentum, which is a length (the lever arm) times linear momentum (mv). Therefore, its dimensions are

$$\underbrace{L}_{[r]} \times \underbrace{M}_{[m]} \times \underbrace{LT^{-1}}_{[v]} = ML^2T^{-1}. \tag{5.90}$$

The $\hbar$ might save hydrogen. It contributes a fourth quantity without a fourth independent dimension. Therefore, it creates an independent dimensionless group.

a_0	L	size
$e^2/4\pi\epsilon_0$	ML^3T^{-2}	electrostatic mess
m_e	M	electron mass
$\hbar$	ML^2T^{-1}	quantum

To find it, look for a dimension appearing in only two quantities. Time appears in $e^2/4\pi\epsilon_0$ as T^{-2} and in $\hbar$ as T^{-1}. Therefore, the group contains the quotient $\hbar^2/(e^2/4\pi\epsilon_0)$. Its dimensions are ML, which can be canceled by dividing by $m_e a_0$. The resulting dimensionless group is

$$\frac{\hbar^2}{m_e a_0 (e^2/4\pi\epsilon_0)}. \tag{5.91}$$

As the only independent dimensionless group, it must be a dimensionless constant. Therefore, the size (as the radius) of hydrogen is given by

$$a_0 \sim \frac{\hbar^2}{m_e (e^2/4\pi\epsilon_0)}. \tag{5.92}$$

Now let's plug in the constants. We could just look up each constant, but that approach gives exponent whiplash; the powers of ten swing up and down and end on a seemingly random value. To gain insight, we'll instead use the numbers for the backs of envelopes (p. xvii). That table deliberately has no entry for $\hbar$ by itself, nor for the electron mass m_e. Both omissions can be resolved with a symmetry operation (Problem 5.35).

Problem 5.35   Shortcuts for atomic calculations

Many atomic problems, such as the size or binding energy of hydrogen, end up in expressions with $\hbar$, the electron mass m_e, and $e^2/4\pi\epsilon_0$. You can avoid remembering those constants by instead remembering the following values:

$$\hbar c \approx 200\,\text{eV nm} = 2000\,\text{eV Å}$$
$$m_e c^2 \sim 0.5 \times 10^6\,\text{eV} \quad \text{(the electron rest energy)}$$
$$\frac{e^2/4\pi\epsilon_0}{\hbar c} \equiv \alpha \approx 1/137 \quad \text{(the fine-structure constant).} \tag{5.93}$$

Use these values and dimensional analysis to find the energy of a photon of green light (which has a wavelength of approximately 0.5 micrometers), expressing your answer in electron volts.

In our expression for the Bohr radius, $\hbar^2/m_e(e^2/4\pi\epsilon_0)$, we can replace $\hbar$ with $\hbar c$ and m_e with $m_e c^2$ simultaneously: Multiply by 1 in the form c^2/c^2; it is a convenient symmetry operation.

$$a_0 \sim \frac{\hbar^2}{m_e(e^2/4\pi\epsilon_0)} \times \frac{c^2}{c^2} = \frac{(\hbar c)^2}{m_e c^2 (e^2/4\pi\epsilon_0)}. \tag{5.94}$$

Now the table provides the needed values: for $\hbar c$, $m_e c^2$, and $e^2/4\pi\epsilon_0$. However, we can make one more simplification because the electrostatic mess $e^2/4\pi\epsilon_0$ is related to a dimensionless constant. To see the relation, multiply and divide by $\hbar c$:

$$\frac{e^2}{4\pi\epsilon_0} = \underbrace{\left(\frac{e^2/4\pi\epsilon_0}{\hbar c}\right)}_{\alpha} \hbar c. \tag{5.95}$$

The factor in parentheses is known as the fine-structure constant α; it is a dimensionless measure of the strength of electrostatics, and its numerical value is close to $1/137$ or roughly 0.7×10^{-2}. Then

$$\frac{e^2}{4\pi\epsilon_0} = \alpha\hbar c. \tag{5.96}$$

This substitution further simplifies the size of hydrogen:

$$a_0 \sim \frac{(\hbar c)^2}{m_e c^2 \times e^2/4\pi\epsilon_0} = \frac{(\hbar c)^2}{m_e c^2 \times \alpha\hbar c} = \frac{\hbar c}{\alpha \times m_e c^2}. \tag{5.97}$$

Only now, having simplified the calculation down to abstractions worth remembering (α, $m_e c^2$, and $\hbar c$), do we plug in the entries from the table.

$$a_0 \sim \frac{\overbrace{200\,\text{eV nm}}^{\hbar c}}{\underbrace{0.7 \times 10^{-2}}_{\alpha} \times \underbrace{5 \times 10^5\,\text{eV}}_{m_e c^2}}. \tag{5.98}$$

This calculation we can do mentally. The units of electron volts cancel, leaving only nanometers (nm). The two powers of ten upstairs (in the numerator) and the three powers of ten downstairs (in the denominator) result in 10^{-1} nanometers or 1 ångström (10^{-10} meters). The remaining factors result in a factor of $1/2$: $2/(0.7 \times 5) \approx 1/2$.

Therefore, the size of hydrogen—the Bohr radius—is about 0.5 ångströms:

$$a_0 \sim 0.5 \times 10^{-10}\,\text{m} = 0.5\,\text{Å}. \tag{5.99}$$

Amazingly, the missing dimensionless prefactor is 1. Thus, atoms are ångström-sized. Indeed, hydrogen is 1 ångström in diameter. All other atoms, which have more electrons and therefore more electron shells, are larger. A useful rule of thumb is that a typical atomic diameter is 3 ångströms.

▷ *What binding energy does this size produce?*

The binding energy is the energy required to disassemble the atom by removing the electron and dragging it to infinity. This energy, denoted E_0, should be roughly the electrostatic energy of a proton and electron separated by the Bohr radius a_0:

$$E_0 \sim \frac{e^2}{4\pi\epsilon_0} \frac{1}{a_0}. \tag{5.100}$$

With our result for a_0, the binding energy becomes

$$E_0 \sim m_e \left(\frac{e^2}{4\pi\epsilon_0} \right)^2 \frac{1}{\hbar^2}. \tag{5.101}$$

The missing dimensionless prefactor is just $1/2$:

$$E_0 = \frac{1}{2} m_e \left(\frac{e^2}{4\pi\epsilon_0} \right)^2 \frac{1}{\hbar^2}. \tag{5.102}$$

Problem 5.36   Shortcut to calculate the binding energy of hydrogen

Use the shortcuts in Problem 5.35 to show that the binding energy of hydrogen is roughly 14 electron volts. From your symbolic expression for the energy, which is also the kinetic energy of the electron, estimate its speed as a fraction of c.

In Problem 5.36, you showed, using the values of $\hbar c$, $m_e c^2$, and α, that this energy is roughly 14 electron volts. For the sake of a round number, let's call the binding energy roughly 10 electron volts.

This energy sets the scale for chemical bonds. In Section 3.2.1, we calculated, by unit conversion, that 1 electron volt per molecule corresponded to roughly 100 kilojoules per mole. Therefore, breaking a chemical bond requires roughly 1 megajoule per mole (of bonds). As a rough estimate, it is not far off. For example, if the molecule consists of a few life-related atoms (carbon, oxygen, and hydrogen), then the molar mass is roughly 50 grams—a small jelly donut. Therefore, burning a jelly donut, as our body does slowly when we eat the donut, should produce roughly 1 megajoule—a useful rule of thumb and way of imagining a megajoule.

Problem 5.37   Adding the speed of light

If we assume that a_0 depends on $e^2/4\pi\epsilon_0$, m_e, and c, what size would dimensional analysis predict for hydrogen? This size is called the classical electron radius and denoted r_0. How does it compare to the actual Bohr radius?

Problem 5.38   Thermal expansion

Estimate a typical thermal-expansion coefficient, also denoted α. It is defined as

$$\alpha \equiv \frac{\text{fractional change in a substance's length}}{\text{change in temperature}}. \tag{5.103}$$

The thermal-expansion coefficient depends on the binding energy E_0. Assuming $E_0 \sim 10\,\text{eV}$, compare your estimate for α with its value for everyday substances.

Problem 5.39 Diffraction in the eye

Light passing through an opening, such as a telescope aperture or the pupil in the eye, diffracts (spreads out). By estimating the diffraction angle θ, we will be able to understand aspects of the design of the eye.

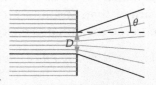

a. What are the valid dimensionless relations connecting the diffraction angle θ, the light wavelength λ, and the pupil diameter D?

b. Diffraction is the result of the photons in the beam acquiring a vertical momentum Δp_y. What is the scaling exponent β in $\theta \propto (\Delta p_y)^\beta$?

c. The Heisenberg uncertainty principle from quantum mechanics says that the uncertainty in the photon's vertical momentum Δp_y is inversely proportional to the pupil diameter D. What is therefore the scaling exponent γ in $\theta \propto D^\gamma$?

d. Therefore find θ as a function of λ and D.

e. Estimate a pupil diameter and the resulting diffraction angle. The light-sensitive cells in the retina that we use in bright light are the cones. They are most dense in the fovea—the central region of the retina that we use for reading and any other task requiring sharp vision. At that location, their density is roughly 0.5×10^7 per square centimeter. Is that what you would predict?

Problem 5.40 Kepler's third law for non-inverse-square force laws

With an inverse-square force, Kepler's third law—the relation between orbit radius and period—is $T \propto r^{3/2}$ (Section 4.2.2). Now generalize the law to forces of the form $F \propto r^n$, using dimensional analysis to find the scaling exponent β in the orbital period $T \propto r^\beta$ as a function of the scaling exponent n in the force law.

Problem 5.41 Ground state energy for general potentials

Just as we can apply dimensional analysis to classical orbits (Problem 5.40), we can apply it to quantum orbits. When the force law is $F \propto r^{-2}$ (electrostatics), or the potential is $V \propto r^{-1}$, we have hydrogen, for which we estimated the binding or ground-state energy. Now generalize to potentials where $V = Cr^\beta$.

The relevant quantities are the ground-state energy E_0, the proportionality constant C in the potential, the quantum constant $\hbar$, and the particle's mass m. These four quantities, built from three dimensions, form one independent dimensionless group. But it is not easy to find. Therefore, use linear algebra to find the exponents γ, δ, and ϵ that make $E_0 C^\gamma \hbar^\delta m^\epsilon$ dimensionless. In the case $\beta = -1$ (electrostatics), check that your result matches our result for hydrogen.

5.5.2 Blackbody radiation

With quantum mechanics and its new constant $\hbar$, we can explain the surface temperatures of planets or even stars. The basis is blackbody radiation: A hot object—a so-called blackbody—radiates energy. (Here, "hot" means warmer than absolute zero, so every object is hot.) Hotter objects radiate more, so the radiated energy flux F depends on the object's temperature T.

▶ *How are the energy flux F and the surface temperature T connected?*

I stated the connection in Section 4.2, but now we know enough dimensional analysis to derive almost the entire result without a detailed study of the quantum theory of radiation. Thus, let's follow the steps of dimensional analysis to find F, and start by listing the relevant quantities.

The list begins with the goal, the energy flux F. Blackbody radiation can be understood properly through the quantum theory of radiation, which is the marriage of quantum mechanics to classical electrodynamics. (A diagram of the connection, which requires the subsequent tool of easy cases, is in Section 8.4.2.) For the purposes of dimensional analysis, this theory supplies two constants of nature: $\hbar$ from quantum mechanics and the speed of light c from classical electrodynamics. The list also includes the object's temperature T, so that dimensional analysis knows how hot the object is; but we'll include it as the thermal energy $k_B T$.

The four quantities built from three independent dimensions produce one independent dimensionless group. Our usual route to finding this group, by looking for dimensions that appear in only one or two variables, fails because mass, like length, appears in three quantities, and time appears in all four. An alternative is to apply linear algebra, as you practiced in Problem 5.41. But it is messy.

F	MT^{-3}	energy flux
$k_B T$	ML^2T^{-2}	thermal energy
$\hbar$	ML^2T^{-1}	quantum
c	LT^{-1}	speed of light

A less brute-force and more physically meaningful alternative is to choose a system of units where $c \equiv 1$ and $\hbar \equiv 1$. Each of these two choices has a physical meaning. Before using the choices, it is worth understanding these meanings. Otherwise we are back to pushing symbols around.

The first choice, $c \equiv 1$, expresses the unity of space and time embedded and embodied in Einstein's theory of special relativity. With $c \equiv 1$, length and time have the same dimensions and are measured in the same units. We can then say that the Sun is 8.3 minutes away from the Earth. That time is equivalent to the distance of 8.3 light minutes (the distance that light travels in 8.3 minutes).

The choice $c \equiv 1$ also expresses the equivalence between mass and energy. We use this choice implicitly when we say that the mass of an electron is

5×10^5 electron volts—which is an energy. The complete statement is that, in the usual units, the electron's rest energy $m_e c^2$ is 5×10^5 electron volts. But when we choose $c \equiv 1$, then 5×10^5 electron volts is also the mass m_e.

The second choice, $\hbar \equiv 1$, expresses the fundamental insight of quantum mechanics, that energy is (angular) frequency. The complete connection between energy and frequency is

$$E = \hbar \omega. \tag{5.104}$$

When $\hbar \equiv 1$, then E is ω.

These two unit choices implicitly incorporate their physical meanings into our analysis. Furthermore, the choices so simplify our table of dimensions that we can find the proportionality between flux and temperature without any linear algebra.

First, choosing $c \equiv 1$ makes the dimensions of length and time equivalent:

$$c \equiv 1 \quad \text{implies} \quad \text{L} \equiv \text{T}. \tag{5.105}$$

Thus, in the table we can replace T with L.

Then adding the choice $\hbar \equiv 1$ contributes the dimensional equation

$$\underbrace{\text{ML}^2\text{T}^{-2}}_{[E]} \equiv \underbrace{\text{T}^{-1}}_{[\omega]} \tag{5.106}$$

It looks like a useless mess; however, replacing T with L (from $c \equiv 1$) simplifies it:

$$\text{M} \equiv \text{L}^{-1}. \tag{5.107}$$

In summary, setting $c \equiv 1$ and $\hbar \equiv 1$ makes length and time equivalent dimensions and makes mass and inverse length equivalent dimensions.

With these equivalences, let's rewrite the table of dimensions, one quantity at a time.

1. *Flux F.* Its dimensions, in the usual system, were MT^{-3}. In the new system, T^{-3} is equivalent to M^3, so the dimensions of flux become M^4.

2. *Thermal energy $k_B T$.* Its dimensions were ML^2T^{-2}. In the new system, T^{-2} is equivalent to L^{-2}, so the dimensions of thermal energy become M. And they should: Choosing $c \equiv 1$ makes energy equivalent to mass.

3. *Quantum constant $\hbar$.* By fiat (when we chose $\hbar \equiv 1$), it is now dimensionless and just 1, so we do not list it.

4. *Speed of light c.* Also by fiat, c is just 1, so we do not list it either.

Our list has become shorter and the dimensions in it

F	M^4	energy flux
k_BT	M	thermal energy

simpler. In the usual system, the list had four quantities (F, k_BT, $\hbar$, and c) and three independent dimensions (M, L, and T). Now the list has only two quantities (F and k_BT) and only one independent dimension (M). In both systems, there is only one dimensionless group.

In the usual system, the Buckingham Pi calculation is as follows:

> 4 quantities
> − 3 independent dimensions
> ———————————————
> 1 independent dimensionless group   (5.108)

In the new system, the calculation is

> 2 quantities
> − 1 independent dimension
> ———————————————
> 1 independent dimensionless group   (5.109)

(If the number of independent dimensionless groups had not been the same, we would know that we had made a mistake in converting to the new system.) In the simpler system, the independent dimensionless group proportional to F is now almost obvious. Because F contains M^4 and k_BT contains M^1, the group is just $F/(k_BT)^4$.

Alas, no lunch is free and no good deed goes unpunished. Now we pay the price for the simplicity: We have to restore c and $\hbar$ in order to find the equivalent group in the usual system of units. Fortunately, the restoration doesn't require any linear algebra.

The first step is to compute the usual dimensions of $F/(k_BT)^4$:

$$\left[\frac{F}{(k_BT)^4}\right] = \frac{MT^{-3}}{\left(ML^2T^{-2}\right)^4} = M^{-3}L^{-8}T^5. \tag{5.110}$$

The rest of the analysis for making this quotient dimensionless just rolls downhill without our needing to make any choices. The only way to get rid of M^{-3} is to multiply by $\hbar^3$: Only $\hbar$ contains a dimension of mass.

The result has dimensions $L^{-2}T^2$:

$$\left[\frac{F}{(k_BT)^4}\hbar^3\right] = \underbrace{M^{-3}L^{-8}T^5}_{[F/(k_BT)^4]} \times \underbrace{\left(ML^2T^{-1}\right)^3}_{[\hbar]} = L^{-2}T^2. \tag{5.111}$$

To clear these dimensions, multiply by c^2. The independent dimensionless group proportional to F is therefore

$$\frac{F\hbar^3 c^2}{(k_B T)^4}. \tag{5.112}$$

As the only independent dimensionless group, it is a constant, so

$$F \sim \frac{(k_B T)^4}{\hbar^3 c^2}. \tag{5.113}$$

Including the dimensionless prefactor hidden in the twiddle ($\sim$),

$$F = \underbrace{\frac{\pi^2}{60} \frac{k_B^4}{\hbar^3 c^2}}_{\sigma} T^4. \tag{5.114}$$

This result is the Stefan–Boltzmann law, and all the constants are lumped into σ, the Stefan–Boltzmann constant:

$$\sigma \equiv \frac{\pi^2}{60} \frac{k_B^4}{\hbar^3 c^2}. \tag{5.115}$$

In Section 4.2.1, we used the scaling exponent in the Stefan–Boltzmann law to estimate the surface temperature of Pluto: We started from the surface temperature of the Earth and used proportional reasoning. Now that we know the full Stefan–Boltzmann law, we can directly calculate a surface temperature. In Problem 5.43, you will estimate the surface temperature of the Earth (and you then try to explain the life-saving discrepancy between prediction and reality). Here, we'll estimate the surface temperature of the Sun using the Stefan–Boltzmann law, the solar flux at the top of the Earth's atmosphere, and proportional reasoning.

Let's work backward from our goal toward what we know. We would like to find the Sun's surface temperature T_{Sun}. If we knew the energy flux F_{Sun} at the Sun's surface, then the Stefan–Boltzmann law would give us the temperature:

$$T_{Sun} = \left(\frac{F_{Sun}}{\sigma} \right)^{1/4}. \tag{5.116}$$

We can find F_{Sun} from the solar flux F at the Earth's orbit by using proportional reasoning. Flux, as we argued in Section 4.2.1, is inversely proportional to distance squared, because the same energy flow (a power) gets smeared over a surface area proportional to the square of the distance from the source. Therefore,

$$\frac{F_{\text{Sun's surface}}}{F} = \left(\frac{r_{\text{Earth's orbit}}}{R_{\text{Sun}}}\right)^2, \tag{5.117}$$

where R_{Sun} is the radius of the Sun.

A related distance ratio is.

$$\frac{D_{\text{Sun}}}{r_{\text{Earth's orbit}}} = \frac{2R_{\text{Sun}}}{r_{\text{Earth's orbit}}}, \tag{5.118}$$

where D_{Sun} is the diameter of the Sun. This ratio is the angular diameter of the Sun, denoted θ_{Sun}. Therefore,

$$\frac{r_{\text{Earth's orbit}}}{R_{\text{Sun}}} = \frac{2}{\theta_{\text{Sun}}}. \tag{5.119}$$

From a home experiment measuring the angular diameter of the Moon, which has the same angular diameter as the Sun, or from the table of constants, the angular diameter θ_{Sun} is approximately 10^{-2}. Therefore, the distance ratio is approximately 200, and the flux ratio is approximately 200^2 or 4×10^4.

Continuing to unwind the reasoning chain, we get the flux at the Sun's surface:

$$F_{\text{Sun's surface}} \approx 4 \times 10^4 \times \underbrace{1300\,\text{W}\,\text{m}^{-2}}_{F} \approx 5 \times 10^7\,\text{W}\,\text{m}^{-2}. \tag{5.120}$$

This flux, from the Stefan–Boltzmann law, corresponds to a surface temperature of roughly 5400 K:

$$T_{\text{Sun}} \approx \left(\frac{5 \times 10^7\,\text{W}\,\text{m}^{-2}}{6 \times 10^{-8}\,\text{W}\,\text{m}^{-2}\,\text{K}^{-4}}\right)^{1/4} \approx 5400\,\text{K}. \tag{5.121}$$

This prediction is quite close to the measured temperature of 5800 K.

Problem 5.42   Redo the blackbody-radiation analysis using linear algebra
Use linear algebra to find the exponents y, z, and w that make the combination $F \times (k_B T)^y \hbar^z c^w$ dimensionless.

Problem 5.43   Blackbody temperature of the Earth
Use the Stefan–Boltzmann law $F = \sigma T^4$ to predict the surface temperature of the Earth. Your prediction should be somewhat colder than reality. How do you explain the (life-saving) difference between prediction and reality?

Problem 5.44   Lengths related by the fine-structure constant
Arrange these important atomic-physics lengths in order of increasing size:

a. the classical electron radius (Problem 5.37) $r_0 \equiv (e^2/4\pi\epsilon_0)/m_e c^2$ (roughly the radius of an electron if its rest energy were entirely electrostatic energy).

b. the Bohr radius a_0 from Section 5.5.1.

c. the reduced wavelength $\lambdabar$ of a photon with energy $2E_0$, where $\lambdabar \equiv \lambda/2\pi$ (analogously to how $\hbar \equiv h/2\pi$ or $f = \omega/2\pi$) and E_0 is the binding energy of hydrogen. (The energy $2E_0$, also the absolute value of the potential energy in hydrogen, is called the Rydberg).

d. the reduced Compton wavelength of the electron, defined as $\hbar/m_e c$.

Place the lengths on a logarithmic scale, and label the gaps (the ratios of successive lengths) in terms of the fine-structure constant α.

5.5.3 Molecular binding energies

We studied hydrogen, which as an element scarcely exists on Earth, mainly to understand chemical bonds. Chemical bonds are formed by attractions between electrons and protons, so the hydrogen atom is the simplest chemical bond. The main defect of this model is that the electron–proton bond in a hydrogen atom is much shorter than in most bonds. A typical chemical bond is roughly 1.5 ångströms, three times larger than the Bohr radius. Because electrostatic energy scales as $E \propto 1/r$, a typical bond energy should be smaller than hydrogen's binding energy by a factor of 3. Hydrogen's binding energy is roughly 14 electron volts, so E_{bond} is roughly 4 electron volts—in agreement with the bond-energy values tabulated in Section 2.1.

Another important bond is the hydrogen bond. These intermolecular bonds, which hold water molecules together, are weaker than the *intra*molecular hydrogen–oxygen bonds within a water molecule. But hydrogen bonds determine important properties of the most important liquid on our planet. For example, the hydrogen-bond energy determines water's heat of vaporization—which determines much of our weather, including the average rainfall on the Earth (as we found in Section 3.4.3).

To estimate the strength of hydrogen bonds, use the proportionality

$$E_{\text{electrostatic}} \propto \frac{q_1 q_2}{r}. \tag{5.122}$$

A hydrogen bond is slightly longer than a typical bond. Instead of the typical 1.5 ångströms, it is roughly 2 ångströms. The bond length is in the denominator, so the greater length reduces the bond energy by a factor of 4/3.

Furthermore, the charges involved, q_1 and q_2, are smaller than the charges in a regular, intramolecular bond. A regular bond is between a full proton and a full electron of charge. A hydrogen bond, however, is between the excess charge on oxygen and the corresponding charge deficit on hydrogen. The excess and deficit are significantly smaller than a full charge. On oxygen the excess may be $0.5e$ (where e is the electron charge). On each hydrogen, because of conservation, it would then be $0.25e$. These reductions in charge contribute factors of $1/2$ and $1/4$ to the hydrogen-bond energy. The resulting energy is roughly 0.4 electron volts:

$$4\,\text{eV} \times \frac{3}{4} \times \frac{1}{2} \times \frac{1}{4} \approx 0.4\,\text{eV}. \tag{5.123}$$

This estimate is not bad considering the rough numbers that it used. Empirically, a typical hydrogen bond is 23 kilojoules per mole or about 0.25 electron volts. Each water molecule forms close to four hydrogen bonds (two from the oxygen to foreign hydrogens, and one from each hydrogen to a foreign oxygen). Thus, each molecule gets credit for almost two hydrogen bonds—one-half of the per-molecule total in order to avoid counting each bond twice. Per water molecule, the result is 0.4 electron volts

Because vaporizing water breaks the hydrogen bonds, but not the intramolecular bonds, the heat of vaporization of water should be approximately 0.4 electron volts per molecule. In macroscopic units, it would be roughly 40 kilojoules per mole—using the conversion factor from Section 3.2.1, that 1 electron volt per molecule is roughly 100 kilojoules per mole.

Because the molar mass of water is 1.8×10^{-2} kilograms, the heat of vaporization is also 2.2 megajoules per kilogram:

$$\frac{40\,\text{kJ}}{\text{mol}} \times \frac{1\,\text{mol}}{1.8 \times 10^{-2}\,\text{kg}} \approx \frac{2.2 \times 10^6\,\text{J}}{\text{kg}}. \tag{5.124}$$

And it is (p. xvii). We used this value in Section 3.4.3 to estimate the average worldwide rainfall. The amount of rain, and so the rate at which many plants grow, is determined by the strength of hydrogen bonds.

Problem 5.45   Rotational energies

The quantum constant $\hbar$ is also the smallest possible angular momentum of a rotating system. Use that information to estimate the rotational energy of a small molecule such as water (in electron volts). To what electromagnetic wavelength does this energy correspond, and where in the electromagnetic spectrum does it lie (for example, radio waves, ultraviolet, gamma rays)?

5.5.4 Stiffness and sound speeds

An important macroscopic consequence of the per-atom and per-molecule energies is the existence of solid matter: substances that resist bending, twisting, squeezing, and stretching. These resistance properties are analogous to spring constants.

However, a spring constant is not the quantity to estimate first, because it is not invariant under simple changes. For example, a thick bar resists stretching more than a thin bar does. Similarly, a shorter bar resists stretching more than a longer bar does. The property independent of the shape or amount of substance—invariant to these changes—is the stiffness or elastic modulus. There are several elastic moduli, of which the Young's modulus is the most broadly useful. It is defined by

$$Y = \frac{\text{stress } \sigma \text{ applied to the substance}}{\text{fractional change in its length } [(\Delta l)/l]}. \tag{5.125}$$

The ratio in the denominator occurs so often that it has its own name and symbol: the strain ϵ.

Stress, like the closely related quantity pressure, is force per area: It is the applied force F divided by the cross-sectional area. The denominator, the  fractional change in length, is the dimensionless ratio $(\Delta l)/l$. Thus, the dimensions of stiffness are the dimensions of pressure—which are also the dimensions of energy density. To see the connection between pressure and energy density, multiply the definition of pressure (force per area) by length over length:

$$\text{pressure} = \frac{\text{force}}{\text{area}} \times \frac{\text{length}}{\text{length}} = \frac{\text{energy}}{\text{volume}}. \tag{5.126}$$

As energy per volume, we can estimate a typical elastic modulus Y. For the numerator, a suitable energy is the typical binding energy per atom—the energy E_{binding} required to remove the atom from the substance by breaking its bonds to surrounding atoms. For the denominator, a suitable volume is a typical atomic volume a^3, where a is a typical interatomic spacing (3 ångströms). The result is

$$Y \sim \frac{E_{\text{binding}}}{a^3}. \tag{5.127}$$

This derivation is slightly incomplete: In multiplying the definition of pressure by length over length, we knew only that the two lengths have the same dimensions, but not whether they have comparable values. If they do not, then the estimate for Y needs a dimensionless prefactor, which might be far from 1. Therefore, before we estimate Y, let's confirm the estimate by using a second method, an analogy—a form of abstraction that we learned in Section 2.4.

The analogy, with which we also began this discussion of stiffness, is between a spring constant (k) and a Young's modulus (Y). There are three physical quantities in the analogy.

1. *Stiffness.* For a single spring, we measure stiffness using k. For a material, we use the Young's modulus Y. Therefore, k and Y are analogous.

2. *Extension.* For a single spring, we measure extension using the absolute length change Δx. For a material, we use the fractional length change $(\Delta l)/l$ (the strain ϵ). Therefore, Δx and $(\Delta l)/l$ are analogous.

3. *Energy.* For a single spring, we measure energy using the energy E directly. For a material, we use an intensive quantity, the energy density $\mathcal{E} \equiv E/V$. Therefore, E and $\mathcal{E}$ are analogous.

Because the energy in a spring is

$$E \sim k(\Delta x)^2, \tag{5.128}$$

by analogy the energy density in a material will be

$$\mathcal{E} \sim Y\left(\frac{\Delta l}{l}\right)^2. \tag{5.129}$$

Let's estimate Y by setting $\Delta l \sim l$. Physically, this choice means stretching or compressing each bond by its own length. This harsh treatment is, as a reasonable assumption, sufficient to break the bond and release the binding energy. Then the right side becomes simply the Young's modulus Y.

The left side, the energy density, becomes simply the bond energy per volume. For the volume, if we use the volume of one atom, roughly a^3, then the bond energy contained in this volume is the binding energy of one atom. Then the Young's modulus becomes

$$Y \sim \frac{E_{\text{binding}}}{a^3}. \tag{5.130}$$

This result is what we predicted based on the dimensions of pressure. However, we now have a physical model of the Young's modulus, based on an analogy between it and spring constant.

Having arrived at this conclusion from dimensions and from an analogy, let's apply it to estimate a typical Young's modulus. To estimate the numerator, the binding energy, start with the typical bond energy of 4 electron volts per bond. With each atom is connected to, say, five or six other atoms, the total energy is roughly 20 electron volts. Because bonds are connections between pairs of atoms, this 20 electron volts counts each bond twice. Therefore, a typical binding energy is 10 electron volts per atom.

Using $a \sim 3$ ångströms, a typical stiffness or Young's modulus is

$$Y \sim \frac{E_{\text{binding}}}{a^3} \sim \frac{10\,\text{eV}}{(3 \times 10^{-10}\,\text{m})^3} \times \frac{1.6 \times 10^{-19}\,\text{J}}{\text{eV}} \sim \frac{1}{2} \times 10^{11}\,\text{J m}^{-3}. \qquad (5.131)$$

For very rough estimates, a convenient value to remember is 10^{11} joules per cubic meter. Because energy density and pressure have the same dimensions, this energy density is also 10^{11} pascals or 10^6 atmospheres. (Because atmospheric pressure is only 1 atmosphere, a factor of 1 million smaller, it has little effect on solids.)

	Y (10^{11} Pa)
diamond	12
steel	2
Cu	1.2
Al	0.7
glass	0.6
granite	0.3
Pb	0.18
concrete	0.17
oak	0.1
ice	0.1

Using a typical stiffness, we can estimate sound speeds in solids. The speed depends not only on the stiffness, but also on the density: denser materials respond more slowly to the forces due to the stiffness, so sound travels more slowly in denser substances. From stiffness Y and density ρ, the only dimensionally correct way to make a speed is $\sqrt{Y/\rho}$. If this speed is the speed of sound, then

$$c_s \sim \sqrt{\frac{Y}{\rho}}. \qquad (5.132)$$

Based on a typical Young's modulus of 0.5×10^{11} pascals and a typical density of 2.5×10^3 kilograms per cubic meter (for example, rock), a typical speed of sound is 5 kilometers per second:

$$c_s \sim \sqrt{\frac{0.5 \times 10^{11}\,\text{Pa}}{2.5 \times 10^3\,\text{kg m}^{-3}}} \approx 5\,\text{km s}^{-1}. \qquad (5.133)$$

This prediction is reasonable for most solids and is exactly correct for steel! With this estimate, we end our dimensional-analysis tour of the properties of materials. Look what dimensional analysis, combined with analogy and proportional reasoning, has given us: the size of atoms, the energies of chemical bonds, the stiffness of materials, and the speed of sound.

5.6 Summary and further problems

A quantity with dimensions has no meaning by itself. As Socrates might have put it, the uncompared quantity is not worth knowing. Using this principle, we learned to rewrite relations in dimensionless form: in terms of combinations of the quantities where the combination has no dimensions. Because the space of dimensionless relations is much smaller than the space of all possible relations, this rewriting simplifies many problems. Like the other two tools in Part II, dimensional analysis discards complexity without loss of information.

Problem 5.46 Oblateness of the Earth

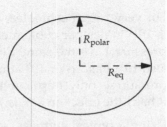

Because the Earth spins on its axis, it is an oblate sphere. You can estimate the oblateness using dimensional analysis and some guessing. Our measure of oblateness will be $\Delta R = R_{eq} - R_{polar}$ (the difference between the polar and equatorial radii). Find two independent dimensionless groups built from ΔR, g, R (the Earth's mean radius), and v (the Earth's rotation speed at the equator). Guess a reasonable relation between them in order to estimate ΔR. Then compare your estimate to the actual value, which is approximately 21.4 kilometers.

Problem 5.47 Huge waves on deep water

One of the highest measured ocean waves was encountered in 1933 by a US Navy oiler, the *USS Ramapo* (a 147-meter-long ship) [34]. The wave period was 14.8 seconds. Find its wavelength using the results of Problem 5.11. Would a wave of this wavelength be dangerous to the ship?

Problem 5.48 Ice skating

For world-record ice skating, estimate the power consumed by sliding friction. (Ice skates sliding on ice have a coefficient of sliding friction around 0.005.) Then make that power meaningful by estimating the ratio

$$\frac{\text{power consumed in sliding friction}}{\text{power consumed by air resistance}}. \tag{5.134}$$

Problem 5.49 Pressure melting during ice skating

Water expands when it freezes. Thus, increasing the pressure on ice should, by Le Chatelier's principle, push it toward becoming water—which lowers its freezing point. Based on the freezing point and the heat of vaporization of water, estimate the change in freezing point caused by ice-skate blades. Is this change large enough to explain why ice-skate blades slip with very low friction on a thin sheet of water?

Problem 5.50   Contact radius

A solid ball of radius R, density ρ, and elastic modulus Y rests on the ground. Using dimensional analysis, how much can you deduce about the contact radius r?

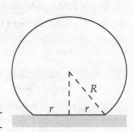

Problem 5.51   Contact time

The ball of Problem 5.50 is dropped from a height, hits a hard table with speed v, and bounces off. Using dimensional analysis, how much can you deduce about the contact time?

Problem 5.52   Floating on water

Some insects can float on water thanks to the surface tension of water. In terms of the bug size l (a length), find the scaling exponents α and β in

$$F_\gamma \propto l^\alpha,$$
$$W \propto l^\beta,$$

(5.135)

where F_γ is the surface-tension force and W is the bug's weight. (Surface tension itself has dimensions of force per length.) Thereby explain why a small-enough bug can float on water.

Problem 5.53   Dimensionless measures of damping

A damped spring–mass system has three parameters: the spring constant k, the mass m, and the damping constant γ. The damping constant determines the damping force through $F_\gamma = -\gamma v$, where v is the velocity of the mass.

a. Use these quantities to make the dimensionless group proportional to γ. Mechanical and structural engineers use this group to the define the dimensionless damping ratio ζ:

$$\zeta \equiv \frac{1}{2} \times \text{the dimensionless group proportional to } \gamma.$$

(5.136)

b. Find the dimensionless group proportional to γ^{-1}. Physicists and electrical engineers, following conventions from the early days of radio, call this group the quality factor Q.

Problem 5.54   Steel cable under its own weight

The stiffness of a material should not be confused with its strength! Strength is the stress (a pressure) at which the substance breaks; it is denoted σ_y. Like stiffness, it is an energy density. The dimensionless ratio σ_y/σ, called the yield strain ϵ_y, has a physical interpretation: the fractional length change at which the substance breaks. For most materials, it lies in the range $10^{-3}\ldots10^{-2}$—with brittle materials (such as rock) toward the lower end. Using the preceding information, estimate the maximum length of a steel cable before it breaks under its own weight.

Problem 5.55   Orbital dynamics

A planet orbits the Sun in an ellipse that can be described
by the distance of closest approach r_{min} and by the fur-
thest distance r_{max}. The length l is their harmonic mean:

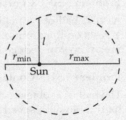

$$l = 2\,\frac{r_{min}r_{max}}{r_{min} + r_{max}} = 2\left(r_{min} \parallel r_{max}\right). \qquad (5.137)$$

(You will meet the harmonic mean again in Problem 8.22,
as an example of a more general kind of mean.)

The table gives r_{min} and r_{max} for the planets, as well as the specific effective potential V, which is the effective potential energy divided by the planet mass m (the
effective potential itself mixes the gravitational potential energy with one component of the kinetic energy). The purpose of this problem is to see how universal
functions organize this seemingly messy data set.

	r_{min} (m)	r_{max} (m)	V (m²s⁻²)
Mercury	4.6001×10^{10}	6.9818×10^{10}	-1.1462×10^{9}
Venus	1.0748×10^{11}	1.0894×10^{11}	-6.1339×10^{8}
Earth	1.4710×10^{11}	1.5210×10^{11}	-4.4369×10^{8}
Mars	2.0666×10^{11}	2.4923×10^{11}	-2.9119×10^{8}
Jupiter	7.4067×10^{11}	8.1601×10^{11}	-8.5277×10^{7}
Saturn	1.3498×10^{12}	1.5036×10^{12}	-4.6523×10^{7}
Uranus	2.7350×10^{12}	3.0063×10^{12}	-2.3122×10^{7}
Neptune	4.4598×10^{12}	4.5370×10^{12}	-1.4755×10^{7}

a. On a graph of V versus r, plot all the data. Each planet provides two data
    points, one for $r = r_{min}$ and one for $r = r_{max}$. The plot should be a mess. But
    you'll straighten it out in the rest of the problem.

b. Now write the relation between V and r in dimensionless form. The relevant
    quantities are V, r, GM_{Sun}, and the length l. Choose your groups so that V
    appears only in one group and r appears only in a separate group.

c. Now use the dimensionless form to replot the data in dimensionless form. All
    the points should lie on one curve. You have found the universal function char-
    acterizing all planetary orbits!

Problem 5.56   Signal propagation speed in coaxial cable

For the coaxial cable of Problem 2.25, estimate the signal propagation speed.

Problem 5.57   Meter stick under pressure

Estimate how much shorter a steel meter stick becomes due to being placed at the
bottom of the ocean. What about a meter stick made of wood?

Problem 5.58   Speed of sound in water

Using the heat of vaporization of water as a measure of the energy density in its weakest bonds, estimate the speed of sound in water.

Problem 5.59   Delta-function potential

A simple potential used as a model to understand molecules is the one-dimensional delta-function potential $V(x) = -E_0 L \delta(x)$, where E_0 is an energy and L is a length (imagine a deep potential of depth E_0 and small width L). Use dimensional analysis to estimate the ground-state energy.

Problem 5.60   Tube flow

In this problem you study fluid flow through a narrow tube. The quantity to predict is Q, the volume flow rate (volume per time). This rate depends on five quantities:

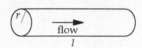

l	the length of the tube
Δp	the pressure difference between the tube ends
r	the radius of the tube
ρ	the density of the fluid
ν	the kinematic viscosity of the fluid

a. Find three independent dimensionless groups G_1, G_2, and G_3 from these six quantities—preparing to write the most general statement as

$$\text{group } 1 = f(\text{group } 2, \text{group } 3). \tag{5.138}$$

Hint: One physically reasonable group is $G_2 = r/l$; to make solving for Q possible, put Q only in group 1 and make this group proportional to Q.

b. Now imagine that the tube is long and thin ($l \gg r$) and that the radius or flow speed is small enough to make the Reynolds number low. Then deduce the form of f using proportional reasoning: First find the scaling exponent β in $Q \propto (\Delta p)^\beta$; then find the scaling exponent γ in $Q \propto l^\gamma$. Hint: If you double Δp and l, what should happen to Q?

Determine the form of f that satisfies all your proportionality requirements. If you get stuck, work backward from the correct result. Look up Poiseuille flow, and use the result to deduce the preceding proportionalities; and then give reasons for why they are that way.

c. The analysis in the preceding parts does not give you the universal, dimensionless constant. Use a syringe and needle to estimate the constant. Compare your estimate with the value that comes from solving the equations of fluid mechanics honestly (by looking up this value).

Problem 5.61   Boiling versus boiling away

Look up the specific heat of water in the table of constants (p. xvii) and estimate the ratio

$$\frac{\text{energy to bring a pot of water to boiling temperature}}{\text{energy to boil away the boiling water}}. \qquad (5.139)$$

Problem 5.62   Kepler's law for elliptical orbits

Kepler's third law connects the orbital period to the minimum and maximum orbital radii r_{min} and r_{max} and to the gravitational strength of the Sun:

$$T = 2\pi \frac{a^{3/2}}{\sqrt{GM_{Sun}}}, \qquad (5.140)$$

where the semimajor axis a is defined as $a \equiv (r_{min} + r_{max})/2$. Write Kepler's third law in dimensionless form, making one independent dimensionless group proportional to T and the other group proportional to r_{min}.

Problem 5.63   Why Mars?

Why did Kepler need data about Mars's orbit to conclude that planets orbit the Sun in an ellipse rather than a circle? Hint: See the data in Problem 5.55.

Problem 5.64   Froude number for a ship's hull speed

For a ship, the hull speed is defined as

$$v \equiv 1.34\sqrt{l}, \qquad (5.141)$$

where v is measured in knots (nautical miles per hour), and l, the waterline length, is measured in feet. The waterline length is, as you might expect, the length of the boat measured at the waterline. The hull speed is a boat's maximum speed before the water drag becomes very large.

Convert this unit-specific formula to an approximate Froude number Fr, the dimensionless number introduced in Section 5.1.1 to estimate the maximum walking speed. For the hull speed, the Froude number is defined as

$$\text{Fr} \equiv \frac{v^2}{gl}. \qquad (5.142)$$

From the approximate Froude number, guess the exact value!

Part III

Discarding complexity
with loss of information

You've organized (Part I); you've discarded complexity without losing information (Part II); yet the phenomenon still resists understanding. When the going gets tough, the tough lower their standards: Approximate first, and worry later. Otherwise you never start, and you can never learn that the approximations would have been accurate enough—if only you had gathered the courage to make them. Helping you make these approximations is the purpose of our final set of tools.

These four tools help us discard complexity while losing information. First, we round or lump complicated numbers and graphs (Chapter 6). Second, we accept that our knowledge is incomplete, and we quantify the uncertainty with the tool of probability (Chapter 7). Third, we study simpler versions of hard problems—the tool of easy cases (Chapter 8). Fourth and finally, by making spring models (Chapter 9), we approximate and can understand many phenomena, including cooking times, sound speeds, and the color of the sky and the sunset.

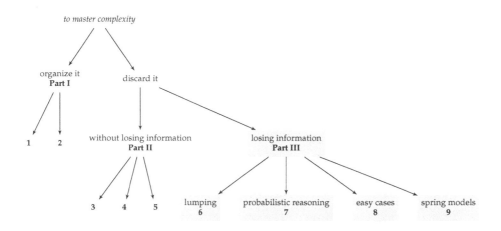

6

Lumping

In 1982, thousands of students in the United States had to estimate 3.04×5.3, choosing 1.6, 16, 160, 1600, or "I don't know." Only 21 percent of 13-year-olds and 37 percent of 17-year-olds chose 16. As Carpenter and colleagues describe [7], the problem is not a lack of calculation skill. On questions testing exact multiplication ("multiply 2.07 by 9.3"), the 13-year-olds scored 57 percent, and the 17-year-olds scored 72 percent correct.

choice	age 13	age 17
1.6	28%	21%
16	**21**	**37**
160	18	17
1600	23	11
don't know	9	12

The problem is a lack of understanding; if you earn roughly $5 per hour for roughly 3 hours, your net worth cannot grow by $1600. The students needed our next tool: rounding or, more generally, lumping.

6.1 Approximate!

Fortunately, rounding is inherent in our perception of quantity: Beyond three items, our perception of "how many" comes with an inherent imprecision. This fuzziness is, for adults, 20 percent: If we briefly see two groups of dots whose totals are within 20 percent, we cannot easily judge which group has more dots. Try it by glancing at the following squares.

In the left pair, one square contains 10 percent more dots than the other
square; in the right pair, one square contains 30 percent more dots than the
other square. In the 10-percent pair, spotting the more numerous square
is difficult. In the 30-percent pair, it is almost obvious at sight. Lumping comes naturally; we just need the courage to do it. We'll develop the
courage first in rounding numbers, the most familiar kind of lumping.

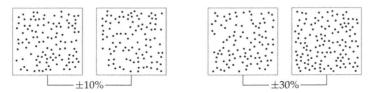

6.2 Rounding on a logarithmic scale

Just as driving to visit a next-door neighbor atrophies our muscles and ability to move around the physical world, asking calculators to do simple arithmetic dulls our ability to navigate the quantitative world. We never develop
an innate sense of sizes and scales in the world. The antidote is to do the
computations ourselves, but approximately—by placing quantities on a logarithmic scale and rounding them to the nearest convenient value.

6.2.1 Rounding to the nearest power of ten

The simplest method of rounding is to round every number to the nearest
power of ten. That simplification turns most calculations into adding and
subtracting integer exponents (the exceptions come from roots, which produce fractional exponents). Here, "nearest" is judged on a logarithmic scale,
where distance is measured not with differences but with ratios or factors.
For example, 50—although closer to 10 than to 100 on a linear scale—is a
factor of 5 greater than 10 but only a factor of 2 smaller than 100. Therefore,
50 is closer to 100 than to 10 and would get rounded to 100.

As practice, let's estimate the number of minutes in a day:

$$1 \text{ day} \times \frac{24 \text{ hours}}{\text{day}} \times \frac{60 \text{ minutes}}{\text{hour}} = 24 \times 60 \text{ minutes}. \tag{6.1}$$

Now we round each factor to the nearest power of 10. Because 24 is only
a factor of 2.4 away from 10, but more than a factor of 4 away from 100, it
gets rounded to 10:

In contrast, 60 is closer to 100 than to 10:

With these approximations, 1 day is approximately 1000 minutes:

$$1 \text{ day} \times \overbrace{\frac{10 \text{ hours}}{\text{day}}}^{\text{from } 24} \times \overbrace{\frac{100 \text{ minutes}}{\text{hour}}}^{\text{from } 60} = 1000 \text{ minutes}. \qquad (6.2)$$

The exact value is 1440 minutes, so the estimate is only 30 percent too small. This error is a reasonable tradeoff to gain a method that requires almost no effort—who needs a calculator to multiply 10 by 100? Furthermore, the accuracy is enough for many calculations, where insight is needed more than accuracy.

Problem 6.1 Rounding to the nearest power of ten
Round these numbers to the nearest power of ten: 200, 0.53, 0.03, and 7.9.

Problem 6.2 Boundary between rounding up or down
We saw that 60 rounded up to 100 but that 24 rounded down to 10. What number is just at the boundary between rounding down to 10 and rounding up to 100?

Problem 6.3 Using rounding to the nearest power of ten
Round the numbers to the nearest power of ten and thereby estimate the products: (a) 27 × 50, (b) 432 × 12, and (c) 3.04 × 5.3.

Problem 6.4 Calculating the bending of light
In Section 5.3.1, we used dimensional analysis to show that the Earth could bend starlight by approximately the angle

$$\theta \sim \frac{\overbrace{6.7 \times 10^{-11} \text{ kg}^{-1} \text{ m}^3 \text{ s}^{-2}}^{G} \times \overbrace{6 \times 10^{24} \text{ kg}}^{m_{\text{Earth}}}}{\underbrace{6.4 \times 10^6 \text{ m}}_{R_{\text{Earth}}} \times \underbrace{10^{17} \text{ m}^2 \text{ s}^{-2}}_{c^2}}. \qquad (6.3)$$

Round each factor to the nearest power of ten in order to estimate the bending angle mentally. How long did making the estimate take?

6.2.2 **Rounding to the nearest half power of ten**

Rounding to the nearest power of ten gives a quick, preliminary estimate. When it is too approximate, we just round more precisely. The next increase in accuracy is to round to the nearest *half* power of ten.

As an illustration, let's estimate the number of seconds in a year. The numerical calculation (without the units) is $365 \times 24 \times 3600$. Now we round each factor by replacing it with a number of the form 10^β. In the previous method, where we rounded to the nearest power of ten, β was an integer. Now β can also be a half-integer (for example, 2.5).

▷ *What is a half power of ten numerically?*

Two half powers of ten multiply to make 10, so each half power is $\sqrt{10}$, or slightly more than 3 (as you found in Problem 6.2). When you need more precision, a half power of ten is 3.2 or 3.16, although that much precision is rarely needed.

In rounding the calculation for the number of seconds, 365 becomes $10^{2.5}$, and, as diagrammed below, 3600 becomes $10^{3.5}$:

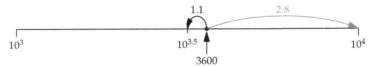

The remaining factor is 24. It is closer to $10^{1.5}$ (roughly 30) than to 10^1:

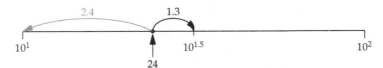

Thus, we replace 24 by $10^{1.5}$. Then the calculation simplifies to

$$\underbrace{10^{2.5}}_{365} \times \underbrace{10^{1.5}}_{24} \times \underbrace{10^{3.5}}_{3600} .$$ (6.4)

Now just add the powers of ten:

$$10^{2.5} \times 10^{1.5} \times 10^{3.5} = 10^{7.5}.$$ (6.5)

Because the 0.5 in the exponent 7.5 contributes a factor of 3, there are about 3×10^7 seconds in a year. I remember this value as $\pi \times 10^7$, which is accurate to 0.5 percent.

Problem 6.5   Earth's orbital speed

Using the estimate of $\pi \times 10^7$ seconds in a year, estimate the Earth's orbital speed around the Sun. Don't use a calculator! (The Earth–Sun distance, 1.5×10^{11} meters, is worth memorizing.)

Problem 6.6   Where does the π come from?

True or false: The π in $\pi \times 10^7$ seconds per year arises because the Earth orbits the Sun in a circle, and a circle has π in its circumference.

Problem 6.7   Only approximately pi

True or false: The π in $\pi \times 10^7$ seconds per year is not exact because the Earth orbits in a slightly noncircular ellipse.

Problem 6.8   Estimating geometric means

In Section 2.3, I talked to my gut to estimate the US annual oil imports. That discussion resulted in the geometric-mean estimate

$$\sqrt{10 \text{ million} \times 1 \text{ trillion}} \text{ barrels per year.} \tag{6.6}$$

Estimate the square root by placing the two quantities 10 million and 1 trillion on a logarithmic scale and finding their midpoint.

6.3   Typical or characteristic values

Lumping not only simplifies numbers, where it is called rounding, it also simplifies complex quantities by creating an abstraction: the typical or characteristic value. We've used this form of lumping implicitly many times; now it's time to discuss it explicitly, in order to appreciate its scope and to develop skill in applying it.

6.3.1   Estimating the population of the United States

Our first explicit example of a typical or characteristic value occurs in the following population estimate. Knowing a population is essential in many estimates about the social world, such as a country's oil imports (Section 1.4), energy consumption, or per-capita land area (Problem 1.14). Here we'll estimate the population of the United States by dividing it into two factors.

$$\text{US population} \sim N_{\text{states}} \times \text{population of a typical US state.} \tag{6.7}$$

The first factor, N_{states}, everyone in America learns in school: $N_{\text{states}} = 50$. The second factor contains the lumping approximation. Rather than using all 50 different state populations, we replace each state with a typical state.

▷ *What is the population of a typical state?*

A large state is California or New York, each of which has a megacity with a population in the several millions. A tiny state is Delaware or Rhode Island. In between is Massachusetts. Because I live there, I know that it has 6 million people. Taking Massachusetts as a typical state, the US population is roughly 50 × 6 million or 300 million. This estimate is more accurate than we deserve: The 2012 population is 314 million.

From this example, you can also see how lumping enhances symmetry reasoning: When there is change, look for what does not change (Section 3.1). Here, each state has its own population, so there's plenty of change among the list of states. Lumping helps us find, or create, a quantity that does not change. We imagine a typical state, one that may not even exist (just as no family has the average of 2.3 children), and replace every state with this state. We've lumped away all the change—throwing away information in exchange for insight into the population of the country.

> **Problem 6.9   German federal states**
> The Federal Republic of Germany has 16 federal states. Pick one at random, multiply its population by 16, and compare that estimate with Germany's population.

6.3.2 Lumping varying physical quantities: How high can animals jump?

Using typical or characteristic values allows us to reason out seemingly impossible questions while sitting in our armchairs. For practice, we'll study how the jump height of an animal depends on its size. For example, should a person be able to jump higher than a locust?

The jump could be a running or a standing high jump. Both types are interesting, but the standing high jump teaches more about lumping. Therefore, imagine that the animal starts from rest and jumps directly upward.

Even with this assumption, the problem looks underspecified. The jump height depends at least on the animal's shape, on how much muscle the animal has, and on the muscle efficiency. It is the kind of problem where lumping, a tool for making assumptions, is most helpful. We'll use lumping and proportional reasoning to find the scaling exponent β in $h \propto m^\beta$, where h is the jump height and m is the animal's mass (our measure of its size).

Finding a scaling exponent usually requires a physical model. You can often build them by imagining an extreme, unrealistic situation, and then asking yourself what physics prevents it from happening. Thus, why can't we jump to the Moon? Because it demands a vast amount of energy, far beyond what our muscles can supply. The point to extract from this thought experiment is that jumping demands energy, which is supplied by muscles.

The appearance of supply and demand suggests, as in the estimate of the number of taxis in Boston (Section 3.4.1), that we equate the demand to the supply. Then we estimate each piece separately—divide and conquer.

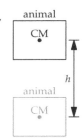

The energy demanded is the gravitational potential energy mgh. Here, g is the gravitational acceleration, and the jump height h is measured as the vertical change in the animal's center of mass (CM). Because all animals feel the same gravity, $E_{\text{demanded}} = mgh$ simplifies to the proportionality $E_{\text{demanded}} \propto mh$.

For the supplied energy, we again divide and conquer:

$$E_{\text{supplied}} \sim \text{muscle mass} \times \text{muscle energy density}, \qquad (6.8)$$

where the muscle energy density is the energy per mass that muscle can supply. This product already contains a lumping assumption: that all the muscles in an animal provide the same energy density. This assumption is reflected in the single approximation sign ($\sim$).

Even so, the result is not simple enough. Each species, and each individual within a species, will have its own muscle mass and energy density. Therefore, let's make the further lumping assumption that all animals, even though they vary in muscle mass, share the same muscle energy density. The assumption is plausible because all muscles use similar biological technology (actin and myosin filaments). Fortunately, making this assumption is an approximation: Lumping throws away actual information, which is how it reduces complexity. With the assumption of a common energy density, the supplied energy becomes the simpler proportionality

$$E_{\text{supplied}} \propto \text{muscle mass}. \qquad (6.9)$$

The simplicity of this result reminds us that an approximate result can be more useful than an exact result.

The muscle mass also varies from animal to animal. Introducing a dimensionless prefactor and making another lumping approximation will manage that complexity. The dimensionless prefactor α will be the fraction of an animal's mass that is muscle:

$$m_{\text{muscle}} = \alpha m. \qquad (6.10)$$

Alas, α varies across species (compare a cheetah and a turtle), within a species, and within the lifetime of an individual—for example, my α is dropping as I sit writing this book. If we account for all these variations, we will be overwhelmed by their complexity. Lumping rescues us: It gives us permission to assume that α is the same for every animal. We replace the diversity of animals with a typical animal. This assumption is not as crazy as it might sound. It doesn't mean that all animals have the same muscle mass. Rather, it means that all animals have the same *fraction* of muscle; as an example, for people, $\alpha \sim 0.4$.

With this lumping assumption, $m_{\mathrm{muscle}} = \alpha m$ becomes the simpler proportionality $m_{\mathrm{muscle}} \propto m$. Because the supplied energy is proportional to the muscle mass, it is proportional to the animal's mass:

$$E_{\mathrm{supplied}} \propto m_{\mathrm{muscle}} \propto m. \qquad (6.11)$$

This result is as simple as we can hope for, and it depends on the right quantity, the animal's mass. Now let's use it to predict how an animal's jump height depend on this mass.

Because the demanded energy and supplied energy are equal, and the demanded energy—the gravitational potential energy—is proportional to mh,

$$mh \propto m. \qquad (6.12)$$

The common factor of m^1 cancels out, leaving h independent of m:

$$h \propto m^0. \qquad (6.13)$$

All (jumping) animals should be able to jump to the same height!

The result always surprises me. My intuition, before doing the calculation, cannot decide how h should behave. On the one hand, small animals seem strong: Ants can lift a mass many times their own body mass, whereas humans can lift only roughly their own body mass. However, my intuition also insists that people should be able to jump higher than locusts.

The data, from *Scaling: Why Animal Size Is So Important* [41, p. 178], contradict my intuition and confirm our lumping and scaling analysis. The mass spans more than eight orders of magnitude (eight decades, or a factor of 10^8), yet the jump height, as a change in the height of the center of mass, varies only by a factor of 3 (half a decade). On a log–log graph, the height-versus-mass line would have a slope less than $1/16$. Our predicted scaling

	m	h
flea	0.5 mg	20 cm
click beetle	40 mg	30 cm
locust	3 g	59 cm
human	70 kg	60 cm

of constant jump height ($h \propto m^0$), which corresponds to a slope of zero, is surprisingly accurate. The outliers, fleas and click beetles, are at the lighter end. (For an explanation, try Problem 6.10.) Furthermore, at the heavier end, locusts and humans, although differing by more than four orders of magnitude in mass, jump to almost exactly the same height.

A moral of this example is that lumping augments proportional reasoning. Proportional reasoning reduces complexity by showing us a notation for ignoring quantities that do not vary. For example, when all animals face the same gravitational field, then $E_{\text{demanded}} = mgh$ simplifies to $E_{\text{demanded}} \propto mh$. Alas, we live in the desert of the real, where "the same" is almost always only an approximation—for example, as with the energy density of muscle in different animals. Lumping rescues us. It gives us permission to replace these changing values with a single, constant, typical value—making the relations amenable to proportional reasoning.

> **Problem 6.10 Jumping fleas**
> The prediction of constant jump height seems to fail at small sizes: Larger animals jump about 60 centimeters, whereas fleas reach only 20 centimeters. In this problem, you evaluate whether air drag can explain the discrepancy.
>
> a. As a lumping approximation, pretend that an animal is a cube with side length l, and assume that it jumps to a height h independent of its mass m. Then find the scaling exponent β in $E_{\text{drag}}/E_{\text{demanded}} \propto l^\beta$, where E_{drag} is the energy consumed by drag and E_{demanded} is the energy needed in the absence of drag.
>
> b. Estimate $E_{\text{drag}}/E_{\text{demanded}}$ for a cubical human that can jump to 60 centimeters. Using the scaling relation, estimate it for a cubical flea. What is your judgment of drag as an explanation for the lower jump heights of fleas?

6.3.3 Period of an ideal spring

A surprising conclusion of dimensional analysis is that the period of a spring, or a small-amplitude pendulum, does not depend on its amplitude (Problem 5.14). However, the mathematical reasoning doesn't give us the why; it doesn't provide a physical insight. That insight comes from lumping, by using characteristic values. We'll try it together by finding the period of a spring; you'll practice by finding the period of a pendulum (Problem 6.11).

This kind of lumping, in exchange for providing physical insight, requires that we make a physical model. Here, a stretched or compressed spring exerts a force on and thereby accelerates the mass. If the spring is stretched to an amplitude x_0, then the force is kx_0 and the acceleration is kx_0/m. This acceleration varies as the mass moves, so analyzing the motion usually requires differential equations. However, this acceleration is also a characteristic acceleration. It sets the scale for the acceleration at all other times. If we replace the changing acceleration with this characteristic acceleration, the complexity vanishes. The problem becomes a constant-acceleration problem where $a \sim kx_0/m$.

With a constant acceleration a for a time comparable to one period T, the mass moves a distance comparable to aT^2, which is kx_0T^2/m. When applying lumping and characteristic values, "comparable" is the verbal translation of the single approximation sign $\sim$. As an equation, we would write

$$\text{distance} \sim aT^2. \tag{6.14}$$

Another useful translation is "of the order of": The distance is of the order of aT^2. Equivalently, the characteristic distance is aT^2.

This characteristic distance must be comparable to the amplitude x_0. Thus,

$$x_0 \sim \frac{kx_0T^2}{m}. \tag{6.15}$$

The amplitude x_0 cancels out! Then the period is $T \sim \sqrt{m/k}$. Lumping thus provides the following explanation for why the period is independent of amplitude: As the amplitude and therefore the travel distance increase, the force and acceleration increase just enough to compensate, leaving the period unchanged.

Problem 6.11   Period of a small-amplitude pendulum using lumping
Use characteristic values to explain why the period of a (small-amplitude) pendulum is independent of the amplitude θ_0.

Problem 6.12   Period of a nonlinear spring
Imagine a nonlinear spring with force law $F \propto x^n$. Use lumping as follows to find how the period T varies with amplitude x_0.

a. Estimate a typical or characteristic acceleration.

b. At this acceleration, roughly how far does the mass travel in one period T?

c. This distance must be comparable to the amplitude x_0. Therefore, find the scaling exponent α in $T \propto x_0^\alpha$ (where α will be a function of the scaling exponent n in the force law). Then check your answer to Problem 5.17.

6.3.4 Lumping derivatives

The preceding analysis of the period of a spring–mass system (Section 6.3.3) illustrates a general simplification: Using characteristic values, we can replace derivatives with algebra. The algebraic expression usually provides a physical model. As an example, let's explain, physically, an acceleration that we derived by dimensional analysis in Section 5.1.1: the inward acceleration of an object moving in a circle.

Acceleration is the derivative of velocity: $a = dv/dt$. Using the definition of derivative,

$$\frac{dv}{dt} \equiv \frac{\text{infinitesimal change in } v}{\text{(infinitesimal) time required to make this change in } v}. \tag{6.16}$$

Infinitesimal changes and times are difficult to picture, so an analysis based on calculus often does not help us see why a result is true.

A lumping approximation, by discarding complexity, can give this insight. A way to remember the lumping approximation, first use $6 = 6$ to cancel the 6s in $16/64$:

$$\frac{1\cancel{6}}{\cancel{6}4} = \frac{1}{4}. \tag{6.17}$$

The result is exact! Although this particular cancellation is dubious, it suggests the analogous lumping approximation $d = d$. The resulting cancellation turns derivatives into algebra:

$$\frac{\cancel{d}v}{\cancel{d}t} \sim \frac{v}{t}. \tag{6.18}$$

▷ *What does v/t mean?*

Lumping replaces "infinitesimal" with "characteristic":

$$\frac{v}{t} \sim \frac{\text{characteristic change in } v}{\text{time required to make this change in } v}. \tag{6.19}$$

The numerator asks us to look at the changes in v and to represent them by a characteristic or typical change. The denominator is, as an abbreviation, often called the characteristic time, or time constant, and denoted τ.

In applying this approximation to circular motion, we have to distinguish the velocity vector **v** from its magnitude v (the speed). The speed, at least in constant-speed circular motion, never changes, so dv/dt itself is zero. We are interested in $|d\mathbf{v}|/dt$: the magnitude of the vector's derivative, rather than the derivative of the vector's magnitude. The lumped acceleration a is

$$a \sim \frac{\left|\text{characteristic change in } \mathbf{v}\right|}{\text{time required to make this change in } \mathbf{v}}. \tag{6.20}$$

The maximum change in $\mathbf{v}$ is reversing direction, from $+\mathbf{v}$ to $-\mathbf{v}$. A characteristic change in $\mathbf{v}$ is $\mathbf{v}$ itself, or any value comparable to it. This range of possibilities is captured by the single approximation sign, which stands for "is comparable to." With that notation,

$$\text{characteristic change in } \mathbf{v} \sim \mathbf{v}. \tag{6.21}$$

Here is an example of such a change, showing the velocity vectors before and after the particle has rotated partway around the circle.

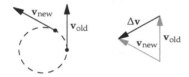

When $\mathbf{v}$ makes its significant change, from $\mathbf{v}_{\text{old}}$ to $\mathbf{v}_{\text{new}}$, it produces a change in $\mathbf{v}$ comparable in magnitude to v. The characteristic time τ—the time required to make this change—is a decent fraction of a period of revolution. Because a full period is $2\pi r/v$, the characteristic time is comparable to r/v. Using τ to represent the characteristic time, and $\sim$ to hide dimensionless prefactors, we write $\tau \sim r/v$.

Here's another way to arrive at $\tau \sim r/v$. If the triangle of $\mathbf{v}_{\text{old}}$, $\mathbf{v}_{\text{new}}$, and $\Delta \mathbf{v}$ is an equilateral triangle, then it corresponds to the particle rotating $60°$ around the circle, which is almost exactly 1 radian. Because 1 radian creates an arc of length r (the same proportionality tells us that 2π radians produces the circumference $2\pi r$), the travel time is r/v, and $\tau \sim r/v$. Then the acceleration is roughly v^2/r:

$$a \sim \frac{v}{\tau} \sim \frac{v}{r/v} = \frac{v^2}{r}. \tag{6.22}$$

This equation encapsulates a physical, proportional-reasoning explanation of $a = v^2/r$. Namely, in circular motion, the velocity vector changes direction significantly in approximately 1 radian of rotation. This motion requires a time $\tau \sim r/v$. Therefore, the circular acceleration a contains two factors of v—one factor from the v itself and one factor from the time in the denominator—and it contains $1/r$, also from the time in the denominator.

6.3.5 Simplifying with characteristic values: Yield of the first atomic blast

In Section 5.2.2, we used dimensional analysis to estimate the yield of an atomic blast. We accurately predicted that the blast energy E is related to the blast radius R and the air density ρ:

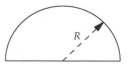

$$E \sim \frac{\rho R^5}{t^2},\qquad(6.23)$$

where t is the time since the blast. However, dimensional analysis, as a mathematical argument, does not give a physical explanation for its results, which can often feel like magic. Lumping explains the magic by helping us analyze a physical model—a model whose analysis would otherwise be absurd in its complexity.

The physical model is that the blast increases the thermal energy of the air molecules, thus increasing the speed of sound—which is the speed at which the blast expands. Using this model exactly requires setting up and solving differential equations, as Taylor did [44]. Lumping, by turning calculus into algebra, simplifies the equations without discarding their physical meaning.

The first step is to estimate the thermal energy. It comes almost entirely from the hot fireball—that is, from the blast energy. This energy is spread unevenly over the blast, with a higher energy density nearer the blast center and a lower density farther away. For the lumping approximation, smear the blast energy E evenly throughout a sphere of radius R. This sphere's volume is comparable to R^3, so the typical or characteristic energy density is $\mathcal{E} \sim E/R^3$.

The next step is to use this energy density to estimate the speed of sound c_s. As we discussed in Section 5.4.1, the speed of sound is comparable to the thermal speed, so

$$c_s \sim \sqrt{\frac{k_B T}{m}},\qquad(6.24)$$

where $k_B T$ is the approximate thermal energy of one air molecule and m is the mass of one air molecule.

To connect this speed to the energy density $\mathcal{E}$, let's convert energy per molecule into energy per volume, by multiplying $k_B T$ by the number density n (the number of air molecules per volume). The result is $n k_B T$, which is the thermal and therefore roughly the blast energy density $\mathcal{E} \sim E/R^3$.

To be fair, we also multiply the denominator m by n. This step converts mass per molecule into mass per volume, which gives the density ρ:

$$\underbrace{\frac{\text{mass}}{\text{molecule}}}_{m} \times \underbrace{\frac{\text{molecules}}{\text{volume}}}_{n} = \underbrace{\frac{\text{mass}}{\text{volume}}}_{\rho}. \tag{6.25}$$

Therefore, the speed of sound is comparable to $\sqrt{E/\rho R^3}$:

$$c_s \sim \sqrt{\frac{nk_BT}{nm}} \sim \sqrt{\frac{E/R^3}{\rho}} = \sqrt{\frac{E}{\rho R^3}}. \tag{6.26}$$

This speed is the rate at which the blast expands.

▶ *Based on this speed, how large should the blast be after time t?*

Because the energy density and therefore the sound speed decreases as R increases, the blast radius is not simply the speed at time t multiplied by the time. But we can make a further lumping approximation: that the typical or characteristic speed, for the entire time t, is $\sqrt{E/\rho R^3}$. In evaluating this speed, we'll use the radius R at time t as the characteristic radius.

$$\underbrace{\text{blast radius}}_{R} \sim \underbrace{\text{characteristic speed}}_{c_s\sim\sqrt{E/\rho R^3}} \times \underbrace{\text{time}}_{t}. \tag{6.27}$$

The solution for the blast energy E is $E \sim \rho R^5/t^2$, as we found using dimensional analysis. Lumping complements that mathematical reasoning with a physical model.

6.4 Applying lumping to shapes

Lumping by replacing varying quantities with their typical, or characteristic, values is close to the next form of lumping: lumping shapes. Our first illustration is explaining a curious fact about everyday materials.

6.4.1 Densities of liquids and solids

Among books teaching the art of approximation, a classic is *Consider a Spherical Cow* [22], so named because a sphere is a much simpler shape than a cow. An even simpler shape is a cube. Thus, a powerful form of lumping is to replace complex shapes by a comparably sized cube. With this idea, we can explain why most solids and liquids have densities between 1 and 10 times the density of water.

Each atom is a complex, ill-defined shape, but pretend that it is a cube. Because the atoms touch, the density of the substance is approximately the density of one approximating cube:

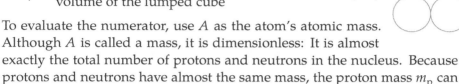

$$\rho \approx \frac{\text{mass of the atom}}{\text{volume of the lumped cube}}. \tag{6.28}$$

To evaluate the numerator, use A as the atom's atomic mass. Although A is called a mass, it is dimensionless: It is almost exactly the total number of protons and neutrons in the nucleus. Because protons and neutrons have almost the same mass, the proton mass m_p can represent the mass of either constituent. Then the mass of one cube is Am_p.

To find the denominator, the cube's volume, make each cube's side be a typical atomic diameter of $a \sim 3$ ångströms. This size is based on our calculation in Section 5.5.1 of the diameter of the smallest atom, hydrogen (whose diameter was roughly 1 ångström). Then the density becomes

$$\rho \sim \frac{Am_p}{(3\,\text{Å})^3}. \tag{6.29}$$

To avoid looking up the proton mass m_p, let's multiply this fraction by N_A/N_A, where N_A is Avogadro's number:

$$\rho \sim \frac{A \times m_p N_A}{(3\,\text{Å})^3 \times N_A}. \tag{6.30}$$

The numerator is A grams per mole, because 1 mole of protons (roughly, of hydrogen) has a mass of 1 gram.

The denominator is roughly 18 cubic centimeters per mole:

$$\underbrace{3 \times 10^{-23}\,\text{cm}^3}_{(3\,\text{Å})^3} \times \underbrace{6 \times 10^{23}\,\text{mol}^{-1}}_{N_A} = \frac{18\,\text{cm}^3}{\text{mol}}. \tag{6.31}$$

A typical solid or liquid density is then related simply to the substance's atomic mass A:

$$\rho \sim \frac{A\,\text{g}\,\text{mol}^{-1}}{18\,\text{cm}^3\,\text{mol}^{-1}} = \frac{A}{18}\frac{\text{g}}{\text{cm}^3}. \tag{6.32}$$

Problem 6.13   Rounding to estimate atomic volume
Use rounding to the nearest half power of ten (Section 6.2.2) to show that a cube with side length 3 ångströms has a volume of approximately 3×10^{-23} cubic centimeters.

Many common elements have atomic masses between 18 and 180, so the densities of many liquids and solids should lie between 1 and 10 grams per cubic centimeter. As shown in the table, the prediction is reasonable. The table even includes a joker: water is not an element! Yet the density estimate is exact.

	A	est.	actual
		$\rho\,(\mathrm{g\,cm^{-3}})$	
Li	7	0.4	0.5
H_2O	18	1.0	1.0
Si	28	1.6	2.4
Fe	56	3.1	7.9
Hg	201	11.2	13.5
U	238	13.3	18.7
Au	197	10.9	19.3

The table also shows why centimeters and grams are, for materials physics, more convenient than are meters and kilograms. A typical solid has a density of a few grams per cubic centimeter. Such modest numbers are easy to remember and handle. In contrast, a density of 3000 kilograms per cubic meter, although mathematically equivalent, is mentally unwieldy. On each use, you have to think, "How many powers of ten were there again?" Therefore, the table gives densities in the mentally friendly units of grams per cubic centimeter.

Problem 6.14   Accounting for discrepancies in the density prediction
In the table of densities, iron (Fe) shows the biggest discrepancy between the actual and predicted densities, by roughly a factor of 2.5. Use that factor to make an improved estimate for the interatomic spacing in iron.

Problem 6.15   Number density of conduction electrons
Estimate the number density of free (conduction) electrons in a copper wire.

Problem 6.16   Typical drift speed in a wire
a. Use your estimate of the number density of conduction electrons in a copper wire (Problem 6.15) to estimate the drift speed of electrons in the wire connecting a typical lamp to the wall socket.

b. Estimate the time required for electrons to travel at this drift speed from the wall socket to the light bulb. Then how does the light bulb turn on right after you flip the switch?

6.4.2   Graph lumping: The number of undergraduate students

A particularly important shape is a graph. When applied to graphs, the idea of lumping shapes simplifies many a problem, making evident its essential features. As an example of graph lumping, we'll estimate the number of US undergraduate students. Such back-of-the-envelope market-size estimates are valuable for business planning and for making public policy.

The first step is to estimate the number of people in the United States who are 18, 19, 20, or 21 years old. This total provides, at least in the United States, the pool from which most undergraduate students come. Because not all 18-to-21-year-olds go to college, at the end we will multiply the total by the fraction of adults who are college graduates.

Finding the exact pool size requires the birth date of every person in the United States. Although these data are collected once every decade by the US Census Bureau, they would only overwhelm us. As an approximation to the voluminous data, the Census Bureau also publishes the number of people at each age. For example, the 1991 data 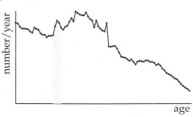 are the wiggly line in the graph. The left side of the graph represents the number of infants and toddlers in 1991, and the right side represents the number of older people (also in 1991). The undergraduate pool size, representing all 18-, 19-, 20-, and 21-year-olds, is the shaded area. (The peak around the ages 30–35 represents the baby boomers, born in the period after World War Two.)

Unfortunately, even this graph depends on the huge resources of the Census Bureau, so it is not suited for back-of-the-envelope estimates. It also provides little insight or transfer value. Insight comes from lumping: from turning the complex, wiggly curve into a rectangle. The rectangle's dimensions can be determined without any information from the Census Bureau.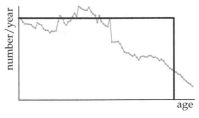

▷ *What are the height and width of this rectangle?*

The rectangle's width is a time, so it must be a characteristic time related to the population. A good guess is the life expectancy, because the age distribution varies significantly over that time. In the United States, the life expectancy is roughly 75 years, which will be the rectangle's width. In 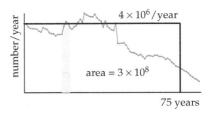 this lumping approximation, everyone lives happily until a sudden death at his or her 75th birthday. This all-or-nothing reasoning is the essential characteristic of lumping, making it such a useful approximation.

The rectangle's height can be computed from its area, which is the US population—approximately 300 million (as we estimated in Section 6.3.1). Therefore, the height is 4 million per year:

$$\text{height} = \frac{\text{area}}{\text{width}} \sim \frac{300 \text{ million}}{75 \text{ years}} = \frac{4 \text{ million}}{\text{year}}. \tag{6.33}$$

Based on this estimate, there should be 16 million people (in the United States) with the four undergraduate ages of 18, 19, 20, or 21 years.

Not everyone in this pool is an undergraduate. Therefore, the final step is to account for the fraction of adults who are college graduates. In the United States, where education has traditionally been widely spread throughout the population, this fraction (or adjustment factor) is high—say, 0.5. The number of US undergraduates should be about 8 million.

For comparison, the 2010 US census data gives 5.361 million enrolled in four-year colleges, and 4.942 million in two-year colleges, for a combined total of almost exactly 10 million. Our estimate, which lies halfway between the four-year and the combined total, is quite good!

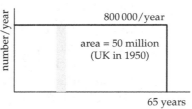

Even had it been terrible, with a large discrepancy between the estimate and the actual number, we would still have learned useful information about a society. As an example, let's use the same method to estimate how many university students graduated in 1950 in the United Kingdom. ("College" in American English and "university" in British English are roughly equivalent.) The UK population at the time was 50 million. With a life expectancy of, say, 65 years, the rectangle's height is roughly 800 000 per year. If, as in the United States of today, 50 percent of the age-eligible population goes to university, 400 000 should graduate each year. However, in 1950, the actual number was roughly 17 000—more than a factor of 20 smaller than the estimate.

Such a huge error probably does not come from the population estimate. Instead, the fraction of 50 percent of university participation must be far too high. Indeed, rather than 50 percent, the actual figure for 1950 is 3.4 percent. The difference between 50- and 3-percent university participation makes the United Kingdom of 1950 a very different society from the United States of 2010 or even from the United Kingdom of 2010, where the fraction going to university is roughly 40 percent.

6.4.3 Cone free-fall time and distance

For our next illustration of graph lumping, let's turn from the social to the physical world, to our falling cones. For the single cone of Section 3.5.2 and the cone race of Section 4.3.1, we assumed that the cones made their entire journey at their terminal speed. But this assumption cannot be exact: Just after the cones are released, their fall speed is zero. Graph lumping will help us evaluate and refine this assumption.

▷ *How far does the cone fall before it reaches its terminal speed?*

Taken literally, the answer is infinity, because no object reaches its terminal speed. As it approaches its terminal speed, the drag and weight more nearly balance, the force and acceleration get closer to zero, and the speed changes ever more slowly. So an object can approach its terminal speed only ever more closely. Let's therefore rephrase the question to ask how far the cone falls before it has reached a significant fraction of its terminal speed. In the "significant," you can see the door through which we will bring in lumping.

Here is a sketch of the actual fall speed versus time. At first, the speed increases rapidly. As the speed increases, so does the drag. The net force and the acceleration fall, so the speed increases more slowly. As a lumping approximation, we'll replace the smooth curve by a slanted and a horizontal segment.

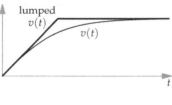

A partly triangular lumping approximation might look different from the population lumping rectangles we constructed in Section 6.4.2. However, it is merely the integral of a rectangle: It is equivalent to replacing the actual, complicated acceleration by a rectangle representing the period of free fall. Then the acceleration drops abruptly to zero.

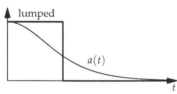

Just as we labeled the population rectangle with its height and width, here we label the velocity graph with speeds, times, and slopes. The graph has two segments. The second, horizontal segment shows the cone's falling at its terminal speed v_{term}. This speed, as we found experimentally in Section 3.5.2, is roughly 1 meter per second.

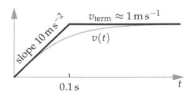

The first, slanted segment represents free fall, as if there were no air resistance. The free-fall acceleration is g, so the free-fall velocity has slope g: In 1 second, the speed increases by 10 meters per second. The slanted and horizontal segments meet where $gt = v_{\text{term}}$. Thus, they meet at approximately 0.1 seconds. Based on the lumping approximation, we expect the cone to reach a significant fraction of its terminal speed by 0.1 seconds—which is only 5 percent of the total fall time of 2 seconds.

▷ *Roughly how far does the cone fall in this time?*

Use lumping again! The distance is the area of the shaded triangle. The triangle's base is 0.1 seconds, and its height is the terminal speed, approximately 1 meter per second. So its area is 5 centimeters:

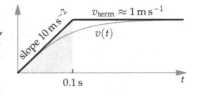

$$\frac{1}{2} \times 0.1\,\text{s} \times 1\,\text{m}\,\text{s}^{-1} = 5\,\text{cm}. \qquad (6.34)$$

After only 2.5 percent of its 2-meter journey, the cone has reached a significant fraction of the terminal speed. The "always at terminal speed" approximation is quite accurate. Thanks to the lumping, we could judge the approximation without setting up or solving differential equations.

Problem 6.17   Raindrop terminal speed
Sketch the fall speed of a large raindrop versus time. Roughly how long and how far does it fall before it reaches (a significant fraction of) its terminal speed?

Problem 6.18   Actual cone fall speed
Set up and solve the differential equations for the fall of a cone with drag proportional to v^2 and terminal speed of 1 meter per second. What is its speed as a fraction of v_{term} after the cone (a) has fallen for 0.1 seconds, or (b) has fallen 5 centimeters?

6.4.4 How viscosity burns up energy

Lumping is particularly useful for gaining insight into fluid flow, a subject where the governing equations, the Navier–Stokes equations, are so complex that almost no problem has an exact solution. These equations were introduced in Section 3.5 to encourage you to use conservation reasoning. Here their specter is invoked to encourage you to use lumping.

In everyday life, an important feature of fluid flow is drag. As we discussed in Section 5.3.2, drag (in steady-state flow) results from viscosity: Without viscosity, there can be no drag. Somehow, viscosity dissipates energy.

Using graph lumping, we can understand the mechanism of energy dissipation without having to understand the detailed physics of viscosity. The essential physical idea is that the viscous force, a force from a neighboring region of fluid, slows down fast pieces of fluid and speeds up slow pieces.

As an example, the arrows in the velocity-profile diagram show a fluid's horizontal speed above a flat boundary—for example, the air speed above a frozen lake. Farther above the boundary, the arrows are longer, indicating that the fluid  moves faster. In a high-viscosity liquid, such as honey, the viscous forces are large, and they force nearby regions to move at nearly the same speed: The flow oozes.

Here is the lumped version of the velocity profile. The varying velocity has been replaced by two rectangles corresponding to two chunks of fluid. The top chunk is moving faster than the bottom chunk; because of the velocity difference, each chunk exerts a viscous force on the other.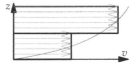

Let's see what consequence this pair of forces has for the total energy of the chunks. To avoid cluttering the essential idea in the analysis with unit algebra, let's give the chunks concrete velocities and masses in a simple unit system. In this unit system, both chunks will have unit mass. The top chunk will move at speed 6 and the bottom chunk at speed 4.

▷ *After a long time, how will viscosity have affected the velocities of the two slabs?*

Because momentum is conserved, the final speeds sum to 10, as they do at the start. Because of the viscous force between the chunks, the top chunk slows down, and the bottom chunk speeds up. The final speeds, once viscosity has done its work, literally and figuratively, are 5 and 5.

The total kinetic energy of the two chunks starts at 26:

$$0.5 \times 1 \times \left(6^2 + 4^2\right) = 26. \tag{6.35}$$

However, their final kinetic energy is only 25:

$$0.5 \times 1 \times \left(5^2 + 5^2\right) = 25. \tag{6.36}$$

Simply because the velocities equalized, the kinetic energy fell! This reduction happens no matter what the initial velocities are (Problem 6.19). The energy difference turns into heat. Our simple physical model, based on lumping, is that viscosity burns up energy by equalizing velocities.

Problem 6.19   Generalizing the argument to other initial velocities
Let the initial velocities of the two chunks be v_1 and v_2, where $v_1 \neq v_2$. Show that
the initial kinetic energy is greater than the final kinetic energy.

Problem 6.20   Shortest-time path
The classic problem in the calculus of variations is to find the
shortest-time path for a mass sliding without friction between
two points. This path is called the brachistochrone. A surpris-
ing conclusion of the full analysis is that this path need not be straight. You can
make this result plausible by using shape lumping: Rather than considering all
possible paths, consider only paths with one corner. Under what conditions is
such a path faster than the straight path (zero corners)?

6.4.5 Mean free path

Lumping smooths out variation. In the famous words of Isaiah 40:4, *the
crooked shall be made straight, and the rough places plain*. We've used this idea
in Section 6.4.2 to simplify a function of time (population versus age) and in
Section 6.4.4 to simplify a function of space (a velocity profile). Now we'll
combine both kinds of lumping to understand and estimate a mean free
path. The mean free path is the average distance that a gas molecule travels
before colliding with another molecule. As we will find in Section 7.3.1 using probabilistic reasoning, mean free paths determine important material
properties, including viscosity and thermal conductivity.

Let's first estimate the mean free path in the simplest model: a spherical
molecule moving in a gas of point molecules. The sphere will have radius r
and the gas molecules a number density n. After analyzing this simplified
situation, we'll replace the point molecules with more realistic spherical
molecules.

The moving molecule has a cross-sectional area $\sigma = \pi r^2$,
and it sweeps out a tube with the same cross-sectional area
(analogous to the tube in our analysis of drag via conservation of energy in Section 3.5.1).

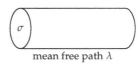

mean free path λ

▷ *How far does the spherical molecule travel down the tube before it hits a point
molecule?*

This distance is the mean free path λ. However, finding it is complicated
because the point molecules, always in motion, keep exiting and entering
the tube. Let's simplify. First, lump in time: Freeze the motion, and pretend

that the point molecules hold those positions. Then lump in space: Rather than spreading the molecules throughout the tube, place them at the far end, at a distance λ from the near end.

The tube, whose length is the mean free path λ, should be just long enough so that the tube contained, before lumping, one point molecule. Then, in the lumped model with that one molecule sitting at the end of the tube, the spherical molecule will hit one molecule after traveling a distance λ.

The number of molecules in the tube is $n\sigma\lambda$:

$$\text{number} = \underbrace{\text{number density}}_{n} \times \underbrace{\text{volume}}_{\sigma\lambda} = n\sigma\lambda. \tag{6.37}$$

Therefore, λ is determined by the requirement that

$$n\sigma\lambda \sim 1. \tag{6.38}$$

For motion through the gas of point molecules, σ is πr^2, so

$$\lambda \sim \frac{1}{n\pi r^2}. \tag{6.39}$$

To make the model more realistic, let's replace the point molecules with spherical molecules. For simplicity, and to model the most frequent case, these molecules will also have radius r.

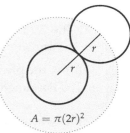

$A = \pi(2r)^2$

▷ *How does this change affect the mean free path?*

Now a collision happens if the centers of two molecules approach within a distance $d = 2r$, the diameter of the sphere. Thus, σ—called the scattering cross section—becomes $\pi(2r)^2$ or πd^2. The mean free path is then a factor of 4 smaller and is

$$\lambda \sim \frac{1}{n\pi d^2}. \tag{6.40}$$

Let's evaluate λ for an air molecule traveling in air. For the diameter d, use the typical atomic diameter of 3 ångströms or 3×10^{-10} meters, which also works for small molecules (like air and water). To estimate the number density n, use the ideal-gas law in the form that, at standard temperature and pressure, 1 mole occupies 22 liters. Therefore,

$$\frac{1}{n} = \frac{22\,\ell}{6 \times 10^{23}\text{ molecules}}, \tag{6.41}$$

and the mean free path is

$$\lambda \sim \frac{2.2 \times 10^{-2}\,\text{m}^3}{6 \times 10^{23}} \times \frac{1}{\pi \times (3 \times 10^{-10}\,\text{m})^2}. \tag{6.42}$$

To evaluate this expression in your head, divide the calculation into three steps—doing the most important first.

1. *Units.* The numerator contains cubic meters; the denominator contains square meters. Their quotient is meters to the first power—as it should be for a mean free path.

2. *Powers of ten.* The numerator contains −2 powers of ten. The denominator contains 3 powers of ten: 23 in Avogadro's number and −20 in $(10^{-10})^2$. The quotient is −5 powers of ten. Along with the units, the expression so far is 10^{-5} meters.

3. *Everything else.* The remaining factors are

$$\frac{2.2}{6 \times \pi \times 3 \times 3}. \tag{6.43}$$

The 2.2/6 is roughly $10^{-0.5}$. The three factors of 3 (one contributed by π) are $10^{1.5}$. Therefore, the remaining factors contribute 10^{-2}.

Putting the pieces together, the mean free path becomes 10^{-7} meters, which is 100 nanometers—quite close to the true value of 68 nanometers.

Problem 6.21   Electric field due to a uniform sheet of charge

In this problem, you investigate a physical model to estimate the electric field above a uniform sheet of charge with areal charge density σ.

a. Explain why the electric field must be vertical.

b. At a height z above the sheet, what region of the sheet contributes significantly to this field?

c. By lumping the sheet into a point charge with the same charge as the significant region, estimate the electric field at a height z, and give the scaling exponent n in $E \propto z^n$.

Confirm your result by comparing it to the prediction from dimensional analysis (Problem 5.34).

Problem 6.22   Electric field inside a spherical shell

Another puzzling electrostatic phenomenon is that the electric field from a uniform shell of charge is zero *everywhere* inside the shell. At the center, the field must be zero by symmetry. However, even away from the center, the field is still zero. Explain this phenomenon using a physical model and lumping.

6.4.6 **Lumping the path of light bent by gravity**

Lumping, as we have seen, replaces a complex, changing process with a simpler, constant process. In the next example, we'll use that simplification to build and analyze a physical model for the bending of starlight by the Sun. In Section 5.3.1, using dimensional analysis and educated guessing, we concluded that the bending angle is roughly Gm/rc^2, where m is the mass of the Sun and r is the distance of closest approach (for a ray grazing the Sun, it is just the radius of the Sun). Lumping provides a physical model for this result; this model will, in Section 8.2.2.2, allow us to predict the effect of extremely strong gravitational fields.

Imagine again a beam (or photon) of light that leaves a distant star. In its journey, it grazes the surface of the Sun and reaches our eye. To estimate the deflection angle using lumping, first identify the changing process—the source of the complexity. Here, the light beam deflects from its original, straight path and the gravitational force from the Sun changes in magnitude and direction as the photon travels. Therefore, calculating the deflection angle requires setting up and evaluating an integral—and carefully checking its trigonometric factors, such as the cosines and secants.

The antidote to complicated integrals is lumping. The lumping approximation simply pretends that the beam bends only near the Sun. In this approximation, only near the Sun does gravity operate. We further assume that, while the photon is near the Sun, its downward acceleration (the acceleration perpendicular to the path) is constant, rather than varying rapidly with position.

The problem then simplifies to estimating the deflection while the beam is near the Sun. As a further lumping approximation, let's define "near" to mean "within r on either side of the location of closest approach." The justification is dimensional: The only length in the problem is the distance of closest approach, which is r; therefore, "near" and "far" are defined 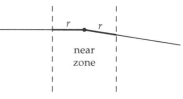 relative to the characteristic distance r.

The geometry is simplest at the point of closest approach. Therefore, let's make the further lumping approximation that, although the light beam tracks how much downward deflection should happen, the total deflection happens only at the point of closest approach.

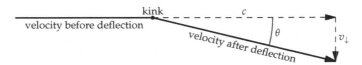

This lumped path, instead of changing direction smoothly, has a kink (a corner) at the point of closest approach.

The deflection angle, in the small-angle approximation, is the ratio of velocity components:

$$\theta \approx \frac{v_\downarrow}{c},\qquad(6.44)$$

where c is the speed of light, which is the forward velocity, and $v_\downarrow$ is the accumulated downward velocity. This downward velocity comes from the downward acceleration due to the Sun's gravity.

In the exact analysis, the downward acceleration varies along the path. As a function of time, it looks like a bell curve centered at the point of closest approach (labeled as $t = 0$, for symmetry). The downward velocity is the integral of (the area under) the entire $a_\downarrow(t)$ curve.

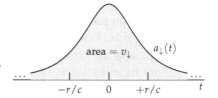

Our lumping analysis replaces the entire area by a rectangle centered around the peak. This rectangle has height Gm/r^2: the characteristic (and peak) downward acceleration. It has width comparable to r/c: the time that the beam spends near the Sun. Therefore, its area, which is the lumping approximation to $v_\downarrow$, is roughly Gm/rc:

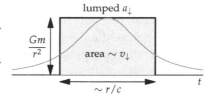

$$\underbrace{v_\downarrow}_{Gm/rc} \sim \underbrace{\text{characteristic downward acceleration}}_{Gm/r^2} \times \underbrace{\text{deflection time.}}_{\sim r/c}\quad(6.45)$$

The deflection angle is therefore comparable to Gm/rc^2:

$$\theta \sim \frac{\overbrace{\frac{Gm/rc}}^{v_\downarrow}}{c} = \frac{Gm}{rc^2}.\qquad(6.46)$$

The lumping argument explains, with a physical model, the deflection angle that we predicted using dimensional analysis and educated guessing (Section 5.3.1). Lumping once again complements dimensional analysis.

Problem 6.23 Sketching the actual and lumped deflection angle
On axes for cumulative deflection θ versus distance along the beam, sketch (a) the actual curve, (b) the lumped curve, assuming that the deflection happens only while the beam is near the Sun, and (c) the lumped curve, assuming, as in the text, that the deflection happens only at the point of closest approach.

6.4.7 All-or-nothing reasoning: Solid mechanics by lumping

For estimating the bending of light, the heart of the lumping analysis was all-or-nothing reasoning: replacing the complex, varying downward acceleration with a simpler curve that was either zero or a nonzero constant. To practice this idea, we'll apply it to an example from solid mechanics, also a subject fraught with differential equations. In particular, we'll estimate the contact radius of a solid ball resting on the ground.

We know a bit about the contact radius: In Problem 5.50, you used dimensional analysis to find that the contact radius r is given by

$$\frac{r}{R} = f\left(\frac{\rho g R}{Y}\right),$$ (6.47)

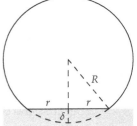

where R is the ball's radius, ρ is its density, Y is its Young's modulus, and f is a dimensionless function. The function f is not determined by dimensional analysis, which is purely mathematical reasoning. Finding f requires a physical model; the easiest way to make and analyze such a model is by making lumping approximations.

Physically, the ground compresses the tip of the ball by a small distance δ, making a flat circle of radius r in contact with the ground. The ball fights back, trying to restore its natural, spherical shape. When the ball rests on the table, the restoring force equals its weight. This constraint will give us enough information to find the dimensionless function f.

The restoring force comes from the stress (or pressure) over the contact surface. To estimate this stress, let's make the lumping approximation that it is constant over the contact surface and equal to a typical or characteristic stress. This approximation is analogous to replacing the varying population curve with a constant value (and making a rectangle). With that approximation, the restoring force is force $\sim$ typical stress $\times$ contact area. (6.48)

We can estimate the stress based on the strain (the fractional compression). The stress and strain are related by the Young's modulus Y:

typical stress $\sim Y \times$ typical strain. (6.49)

The strain, like the stress, varies throughout the ball. In the lumping or all-or-nothing approximation, the strain becomes a characteristic or typical strain in a region around the contact surface, and zero outside this region. The typical strain is a fractional length change:

$$\text{strain} = \frac{\text{length change}}{\text{length}}. \qquad (6.50)$$

The numerator is δ: the amount by which the tip of the ball is compressed. The denominator is the size of the mysterious compressed region.

▶ *How large is the compressed region?*

Because the ball's radius or diameter changes, the compressed region might be the whole ball. This tempting reasoning turns out to be incorrect. To see why, imagine an extreme case: compressing a huge, 10-meter cube of rubber (rubber, because one can imagine compressing it). By pressing your finger onto the center of one face, you'll indent the face by, say, 1 millimeter over an area of 1 square centimeter. In the notation that we use for the ball, $\delta \sim 1$ millimeter, $r \sim 1$ centimeter, and $R \sim 10$ meters.

The strained region is not the whole cube, nor any significant fraction of it! Rather, its radius is comparable to the radius of the contact region—a fingertip ($r \sim 1$ centimeter). From this thought experiment, we learn that when the object is large enough ($R \gg r$), the strained volume is related not to the size of the object, but rather to the size of contact region (r).

Therefore, in the estimate for the typical strain, the length in the denominator is r. The typical strain ϵ is then δ/r.

Because the compression δ, in contrast to the contact radius r and the ball's radius R, is not easily visible, let's rewrite δ in terms of r and R. Amazingly, they are related by a geometric mean, because their geometry reproduces the geometry of the horizon distance (Section 2.3). The compression δ is analogous to one's height above sea level. The contact radius r is analogous to the horizon distance. And the ball's radius R is analogous to the Earth's radius.

Just as the horizon distance is roughly the geometric mean of the two surrounding lengths, so should the contact radius be roughly the geometric mean of the compression length and the ball's radius. On a logarithmic scale, r is roughly halfway between δ and R. (Like the horizon distance, r is actually halfway between δ and the diameter $2R$. However, the factor of 2 does not matter to this lumping analysis.)

tip compression	contact radius	ball's radius
δ	r	R

Therefore, δ/r, which is the typical strain, is also roughly r/R.

The restoring force is therefore comparable to Yr^3/R:

$$\underbrace{\text{restoring force}}_{Yr^3/R} \sim Y \times \overbrace{\underbrace{\text{typical strain}}_{r/R}}^{\text{typical stress}} \times \underbrace{\text{contact area}}_{r^2}. \qquad (6.51)$$

This force balances the weight, which is comparable to $\rho g R^3$:

$$\frac{Yr^3}{R} \sim \rho g R^3. \qquad (6.52)$$

Therefore, in dimensionless form, the contact radius is given by

$$\frac{r}{R} \sim \left(\frac{\rho g R}{Y}\right)^{1/3}, \qquad (6.53)$$

and the dimensionless function f in

$$\frac{r}{R} = f\left(\frac{\rho g R}{Y}\right) \qquad (6.54)$$

is a cube root (multiplied by a dimensionless constant).

▷ *How large is the contact radius r in practice?*

Let's plug in numbers for a superball (a small, highly elastic rubber ball) resting on the ground. The density of rubber is roughly the density of water. A superball is small—say, $R \sim 1$ centimeter. Its elastic modulus is roughly $Y \sim 3 \times 10^7$ pascals. (This elastic modulus is a factor of 300 smaller than oak's and almost a factor of 10^4 smaller than steel's.) Then

$$\frac{r}{R} \sim \left(\frac{\overbrace{10^3\,\text{kg/m}^3}^{\rho} \times \overbrace{10\,\text{m s}^{-2}}^{g} \times \overbrace{10^{-2}\,\text{m}}^{R}}{\underbrace{3 \times 10^7\,\text{Pa}}_{Y}}\right)^{1/3}. \qquad (6.55)$$

The quotient inside the parentheses is $10^{-5.5}$. Its cube root is roughly 10^{-2}, so r/R is roughly 10^{-2}, and r is roughly 0.1 millimeters (a sheet of paper).

From this example, we see how lumping builds on dimensional analysis. Dimensional analysis gave the form of the relation between r/R and $\rho g R/Y$. Lumping provided a physical model that specified the particular form.

Problem 6.24   Compression of the tip

For the superball, estimate δ, the compression of its tip when it is resting on the ground.

Problem 6.25   Contact radius of a marble

Estimate the contact radius of a small glass marble sitting on a (very) hard table.

Problem 6.26   Energy argument

The potential-energy density due to deforming a solid is comparable to $Y\epsilon^2$. Using this relation, find a second explanation for why r/R is comparable to $(\rho g R/Y)^{1/3}$.

Problem 6.27   Mountain heights

On Earth, the tallest mountain (Everest) is 9 kilometers high. On Mars, the tallest mountain (Olympus) is 27 kilometers high. Make a lumped model of a mountain to explain the factor-of-3 difference. If the same reasoning applied to the Moon, how high would its tallest mountain be? Why is it so much shorter?

Problem 6.28   Asteroid shapes

The tallest mountain on Earth is much smaller than the Earth. Using the mountain-height data of Problem 6.27, estimate the maximum radius of a planetary body made of rock (like the Earth or Mars) that has mountains comparable in size to the body. What consequence does this size have for the shapes of asteroids?

Problem 6.29   Contact time

Imagine a ball dropped from a height, hitting a hard table with impact speed v. Find the scaling exponent β in

$$\frac{\tau c_s}{R} \sim \left(\frac{v}{c_s}\right)^{\beta}, \tag{6.56}$$

where τ is the contact time and c_s is the speed of sound in the ball. Then write τ in the form $\tau \sim R/v_{\text{effective}}$, where $v_{\text{effective}}$ is a weighted geometric mean of the impact speed v and the sound speed c_s.

Problem 6.30   Contact force

For a small steel ball bouncing from a steel table with impact speed 1 meter per second (Problem 6.29), how large is the contact force compared to the ball's weight?

6.5 Quantum mechanics

When we introduced quantum mechanics to estimate the size of hydrogen (Section 5.5.1), it was an actor in dimensional analysis. There, quantum mechanics merely contributed a new constant of nature $\hbar$. This contribution was mathematical. By using lumping to introduce quantum mechanics, we will gain a physical intuition for the effect of quantum mechanics.

6.5.1 Particle in a box: Size of neutron stars

In mechanics, the simplest useful model is motion in a straight line at constant acceleration (which includes constant velocity). This model underlies numerous analyses—for example, the lumping analysis of a pendulum period (Section 6.3.3). In quantum mechanics, the simplest useful model is a particle confined to a box. Let's give the particle a mass m and the box a width a.

box

$\bullet\, m$

width a

▷ *What is the lowest possible energy of the particle—that is, its ground-state energy?*

Because this box can be a lumping model for more complex problems, this energy will help us explain the binding energy of hydrogen. A lumping analysis starts with Heisenberg's uncertainty principle:

$$\Delta p \Delta x \sim \hbar. \tag{6.57}$$

Here, Δp is the spread (the uncertainty) in the particle's momentum, Δx is the spread in its position, and $\hbar$ is the quantum constant. If we know the position precisely (that is, if the uncertainty Δx is tiny), then the momentum uncertainty Δp must be large in order to make the product $\Delta p \Delta x$ comparable to $\hbar$. In contrast, if we hardly know the position (Δx is large), then the momentum uncertainty Δp can be small. This relation is the physical contribution of quantum mechanics.

Let's apply it to the particle in the box. The particle could be anywhere in the box, so its position uncertainty Δx is comparable to the box width a:

$$\Delta x \sim a. \tag{6.58}$$

Because of the uncertainty principle, the consequence of confining the particle to the box implies a momentum uncertainty $\hbar/a$:

$$\Delta p \sim \frac{\hbar}{\Delta x} \sim \frac{\hbar}{a}. \tag{6.59}$$

This momentum corresponds to a kinetic energy $E \sim (\Delta p)^2/m$. As a consequence, the particle acquires an energy, called the confinement energy:

$$E \sim \frac{\hbar^2}{ma^2}. \tag{6.60}$$

This simple result enables us to estimate the radius of a neutron star: a star without any nuclear fusion (which normally resists gravity) and heavy enough that gravity has collapsed the protons and electrons into neutrons. The atomic structure is gone, and the star is one giant, neutral nucleus (except in the crust, where gravity is not strong enough to make neutrons).

> **Problem 6.31   Dimensional analysis for the size of a neutron star**
> Use dimensional analysis to find the radius R of a neutron star of mass M.

Two physical effects compete: Gravity tries to crush the star; and quantum mechanics, armed with the uncertainty principle, resists by penalizing confinement. To begin quantifying these effects, let the star have radius R and contain N neutrons, so it has mass $M = Nm_n$.

Gravity's contribution to crushing the star is reflected in the star's gravitational potential energy

$$E_{\text{potential}} \sim \frac{GM^2}{R}. \tag{6.61}$$

This energy is negative: Gravity is happier (the energy is lower) when R is smaller. Here, the minus sign is incorporated into the single approximation sign $\sim$.

For the other side of the competition, we confine each neutron to a box. To find the box width a, imagine the star as a three-dimensional cubic lattice of neutrons. Because the star contains N neutrons, the lattice is $N^{1/3} \times N^{1/3} \times N^{1/3}$. Each side has length comparable to R, so $aN^{1/3} \sim R$ and $a \sim R/N^{1/3}$.

Each neutron gets a confinement (kinetic) energy $\hbar^2/m_n a^2$. For all the N neutrons together, and using $a \sim R/N^{1/3}$,

$$E_{\text{kinetic}} \sim N\frac{\hbar^2}{m_n a^2} \sim N^{5/3}\frac{\hbar^2}{m_n R^2}. \tag{6.62}$$

In terms of the known masses M and m_n (instead of N), the total confinement energy is

$$E_{\text{kinetic}} \sim \frac{M^{5/3}}{m_n^{8/3}}\frac{\hbar^2}{R^2}. \tag{6.63}$$

The kinetic energy is proportional to R^{-2}, whereas the potential energy is proportional to R^{-1}. On log–log axes of energy versus radius, the kinetic energy is a straight line with a -2 slope, and the potential energy is a straight line with a -1 slope. Because the slopes differ, the lines cross. The radius where they cross, which is roughly the radius that minimizes the total energy, is also roughly the radius of the neutron star. (We used the same reasoning in Section 4.6.1, to find the speed that minimized the energy required to fly.)

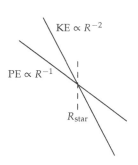

With all the constants included, the crossing happens when R satisfies

$$\frac{GM^2}{R} \sim \frac{M^{5/3}}{m_{\text{n}}^{8/3}} \frac{\hbar^2}{R^2}. \tag{6.64}$$

The radius of the neutron star is then

$$R \sim \frac{\hbar^2}{GM^{1/3}m_{\text{n}}^{8/3}}. \tag{6.65}$$

If the Sun, with a mass of 2×10^{30} kilograms, became a neutron star, it would have a radius of roughly 3 kilometers:

$$R \sim \frac{\left(10^{-34}\,\text{kg}\,\text{m}^2\,\text{s}^{-1}\right)^2}{7 \times 10^{-11}\,\text{kg}^{-1}\,\text{m}^3\,\text{s}^{-2} \times \left(2 \times 10^{30}\,\text{kg}\right)^{1/3} \times \left(1.6 \times 10^{-27}\,\text{kg}\right)^{8/3}} \tag{6.66}$$

$$\sim 3\,\text{km}.$$

The true neutron-star radius for the Sun, after including the complicated physics of a varying density and pressure, is approximately 10 kilometers. (In reality, the Sun's gravity wouldn't be strong enough to crush the electrons and protons into neutrons, and it would end up as a white dwarf rather than a neutron star. However, Sirius, the subject of Problem 6.32, is massive enough.)

Here is a friendly numerical form that incorporates the 10-kilometer information. It makes explicit the scaling relation between the radius and the mass, and it measures mass in the convenient units of the solar mass.

$$R \approx 10\,\text{km} \times \left(\frac{M_{\text{Sun}}}{M}\right)^{1/3}. \tag{6.67}$$

Our analysis, which used several lumping approximations and neglected many dimensionless constants throughout, predicted a prefactor of 3 kilometers instead of 10 kilometers. An error of a factor of 3 is a worthwhile tradeoff. We avoid the complicated analysis of a quantum fluid of electrons, protons, and neutrons, and still gain insight into the essential physical idea: The size of a neutron star results from a competition between gravity and quantum mechanics. (You will find many insightful astrophysical scalings in an article on the "Astronomical reach of fundamental physics" [5].)

> **Problem 6.32   Sirius as a neutron star**
> Sirius, the brightest star in the night sky, has a mass of 4×10^{30} kilograms. What would be its radius as a neutron star?

6.5.2   Hydrogen by lumping

Having practiced quantum mechanics and lumping in the large (a neutron star), let's go to the small: We'll use lumping to complement the dimensional-analysis estimate of the size of hydrogen in Section 5.5.1.

Its size is the radius of the orbit with the lowest energy. Because the energy is a sum of the potential energy from the electrostatic attraction and the kinetic energy from quantum mechanics, hydrogen is analogous to a neutron star, where quantum mechanics competes instead against gravitation. Partly because electrostatics is much stronger than gravity (Problem 6.34), hydrogen is very tiny.

For now, let the hydrogen atom have an unknown radius r. As a lumping approximation, the complicated electrostatic potential becomes a box of width r. Then the electron's momentum uncertainty is $\Delta p \sim \hbar/r$ and its confinement energy is $\hbar^2/m_e r^2$:

$$E_{\text{kinetic}} \sim \frac{(\Delta p)^2}{m_e} \sim \frac{\hbar^2}{m_e r^2}. \tag{6.68}$$

This energy competes against the electrostatic energy, whose estimate also requires a lumping approximation. For in quantum mechanics, an electron is not at any definite location. Depending on your interpretation of quantum mechanics, and speaking roughly, it is either smeared over the box, or it has a probability of being anywhere in the box. On either interpretation, calculating the potential energy requires an integral. However, we can approximate it by using lumping: The electron's typical, or characteristic, distance from the proton is simply r, the radius of hydrogen (which is the size of the lumping box).

Then the potential energy is the electrostatic energy of a proton and electron separated by r:

$$E_{\text{potential}} \sim -\frac{e^2}{4\pi\epsilon_0 r}. \tag{6.69}$$

The total energy is their sum

$$E = E_{\text{potential}} + E_{\text{kinetic}} \sim -\frac{e^2}{4\pi\epsilon_0 r} + \frac{\hbar^2}{m_e r^2}. \tag{6.70}$$

The electron adjusts r to minimize this energy. As we learned in the analysis of lift (Section 4.6.1), when two terms with different scaling exponents compete, the minimum occurs where the terms are comparable. (In Section 8.3.2.2, we'll return to this example with the additional tool of easy cases and also sketch the energies on log–log axes.) Using the Bohr radius a_0 as the separation of minimum energy, comparability means

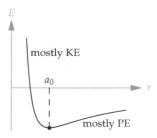

$$\underbrace{\frac{e^2}{4\pi\epsilon_0 a_0}}_{E_{\text{potential}}} \sim \underbrace{\frac{\hbar^2}{m_e a_0^2}}_{E_{\text{kinetic}}}. \tag{6.71}$$

The Bohr radius is therefore

$$a_0 \sim \frac{\hbar^2}{m_e(e^2/4\pi\epsilon_0)}, \tag{6.72}$$

as we found in Section 5.5.1 using dimensional analysis. Now we also get a physical model: The size of hydrogen results from competition between electrostatics and quantum mechanics.

Problem 6.33 Minimizing the energy in hydrogen using symmetry

Use symmetry to find the minimum-energy separation in hydrogen, where the total energy has the form

$$\frac{A}{r^2} - \frac{B}{r}. \tag{6.73}$$

(See Section 3.2.3 for related examples.)

Problem 6.34 Electrostatics versus gravitation

Estimate the ratio of the gravitational to the electrostatic force in hydrogen. What are the similarities and differences between this estimate and the estimate that you made in Problem 2.15?

6.6 Summary and further problems

Lumping is our first tool for discarding complexity with loss of information. By doing so, it simplifies complicated problems where our previous set of tools could not. Curves become straight lines, calculus becomes algebra, and even quantum mechanics becomes comprehensible.

Problem 6.35   Precession of the equinoxes

Because the Earth is oblate (Problem 5.46), and thus has a bit of fat near the equator, and because its rotation axis is tilted relative to the Earth's orbital plane, the Sun and the Moon each exert a slight torque on the Earth. This torque slowly rotates (precesses) the Earth's axis of rotation (so the north star will not always be the north star).

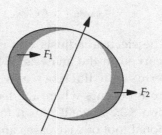

a. Explain why F_1 and F_2, the gravitational forces on the two bulges, are almost exactly equal in direction but not equal in magnitude.

b. Use lumping to find the scaling exponents x and y in

$$\text{torque} \propto m^x l^y, \tag{6.74}$$

where m is the mass of the object (either the Sun or Moon) and l is its distance from the Earth. Then estimate the ratio

$$\frac{\text{torque from the Sun}}{\text{torque from the Moon}}. \tag{6.75}$$

c. By including the constants of proportionality, estimate the total torque and the precession rate.

Problem 6.36   Graph lumping in reverse

In a constant-temperature (isothermal) atmosphere, which isn't a terrible approximation to the actual atmosphere, the air density falls exponentially with height:

$$\rho = \rho_0 e^{-z/H}, \tag{6.76}$$

where z is the height above sea level, ρ_0 is the density at sea level, and H is the scale height of the atmosphere. We estimated H in Section 5.4.1 using dimensional analysis; in this problem, you'll estimate it by using lumping in reverse.

a. Sketch $\rho(z)$ versus z as given above. On the same graph, sketch a lumped $\rho(z)$ that represents a sea-level-density atmosphere for $z < H$ and zero density for $z \geq H$. Ensure that your lumping rectangle has the same area as the exponentially decaying $\rho(z)$ curve.

b. Use your lumping rectangle and the sea-level pressure p_0 to estimate H. Then estimate the relative density at the top of Mount Everest (roughly 9 kilometers). Check your estimate by looking up the actual air density on Mount Everest.

7
Probabilistic reasoning

Our previous tool, lumping, helps us simplify by discarding less important information. Our next tool, probabilistic reasoning, helps us when our information is already incomplete—when we've discarded even the chance or the wish to collect the missing information.

7.1 Probability as degree of belief: Bayesian probability

The essential concept in using probability to simplify the world is that probability is a degree of belief. Therefore, a probability is based on our knowledge, and it changes when our knowledge changes.

7.1.1 Is it my telephone number?

Here is an example from soon after I had moved to England. I was talking to a friend on the phone, of the old-fashioned variety with wires connecting it to the wall. David needed to call me back. However, having just moved to the apartment, I was unsure of my phone number; plus, for anyone used to American phone numbers, British phone numbers have a strange and hard-to-remember format. I had a reasonably likely guess, which I gave David so that he could call me back. After I hung up, I tested my guess by picking up my phone and dialing my guess—and got a busy signal.

▷ *Given this experimental evidence, how sure am I that the candidate number is my phone number? Quantitatively, what odds should I give?*

This question makes no sense if probability is seen as long-run frequency. In that view, the probability of a coin turning up heads is $1/2$ because $1/2$ is the limiting proportion of heads in an ever-longer series of tosses. However, for evaluating the plausibility of the phone number, this interpretation—called the frequentist interpretation—cannot apply, because there is no repeated experiment.

The frequentist interpretation gets stuck because it places probability in the physical system itself. The alternative—that probability reflects the incompleteness of our knowledge—is known as the Bayesian interpretation of probability. It is the interpretation suited for mastering complexity. A book-length discussion and application of this fundamental point is Edwin Jaynes's *Probability Theory: The Logic of Science* [26].

The Bayesian interpretation is based on one simple idea: A probability reflects our degree of belief in a hypothesis. Probabilities are therefore *subjective:* Someone with different knowledge will have different probabilities. Thus, by collecting evidence, our degrees of belief change. Evidence changes probabilities.

▷ *In the phone-number problem, what is the hypothesis and what is the evidence?*

The hypothesis—often denoted H—is the statement about the world whose credibility we would like to judge. Here,

$$H \equiv \text{My phone-number guess is correct.} \tag{7.1}$$

The evidence—often denoted E or D (for data)—is the information that we collect, obtain, or learn and then use to judge the hypothesis. It augments our knowledge. Here, E is the result of the experiment:

$$E \equiv \text{Dialing my guess gave a busy signal.} \tag{7.2}$$

Any hypothesis has an initial probability $\Pr(H)$. This probability is called the prior probability, because it is the probability prior to, or before, incorporating the evidence. After learning the evidence E, the hypothesis has a new probability $\Pr(H|E)$: the probability of the hypothesis H given—that is, upon assuming—the evidence E. This probability is called the posterior probability, because it is the probability, or degree of belief, after including the evidence.

The recipe for using evidence to update probabilities is Bayes' theorem:

$$\Pr(H|E) \propto \Pr(H) \times \Pr(E|H). \tag{7.3}$$

The new factor, the probability $\Pr(E|H)$—the probability of the evidence given the hypothesis—is called the likelihood. It measures how well the candidate theory (the hypothesis) explains the evidence. Bayes' theorem then says that

$$\underbrace{\text{posterior probability}}_{\Pr(H|E)} \propto \underbrace{\text{prior probability}}_{\Pr(H)} \times \underbrace{\text{explanatory power}}_{\Pr(E|H)}. \tag{7.4}$$

(The constant of proportionality is chosen so that the posterior probabilities for all the competing hypotheses add to 1.) Both probabilities on the right are necessary. Without the likelihood, we could not change our probabilities. Without the prior probability, we would always prefer the hypothesis with the maximum likelihood, no matter how contrived or post hoc.

In a frequent use of Bayes' theorem, there are only two hypotheses, H and its negation $\overline{H}$. In this problem, $\overline{H}$ is the statement that my guess is wrong. With only two hypotheses, a compact form of Bayes' theorem uses odds instead of probabilities, thereby avoiding the constant of proportionality:

$$\underbrace{\text{posterior odds}}_{O(H|E)} = \underbrace{\text{prior odds}}_{O(H)} \times \frac{\Pr(E|H)}{\Pr(E|\overline{H})}. \tag{7.5}$$

The odds O are related to the probability p by $O = p/(1-p)$. For example, a probability of $p = 2/3$ corresponds to an odds of 2—often written as 2:1 and read as "2-to-1 odds."

Problem 7.1   Converting probabilities to odds
Convert the following probabilities to odds: (a) 0.01, (b) 0.9, (c) 0.75, and (d) 0.3.

Problem 7.2   Converting odds to probabilities
Convert the following odds to probabilities: (a) 3, (b) 1/3, (c) 1:9, and (d) 4-to-1.

The ratio $\Pr(E|H)/\Pr(E|\overline{H})$ is called the likelihood ratio. Its numerator measures how well the hypothesis H explains the evidence E; its denominator measures how well the contrary hypothesis $\overline{H}$ explains the same evidence. So their ratio measures the relative explanatory power of the two hypotheses. Bayes' theorem, in the odds form, is simple:

$$\text{updated odds} = \text{initial odds} \times \text{relative explanatory power}. \tag{7.6}$$

Let's use Bayes' theorem to judge my phone-number guess. Before the experiment, I was not too sure of the phone number; $\Pr(H)$ is perhaps $1/2$, making $O(H) = 1$. In the likelihood ratio, the numerator $\Pr(E|H)$ is the probability of getting a busy signal assuming ("given") that my guess is correct. Because I would be dialing my own phone using my phone, I would definitely get a busy signal. Thus, $\Pr(E|H) = 1$: The hypothesis of a correct guess (H) explains the data as well as possible.

The trickier estimate is the denominator $\Pr(E|\overline{H})$: the probability of getting a busy signal assuming that my guess is incorrect. I'll assume that my guess is still a valid phone number (I nowadays rarely get the recorded message saying that I have dialed an invalid number). Then I would be dialing a random person's phone. Thus, $\Pr(E|\overline{H})$ is the probability that a random valid phone is busy. It is probably similar to the fraction of the day that my own phone is busy. In my household, the phone is in use for 0.5 hours in a 24-hour day, and the busy fraction could be $0.5/24$.

However, that estimate uses an overly long time, 24 hours, for the denominator. If I do the experiment at 3 am and my guess is wrong, I would wake up an innocent bystander. Furthermore, I am not often on the phone at 3 am. A more reasonable denominator is 10 hours (9 am to 7 pm), making the busy fraction and the likelihood $\Pr(E|\overline{H})$ roughly 0.05. An incorrect guess ($\overline{H}$) is a lousy explanation for the data.

The relative explanatory power of H and $\overline{H}$, which is measured by the likelihood ratio, is roughly 20:

$$\frac{\Pr(E|H)}{\Pr(E|\overline{H})} \sim \frac{1}{0.05} = 20. \tag{7.7}$$

Because the prior odds were 1 to 1, the updated, posterior odds are 20 to 1:

$$\underbrace{\text{posterior odds}}_{O(H|E)\sim 20} = \underbrace{\text{prior odds}}_{O(H)\sim 1} \times \underbrace{\text{likelihood ratio}}_{\Pr(E|H)\,/\,\Pr(E|\overline{H})\,\sim\,20} \sim 20. \tag{7.8}$$

My guess has become very likely—and it turned out to be correct.

Problem 7.3   PKU testing

In most American states and many countries, newborn babies are tested for the metabolic defect phenylketonuria (PKU). The prior odds of having PKU are about 1 in 10 000. The test gives a false-positive result 0.23 percent of the time; it gives a false-negative result 0.3 percent of the time. What are $\Pr(\text{PKU}|\text{positive test})$ and $\Pr(\text{PKU}|\text{negative test})$?

7.2 Plausible ranges: Why divide and conquer works

The Bayesian understanding of probability as degree of belief will show us why divide-and-conquer reasoning (Chapter 1) works. We'll see how it increases our confidence in an estimate and decreases our uncertainty as we analyze a divide-and-conquer estimate in slow motion.

7.2.1 Land area of the United Kingdom

The estimate will be the land area of the United Kingdom, which is where I was born and later spent many years. So I have some implicit knowledge of the area, but I don't know it explicitly. The estimate therefore serves as a model for the common situation where we know more than we think we do and need to bring out and use that knowledge.

To make the initial estimate, the baseline against which to compare the divide-and-conquer estimate, I talk to my gut using the rubric of Section 1.6. For the lower end of my range, 10^4 square kilometers feels right: I would be fairly surprised if the area were smaller. For the upper end, 10^7 square kilometers feels right: I would be fairly surprised if the area were larger. Combining the endpoints, I would be mildly surprised if the area were smaller than 10^4 square kilometers or greater than 10^7 square kilometers. The wide range, spanning three orders of magnitude, reflects the difficulty of estimating an area without using divide-and-conquer reasoning.

My confidence in the gut estimate is my probability for the hypothesis H:

$H \equiv$ The UK's land area lies in the range $10^4 \ldots 10^7$ km^2.   (7.9)

This probability assumes, or is based on, my background knowledge K:

$K \equiv$ what I know about the area *before* using divide and conquer. (7.10)

My confidence or degree of belief in the guess is the conditional probability $\Pr(H|K)$: the probability that the area lies in the range, based on my knowledge before applying divide-and-conquer reasoning. Alas, no algorithm is known for computing a probability based on such complicated background information. The best that we can do is to introspect: to hold a further gut discussion. This discussion concerns not the area itself, but rather the degree of belief about the range $10^4 \ldots 10^7$ square kilometers.

My gut chose the range for which I would feel mild surprise, but not shock, to learn that the area lies outside it. The surprise implies that the probability $\Pr(H|K)$ is larger than 1/2: If $\Pr(H|K)$ were less than 1/2, I would be surprised to find the area *inside* the range. The mildness of the surprise suggests that $\Pr(H\,|\,K)$ is not much larger than $1/2$. The probability feels like $2/3$. The corresponding odds are 2:1; I'd give 2-to-1 odds that the area lies within the plausible range. With a further assumption of symmetry, so that the area's falling below and above the range are equally likely, the plausible range represents the following probabilities:

$$10^4\,\text{km}^2 \qquad\qquad \text{plausible range} \qquad\qquad 10^7\,\text{km}^2$$

| $p \approx 1/6$ | $p \approx 2/3$ | $p \approx 1/6$ |

where the wavy lines at the ends indicate that the left and right ranges extend down to zero and up to infinity, respectively.

Now let's see how a divide-and-conquer estimate changes the plausible range. To make this estimate, lump the area into a rectangle with the same area and aspect ratio as the United Kingdom. My own best-guess rectangle is superimposed on the map outline of the United Kingdom. Its area is the product of its width and height. Then the area estimate divides into two simpler estimates.

1. *Lumped width.* Before I ask my gut for the plausible range for the width, I prepare by reviewing my knowledge of the width. Crossing the United Kingdom in the south, say from London to Cornwall, takes maybe 4 hours by car. But much of the United Kingdom is thinner, so the average, or lumped, width corresponds to maybe 3 hours of driving. My gut is content with the range 150...250 miles or 240...400 kilometers:

$$240\,\text{km} \qquad\qquad \text{plausible range} \qquad\qquad 400\,\text{km}$$

| $p \approx 1/6$ | $p \approx 2/3$ | $p \approx 1/6$ |

On a logarithmic scale, which is the correct scale for positive quantities such as the width and height, the midpoint of the range is $\sqrt{240 \times 400}$ or 310 kilometers. It is my best estimate of the width.

2. *Lumped height.* The train journey from London in the south of England to Edinburgh in Scotland, in the northern part of the United Kingdom, takes about 5 hours at, say, 80 miles per hour. In addition to these 400 miles, there's more latitude in Scotland north of Edinburgh and some in England south of London. Thus, my gut estimate for the height of

the lumping rectangle is 500 miles. This distance is the midpoint of my plausible range for the height.

The estimate feels accurate to plus or minus 20 percent (roughly ±100 miles). On a logarithmic (or multiplicative) scale, the range would be roughly a factor of 1.2 on either side of midpoint:

$$\underbrace{420}_{500/1.2} \ldots \underbrace{600}_{500\times1.2} \text{ miles.} \qquad (7.11)$$

(A more accurate value for 500/1.2 is 417, but there's little reason to strive for such accuracy when these endpoints are rough estimates anyway.) In metric units, the range is 670...960 kilometers, with a midpoint of 800 kilometers. Here is its probability interpretation:

670 km	plausible range	960 km
$p \approx 1/6$	$p \approx 2/3$	$p \approx 1/6$

The next step is to combine the plausible ranges for the height and the width in order to make the plausible range for the area. A first approach, because the area is the product of the width and height, is simply to multiply the endpoints of the width and height ranges:

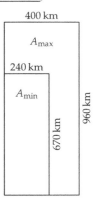

$$A_{\min} \approx 240 \, \text{km} \times 670 \, \text{km} \approx 160\,000 \, \text{km}^2;$$
$$A_{\max} \approx 400 \, \text{km} \times 960 \, \text{km} \approx 380\,000 \, \text{km}^2. \qquad (7.12)$$

The geometric mean (the midpoint) of these endpoints is 250 000 square kilometers.

Although reasonable, this approach overestimates the width of the plausible range—a mistake that we'll correct shortly. However, even this overestimated range spans only a factor of 2.4, whereas my starting range of $10^4...10^7$ square kilometers spans a factor of 1000. Divide-and-conquer reasoning has significantly narrowed my plausible range by replacing a quantity about which I have vague knowledge, namely the area, with quantities about which I have more precise knowledge.

The second bonus is that subdividing into many quantities carries only a small penalty, smaller than suggested by simply multiplying the endpoints. Multiplying the endpoints produces a range whose width is the product of the two widths. But this width assumes the worst. To see how, imagine an extreme case: estimating a quantity that is the product of ten independent factors, each of which you know to within a factor of 2 (in other words, each plausible range spans a factor of 4). Does the plausible range for the final quantity span a factor of 4^{10} (approximately 10^6)? That conclusion is terribly pessimistic. More likely, several of the ten estimates will be too large and several will be too small, allowing many errors to cancel.

Correctly computing the plausible range for the area requires a complete probabilistic description of the plausible ranges for the width and height. With it we could compute the probability of each possible product. That description is, somewhere, available to the person giving the range. But no one knows how to deduce such a complete set of probabilities based on the complex, diffuse, and seemingly contradictory information lodged in a human mind.

A simple solution is to specify a reasonable probability distribution. We'll use the log-normal distribution. This choice means that, on a logarithmic scale, the probability distribution is a normal distribution, also known as a Gaussian distribution.

Here's the log-normal distribution for my gut estimate of the height of the UK lumping rectangle. The horizontal axis is logarithmic: Distances correspond to ratios rather than differences. Therefore, 800 rather than 815 kilometers lies halfway between 670 and 960 kilometers. These endpoints are a factor of 1.2 smaller or larger than 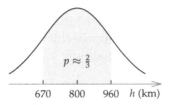 the midpoint. The peak at 800 kilometers reflects my belief that 800 kilometers is the best guess. The shaded area of 2/3 quantifies my confidence that the true height lies in the range 670...960 kilometers.

We use the log-normal distribution for several reasons. First, our mental hardware compares quantities using ratio rather than absolute difference. (By "quantity," I mean inherently positive values, such as distance, rather than signed values such as position.) In short, our hardware places quantities on a logarithmic scale. To represent our thinking, we therefore place the distribution on a logarithmic scale.

Second, the normal distribution, which has only two parameters (midpoint and width), is simple to describe. This simplicity helps us when we translate our internal gut knowledge into the distribution's parameters. From our gut estimates of the lower and upper endpoints, we just find the corresponding midpoint and width (on a logarithmic scale!): The midpoint is the geometric mean of the two endpoints, and the width is the square root of the ratio between the upper and lower endpoints.

Third, normal distributions combine simply. When we add two quantities represented by a normal distribution, their sum is also represented by a normal distribution.

On the other hand, the normal distribution does not represent all aspects of our gut knowledge. In particular, the tails of the normal distribution are too thin, reflecting an unrealistically high confidence that our estimate does not contain a huge error. Fortunately, for our analyses, this problem is not so significant, because we will concern ourselves with the location and width of the central region, and not take the thin tails too seriously. With that caveat, we'll represent our plausible range as a normal distribution on a logarithmic scale: a log-normal distribution.

Here is the other log-normal distribution, for the width of the UK lumping rectangle. The shaded range is the so-called one-sigma range $\mu - \sigma$ to $\mu + \sigma$, where μ is the midpoint (here, 310 kilometers) and σ is the width, measured as a distance on a logarithmic scale. In a normal distribution, the one-sigma range contains 68 percent

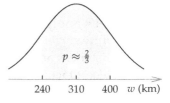

of the probability—conveniently close to 2/3. When we ask our plausible ranges to contain a 2/3 probability, we are estimating a one-sigma range.

The two log-normal distributions supply the probabilistic description required to combine the plausible ranges. The rules of probability theory (Problem 7.5) produce the following two-part recipe.

1. The midpoint of the plausible range for the area A is the product of the midpoint of the plausible ranges for h and w. Here, the height midpoint is 800 kilometers and the width midpoint is 310 kilometers, so the area midpoint is roughly 250 000 square kilometers:

$$\underbrace{800\,\text{km}}_{h} \times \underbrace{310\,\text{km}}_{w} \approx 250\,000\,\text{km}^2. \tag{7.13}$$

2. To compute the plausible range's width or half width, first express the individual half widths (the σ values) in logarithmic units. Convenient units include factors of 10, also known as bels, or the even-more-convenient decibels. A decibel, whose abbreviation is dB, is one-tenth of a factor of 10. Here is the conversion between a factor f and decibels:

$$\text{number of decibels} = 10 \log_{10} f. \tag{7.14}$$

For example, a factor of 3 is close to 5 decibels (because 3 is almost one-half of a power of ten), and a factor of 2 is almost exactly 3 decibels.

(These decibels are slightly more general than the acoustic decibels introduced in Problem 3.10: Acoustic decibels measure energy flux relative to a reference value, usually 10^{-12} watts per square meter. Both kinds of decibels measure factors of 10, but the decibels here have no implicit reference value.)

In decibels, bels, or any logarithmic unit, the half width (the σ) of the product's range is the Pythagorean sum of the individual half widths (the σ values). Using σ_x to represent the half width of the plausible range for the quantity x, the recipe is

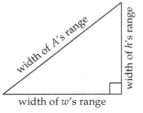

$$\sigma_A = \sqrt{\sigma_h^2 + \sigma_w^2}. \qquad (7.15)$$

width of w's range

Let's apply this recipe to our example. The plausible range for the height (h) was 800 kilometers give or take a factor of 1.2. On a logarithmic scale, distances are measured by ratios or factors, so think of a range as "give or take a factor of" rather than as "plus or minus" (a description that would be appropriate on a linear scale). A factor of 1.2 is about ±0.8 decibels:

$$10 \log_{10} 1.2 \approx 0.8. \qquad (7.16)$$

Therefore, $\sigma_h \approx 0.8$ decibels.

The plausible range for the width (w) was roughly 310 kilometers give or take a factor of 1.3. A factor of 1.3 is ±1.1 decibels:

$$10 \log_{10} 1.3 \approx 1.1. \qquad (7.17)$$

Therefore, $\sigma_w \approx 1.1$ decibels.

The Pythagorean sum of σ_h and σ_w is approximately 1.4 decibels:

$$\sqrt{0.8^2 + 1.1^2} \approx 1.4. \qquad (7.18)$$

As a factor, 1.4 decibels is, coincidentally, approximately a factor of 1.4:

$$10^{1.4/10} \approx 1.4. \qquad (7.19)$$

Because the midpoint of the plausible range is 250 000 kilometers, the UK land area should be 250 000 square kilometers give or take a factor of 1.4. Retaining a bit more accuracy, it is a factor of 1.37.

$$\underbrace{180\,000}_{/1.37} \ldots \underbrace{250\,000}_{\text{midpoint}} \ldots \underbrace{340\,000}_{\times 1.37} \; \text{km}^2. \qquad (7.20)$$

As a probability bar, the range is

180 000 km²	250 000 km²	340 000 km²
$p \approx 1/6$	$p \approx 2/3$	$p \approx 1/6$

The true area is 243 610 square kilometers. This area is comfortably in my predicted range and surprisingly close to the midpoint.

▶ *How surprising is the accuracy of the estimate?*

The surprise can be quantified with a probability: the probability that the true value would be closer to the midpoint (the best estimate) than 243 610 square kilometers is. Here, 243 610 is 2.6 percent or a factor of 1.026 smaller than 250 000. The probability that the true value would be within a factor of 1.026 (on either side) of the midpoint is the tiny shaded region of the log-normal distribution.

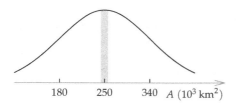

$$180 \qquad 250 \qquad 340 \quad A \ (10^3 \ \text{km}^2)$$

The region is almost exactly a rectangle, so its area is approximately its height multiplied by its width. The height is the peak height of a normal distribution. In $\sigma = 1$ units (in dimensionless form), this height is $1/\sqrt{2\pi}$.

The width of the shaded region, also in $\sigma = 1$ units, is the ratio

$$2 \times \frac{\text{dB equivalent for a factor of 1.026}}{\text{dB equivalent for a factor of 1.37}},$$
(7.21)

which is approximately 0.16. (The factor of 2 arises because the region extends equally on both sides of the peak.) Therefore, the shaded probability is approximately $0.16/\sqrt{2\pi}$ or only 0.07. I am surprised, but encouraged, by the high accuracy of my estimate for the UK's land area. Once again, many individual errors—for example, in estimating journey times and speeds—have canceled out.

Problem 7.4   Volume of a room
Estimate the volume of your favorite room, comparing your plausible ranges before and after using divide-and-conquer reasoning.

Problem 7.5   Justifying the recipe for combining ranges
Use Bayes' theorem to justify the recipe for combining plausible ranges.

Problem 7.6   Practice combining plausible ranges

You are trying to estimate the area of a rectangular field. Your plausible ranges for its width and length are 1...10 meters 10...100 meters, respectively.

a. What are the midpoints of the two plausible ranges?

b. What is the midpoint of the plausible range for the area?

c. What is the too-pessimistic range for the area, obtained by multiplying the corresponding endpoints?

d. What is the actual plausible range for the area, based on combining log-normal distributions? This range should be narrower than the pessimistic range in part (c)!

e. How do the results change if the ranges are instead 2...20 meters for the width and 20...200 meters for the length?

Problem 7.7   Area of A4 paper

If you have a sheet of standard European (A4) paper handy, either in reality or mentally, find your plausible range for its area A by gut estimating its length and width (without using a ruler). Then compare your best estimate (the midpoint of your range) to the official area of An paper, which is 2^{-n} square meters.

Problem 7.8   Estimating a mass

In trying to estimate the mass of an object, your plausible range for its density is 1...5 grams per cubic centimeter and for its volume is 10...50 cubic centimeters. What is (roughly) your plausible range for its mass?

Problem 7.9   Which is the wider range?

Suppose that your knowledge of the quantities a, b, and c is given by these plausible ranges:

$$a = 1...10$$
$$b = 1...10 \tag{7.22}$$
$$c = 1...10.$$

Which quantity—abc or a^2b—has the wider plausible range?

Problem 7.10   Handling division

If a quantity a has the plausible range 1...4, and the quantity b has the plausible range 10...40, what are the plausible ranges for ab and for a/b?

7.2.2   Finding one-sigma endpoints before the midpoint

With our understanding of probability, we can explain two curious and seemingly arbitrary features of how we make gut estimates (Section 1.6).

First, we learned not to ask our gut about its best estimate. Rather, we ask about its lower and upper endpoints, from which we find the best estimate as their midpoint. Second, in finding the endpoints, the standard we use is "mildly surprised": We should be mildly surprised if the true value lies outside the endpoints. Quantitatively, "mild surprise" now means that the probability should 2/3 that the true value lies between the endpoints.

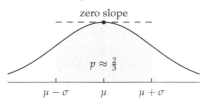

The first feature, that we estimate the endpoints rather than the midpoint directly, is explained by the shape of the log-normal distribution. Imagine trying to locate its midpoint by offering your gut candidate midpoints. The distribution is flat at the midpoint, so the probability hardly varies as the candidates change. Thus, your gut will answer with almost identical sensations of ease over a wide range around the midpoint. As a result, you cannot easily extract a decent midpoint estimate.

The solution to this problem explains the second curious feature: why the plausible range should enclose a probability of 2/3. The solution is to estimate the location of the steepest places on the curve. Those spots are the points of maximum slope (in absolute value). Around these values, the probability changes most rapidly and so does the sensation of ease.

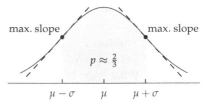

If we use the most convenient logarithmic units, where the midpoint μ is 0 and the half width σ is 1, then the log-normal distribution becomes the reasonably simple form

$$p(x) = \frac{1}{\sqrt{2\pi}} e^{-x^2/2}, \tag{7.23}$$

where x measures half widths away from the peak. Its slope $p'(x)$ is a maximum where the derivative of $p'(x)$, namely $p''(x)$, is zero. These points are also called the inflection or zero-curvature points. (Where the curvature is zero, the curve is straight, so the dashed tangent lines pass through it.)

Ignoring dimensionless prefactors,

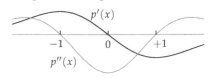

$$p'(x) \sim -xe^{-x^2/2};$$
$$p''(x) \sim (x^2 - 1)e^{-x^2/2}. \tag{7.24}$$

The second derivative is zero when $x = \pm 1$. This value is expressed in the $\mu = 0$, $\sigma = 1$ system of units. In the usual system,

$x = \pm 1$ means the one-sigma points $\mu \pm \sigma$. So the points of maximum (absolute) slope—the points that our gut can most accurately estimate—are the one-sigma points! We find them first and then find their midpoint.

When we estimate a 2/3-probability range, we are finding a one-sigma range almost exactly: In a normal or log-normal distribution, the one-sigma range contains approximately 68 percent of the probability, which is almost exactly 2/3. For comparison, the two-sigma range contains approximately 95 percent of the probability, which is a popular number in statistical analysis. Therefore, you may also want to find your two-sigma range (Problem 7.11). However, the slope at the two-sigma points is approximately a factor of 2.2 smaller than the slope at the one-sigma endpoints, so the two-sigma range is somewhat harder to estimate than is the one-sigma range. To find the two-sigma range, first estimate the one-sigma range, and then double its width (on a log scale).

This analysis of plausible ranges concludes our introduction to probability and the probabilistic basis of divide-and-conquer reasoning. We have learned that probabilities result from the incompleteness of our knowledge, and how acquiring knowledge changes our probabilities. In the next sections, we will use probability to master the complexity of systems with vast numbers of atoms and molecules, where complete knowledge would be impossible. The analysis begins with a special walk, the random walk.

Problem 7.11   Two-sigma range
My one-sigma range for the UK's land area is 180 000...340 000 square kilometers. What is the two-sigma range?

Problem 7.12   Gold or banknotes?
Having broken into a bank vault, do you take the banknotes or the gold? Assume that your capacity to carry loot is limited by mass rather than by volume.

a. Estimate gold's value density (monetary value per mass)—for example, in dollars per gram. Give plausible ranges for your subestimates and find the resulting plausible range for the value density.

b. For your favorite banknote, give your plausible range for its value density and for the ratio

$$\frac{\text{value density of gold}}{\text{value density of the banknote}}. \tag{7.25}$$

c. Should you take the gold or the banknotes? Use a table of the normal distribution to evaluate the probability that your choice is correct.

7.3 Random walks: Viscosity and heat flow

The great mathematician George Pólya, later to become author of *How to Solve It* [38], was staying at a bed and breakfast in his adopted country of Switzerland and walking daily in the garden between its tall hedge rows. He kept running into a newlywed couple taking their walk. Were they following him, or was it a mathematical necessity? From this question was born the study of random walks.

Random walks are everywhere. In the card game War, cards wander between two players, until one player gets the whole deck. How long does the game last, on average? A molecule of neurotransmitter is released from a synaptic vesicle. It wanders in the 20-nanometer gap, the synaptic cleft, until it binds to a muscle

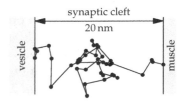

cell; then your leg muscle twitches. How long does the molecule's journey take? On a winter day, you stand outside wearing only a thin layer of clothing, and your body heat wanders through the clothing. How much heat do you lose? And why do large organisms have circulatory systems? Answering these questions requires understanding random walks.

7.3.1 Behavior of random walks: Lumping and probabilistic reasoning

For our first random walk, imagine a perfume molecule wandering in a room, moving in a straight line until collisions with air molecules deflect it in a random direction. This randomness reflects *our* incomplete knowledge: Knowing the complete state of the colliding molecules, we could calculate their paths after the collision (at least, in classical physics). However, we do not have that knowledge and do not want it!

Even without that information, the random motion of one molecule is still complicated. The complexity arises from the generality—that the direction of travel and the distance between collisions can have any value. To simplify, we'll lump in several ways.

Distance. Let's assume that the molecule travels a typical, fixed distance between collisions. This distance is the mean free path λ.

Direction. Let's assume that the molecule travels only along coordinate axes. Let's also study only one-dimensional motion; thus, the molecule moves either left or right (with equal probability).

Time. Let's assume that the molecule travels at a typical, fixed speed v and that every collision happens at regularly spaced clock ticks. These ticks are therefore separated by a characteristic time $\tau = \lambda/v$. This time is called the mean free time.

In this heavily lumped, one-dimensional model, a molecule starts at an origin ($x = 0$) and wanders along a line. At each tick it moves left or right with probability $1/2$ for each direction.

As time passes, the molecule spreads out. Actually, the molecule itself does not spread out! It has a particular position, but we just don't know it. What spreads out is our belief about the position. In the notation of probability theory, this belief is a set of probabilities—a probability distribution—based upon our knowledge of the molecule's starting position:

$$\Pr\left(\text{molecule is at position } x \text{ at time } t \,\middle|\, \text{it was at } x = 0 \text{ at } t = 0\right). \quad (7.26)$$

The changing beliefs are represented by a sequence of probability distributions, one for each time step. For example, at 2τ (after two ticks), the molecule has probability $1/2$ of being at the origin, by either going right then left or left then right. At 3τ, it has probability 0 of being at the origin (why?), but probability $3/8$ of being at $x = +\lambda$. To quantify the spread, which is shown in the following figure, we need an abstraction and a notation.

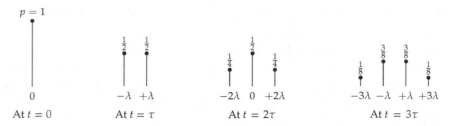

The position of the molecule will be x. Its expected position will be $\langle x \rangle$. The expected position is the weighted average of the possible positions, weighted by their probabilities. Both x and $\langle x \rangle$ are functions of time or, equivalently, of the number of ticks. However, because motion in each direction is equally likely, the expected position does not change (by symmetry). So $\langle x \rangle$, which starts at zero, remains zero.

A useful measure is the squared position x^2—more useful because it is never negative, making moot the symmetry argument that made $\langle x \rangle = 0$. Analogous to $\langle x \rangle$, the expected or mean squared position $\langle x^2 \rangle$ is the average of the possible values of x^2, weighted by their probabilities.

Let's see how $\langle x^2 \rangle$ changes with time. At $t = 0$, the only possibility is $x = 0$, so $\langle x^2 \rangle_{t=0} = 0$. After one clock tick, at $t = \tau$, the possibilities are also limited: $x = +\lambda$ or $-\lambda$. In either case, $x^2 = \lambda^2$. Therefore, $\langle x^2 \rangle_{t=\tau} = \lambda^2$.

It may be the mark of a savage, in Pólya's phrase [37], to generalize a pattern from only two data points. So let's find $t = 2\tau$ and then guess a pattern. At $t = 2\tau$, the position x could be -2λ, 0, or 2λ, with probabilities $1/4$, $1/2$, and $1/4$, respectively. The weighted average $\langle x^2 \rangle$ is $2\lambda^2$:

$$\langle x^2 \rangle = \frac{1}{4} \times (-2\lambda)^2 + \frac{1}{2} \times (0\lambda)^2 + \frac{1}{4} \times (+2\lambda)^2 = 2\lambda^2. \tag{7.27}$$

The pattern seems to be

$$\langle x^2 \rangle_{t=n\tau} = n\lambda^2. \tag{7.28}$$

This conjecture is correct. (You can test it at $t = 3\tau$ and 4τ in Problem 7.13.) Each step contributes the squared step size λ^2 to the squared spread $\langle x^2 \rangle$. We saw this pattern in Section 7.2.1, when we combined plausible ranges. The half width of the plausible range for an area $A = hw$ was given by

$$\sigma_A^2 = \sigma_h^2 + \sigma_w^2, \tag{7.29}$$

where σ_x is the half width of the plausible range for the quantity x. The half widths are step sizes in a random walk—random because the estimate is equally like to be an underestimate or an overestimate (representing stepping left or right, respectively). Therefore, the half widths, like the step sizes in a random walk, add via their squares ("adding in quadrature").

The number of ticks is $n = t/\tau$, so $\langle x^2 \rangle$, which is $n\lambda^2$, is also $t\lambda^2/\tau$. Thus,

$$\frac{\langle x^2 \rangle}{t} = \frac{\lambda^2}{\tau}. \tag{7.30}$$

As time marches on, $\langle x^2 \rangle / t$ remains λ^2/τ! The invariant λ^2/τ is all that we need to know about the details of a random walk.

This abstraction is known as the diffusion constant. It is usually denoted D and has dimensions of $L^2 T^{-1}$. The table gives useful approximate diffusion constants for a particle wandering in three dimensions. In d dimensions, the diffusion constant is defined with a dimensionless prefactor:

	$D\,(\mathrm{m}^2\,\mathrm{s}^{-1})$
air molecules in air	1.5×10^{-5}
perfume molecules in air	10^{-6}
small molecules in water	10^{-9}
large molecules in water	10^{-10}

$$D = \frac{1}{d}\frac{\lambda^2}{\tau}. \tag{7.31}$$

Because λ/τ is the speed v of the molecule between its randomizing collisions, a useful alternative form for estimating D is

$$D = \frac{1}{d}\lambda v. \qquad (7.32)$$

Problem 7.13   Testing the random-walk dispersion conjecture
Make the probability distribution for the particle at $t = 3\tau$ and $t = 4\tau$, and compute each $\langle x^2 \rangle$. Do your results confirm that $\langle x^2 \rangle_{t=n\tau} = n\lambda^2$?

Problem 7.14   Higher dimensions
For a molecule starting at the origin and wandering in two dimensions, $\langle r^2 \rangle = n\lambda^2$, where $r^2 = x^2 + y^2$. Confirm this statement for $t = 0...3\tau$.

A random and a regular walk are analogous in having an invariant. For a regular walk, it is $\langle x \rangle/t$: the speed. For a random walk, it is $\langle x^2 \rangle/t$: the diffusion constant. (The diffusion constant is to a random walk as the speed is to a regular walk.) However, the two walks differ in a scaling exponent. For a regular walk, $\langle x \rangle \propto t$: The scaling exponent connecting position and time is 1. For a random walk, we'll use the rms (root-mean-square) position

$$x_{\text{rms}} \equiv \sqrt{\langle x^2 \rangle} \qquad (7.33)$$

as a measure analogous to $\langle x \rangle$ for a regular walk. Because $\langle x^2 \rangle \propto t$, the rms position x_{rms} is proportional to $t^{1/2}$: In a random walk, the scaling exponent connecting position and time is only 1/2. This scaling exponent has profound effects on heat, drag, and diffusion.

As an example of the effect, let's apply our knowledge of diffusion and random walks to a familiar situation. Across a room someone opens a bottle of perfume or, if your taste in problems is pessimistic, a lunch of leftover fish.

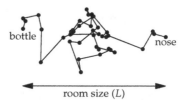

▶ *How long until odor molecules reach your nose?*

As time passes, the molecule wanders farther afield, with its rms position growing proportional to $t^{1/2}$. As a lumping approximation, imagine that the molecule is equally likely to be anywhere within a distance x_{rms} of the source (the perfume bottle or leftover fish). For the molecule to have a significant probability to be at your nose, x_{rms} should be comparable to the room size L. Because $\langle x^2 \rangle = Dt$, the condition is $L^2 \sim Dt$, and the required diffusion time is $t \sim L^2/D$. (For another derivation, see Problem 7.16.)

For perfume molecules diffusing in air, D is roughly 10^{-6} square meters per second. For a 3-meter room, the diffusion time is roughly 4 months:

$$t \sim \frac{L^2}{D} \sim \frac{(3\,\mathrm{m})^2}{10^{-6}\,\mathrm{m^2\,s^{-1}}} \approx 10^7\,\mathrm{s} \approx 4\,\mathrm{months}. \tag{7.34}$$

This estimate does not agree with experiment! After perhaps a minute, you'll notice the aroma, whether perfume or fish. Diffusion is too slow to explain why the odor molecules arrive so quickly. In reality, they make most of the journey using a regular walk: The small, unavoidable air currents in the room transport the molecules much farther and faster than diffusion can. The speedup is a consequence of the change in scaling exponent, from 1/2 (for random walks) to 1 (for regular walks).

Problem 7.15   Diffusion constant of air
Estimate the diffusion constant for air molecules diffusing in air by using

$$D \sim \frac{1}{3} \times \text{mean free path} \times \text{travel speed} \tag{7.35}$$

and the mean free path of air molecules (Section 6.4.5). This value is also its thermal diffusivity κ_{air} and its kinematic viscosity ν_{air}!

Problem 7.16   Dimensional analysis for the diffusion time
Use dimensional analysis to estimate the diffusion time t based on L (the relevant characteristic of the room) and D (the characteristic of the random walk).

A similar estimate explains the existence of circulatory systems. Imagine an oxygen molecule diffusing through our body to a muscle cell, where its services are needed to burn glucose and produce energy. The diffusion distance (our body size) is $L \sim 1$ meter. The diffusion constant for an oxygen molecule in water (a small molecule in water) is roughly 10^{-9} square meters per second. The diffusion time is roughly 10^9 seconds or 30 years:

$$t \sim \frac{L^2}{D} \sim \frac{(1\,\mathrm{m})^2}{10^{-9}\,\mathrm{m^2\,s^{-1}}} = 10^9\,\mathrm{s} \approx 30\,\mathrm{years}. \tag{7.36}$$

Over long distances—long compared to the mean free path λ—diffusion is a slow method of transport! Large organisms, especially warm-blooded organisms with high metabolic rates, need another solution: a circulatory system. It transports oxygen much more efficiently than diffusion can, just as air currents do for perfume. The circulatory system, a branching network of ever-smaller capillaries, ends once the typical distance between the smallest capillaries and a cell is small enough for diffusion to be efficient.

Another biological example of short-distance diffusion is the gap between two neighboring neurons, called the synaptic cleft. Its width is only $L \sim$ 20 nanometers. Signals between neighboring nerve cells, or between a neuron and a muscle cell, travel chemically, as neurotransmitter molecules.

Let's estimate the diffusion time for a neurotransmitter molecule. A neurotransmitter is a large molecule, and it is diffusing in water, so $D \sim 10^{-10}$ square meters per second. Its diffusion time is 4 microseconds:

$$t \sim \frac{L^2}{D} \sim \frac{(2 \times 10^{-8}\,\mathrm{m})^2}{10^{-10}\,\mathrm{m^2\,s^{-1}}} = 4 \times 10^{-6}\,\mathrm{s}. \tag{7.37}$$

Without a comparison, this time means little: We cannot say right away whether the time is long or short. However, because it is much smaller than the spike-timing accuracy of neurons (about 100 microseconds), the time to cross the synaptic cleft is small enough not to affect nerve-signal propagation. For transmitting signals between neighboring neurons, diffusion is an efficient and simple solution.

Problem 7.17   Probability of being at the origin
Pólya's analysis of his encounters with the newlywed couple required first finding the probability p_n that a random walker is at the origin after n clock ticks ($p_0 = 1$). For a d-dimensional random walk, find the scaling exponent $\beta(d)$ in $p_n \propto n^{\beta(d)}$.

Problem 7.18   Expected number of visits to the origin
Using your result from Problem 7.17, estimate the expected number of visits a random walker in one and two dimensions makes to the origin (summed over all ticks $n \geq 0$). Thereby explain Pólya's theorem [36], that a random walker in one or two dimensions always returns to the origin. What makes a three-dimensional random walk different from one or two dimensions?

7.3.2 Types of diffusion constants

Because random walks are everywhere, there are several kinds of diffusion constants. They are named differently depending on what is diffusing, but they share the mathematics of the random walk. Therefore, they all have dimensions of length squared per time ($\mathrm{L^2 T^{-1}}$).

what is diffusing	name of diffusion constant	symbol
particles	diffusion constant	D
energy (heat)	thermal diffusivity	κ
momentum	kinematic viscosity	ν

A handful of useful diffusion constants (for particles) were tabulated on page 251. To complement that table, here are a few useful, approximate thermal diffusivities and kinematic viscosities.

κ_{air}	heat diffusing in air	$1.5 \times 10^{-5} \, \text{m}^2 \, \text{s}^{-1}$
ν_{air}	momentum diffusing in air	1.5×10^{-5}
κ_{water}	heat diffusing in water	1.5×10^{-7}
ν_{water}	momentum diffusing in water	1.0×10^{-6}

In air, all three diffusion constants—D for molecules, κ for energy, and ν for momentum—are roughly 1.5×10^{-5} square meters per second. Their similarity is no coincidence. The same mechanism (diffusion of air molecules) transports molecules, energy, and momentum.

In water, however, the molecular diffusion constant D is several orders of magnitude smaller than the heat- and momentum-diffusion constants (κ and ν, respectively). Even the momentum- and heat-diffusion constants differ roughly by a factor of 7. This dimensionless ratio, ν/κ, is the Prandtl number Pr. For water, I usually remember ν, because it is just a power of ten in SI units, and remember the Prandtl number—the lucky number 7—and use these values to reconstruct κ.

7.3.3 Thermal diffusivities of liquids and solids

The large discrepancy between the molecular and thermal diffusion constants in water indicates that our model for diffusion in water is not complete. The problem is not limited to water. If we had made a similar comparison for any solid, comparing the molecular and thermal diffusion constants (D and κ), the discrepancy would have been even larger.

Indeed, in liquids and solids, in contrast to gases, heat is not transported by molecular motion. In a solid, the molecules sit at their sites in the lattice. They vibrate but scarcely wander. In a liquid, molecules wander but only slowly. Their tight packing keeps the mean free path short and the diffusion constant small. Yet, as everyday experience and the large κ/D ratio suggest, heat can travel quickly in liquids and solids. The reason is that heat is transported by miniature sound waves rather than by molecular motion. The sound waves are called phonons.

By analogy to photons, which represent the vibrations of the electromagnetic field, phonons represent the vibrations of the lattice entities (the atoms or molecules of the liquid or solid). One molecule vibrates, shaking the next molecule, which shakes the next molecule. This chain is the motion of a phonon. Phonons act like particles: They travel through the lattice, bouncing off impurities and other phonons. Like ordinary particles, they have a mean free path and a propagation speed. These properties of their random walk determine the thermal diffusivity:

$$\kappa \approx \frac{1}{3} \times \text{phonon mean free path} \times \text{propagation speed.} \tag{7.38}$$

The propagation speed is the speed of sound c_s, because phonons are tiny sound waves (familiar sound waves contain many phonons, just as light beams contain many photons). Sound speeds in liquids and solids are much higher than the thermal speeds, so we already see one reason why κ, the diffusion constant for heat, is larger than D, the diffusion constant for particles.

The mean free path λ measures how far a phonon travels before bouncing (or scattering) and heading off in a random direction. Let's write $\lambda = \beta a$, where a is the typical lattice spacing (3 ångströms) and β is the number of lattice spacings that the phonon survives.

Then the thermal diffusivity becomes

$$\kappa = \frac{1}{3}c_s\beta a. \tag{7.39}$$

For water, our favorite substance, $c_s \sim 1.5$ kilometers per second (as you estimated in Problem 5.58). Then the predicted thermal diffusivity becomes

$$\kappa_{\text{water}} \sim \beta \times 1.5 \times 10^{-7}\,\text{m}^2\,\text{s}^{-1}. \tag{7.40}$$

Because the actual thermal diffusivity of water is 1.5×10^{-7} square meters per second, our estimate is exact if we use $\beta = 1$. This choice is easy to interpret and remember: In water, the phonons travel roughly one lattice spacing before scattering in a random direction. This distance is so short because water molecules do not sit in an ordered lattice. Their disorder provides irregularities that scatter the phonons. (At the same time, this mean free path is much larger than the mean free path of tightly packed atoms or molecules, which move a fraction of an ångström before getting significantly deflected. Therefore, even in a liquid, κ is much larger than the molecular diffusion constant D.)

To estimate κ for a solid, let's use κ_{water} along with the scaling relation

$$\kappa \propto \lambda c_s. \tag{7.41}$$

In a typical solid, the sound speed c_s is 5 kilometers per second—roughly a factor of 3 faster than in water. The mean free path λ is also longer than in the totally disordered lattice of a liquid. In a solid without too many lattice defects, and at room temperature, a phonon travels a few lattice spacings before scattering—compared to just one lattice spacing in water.

These two differences, each contributing a factor of 3, make the typical thermal diffusivity of a solid a factor of 10 larger than that of water:

$$\kappa_{\text{solid}} \sim \kappa_{\text{water}} \times 10 \approx 1.5 \times 10^{-6}\, \text{m}^2\,\text{s}^{-1}. \tag{7.42}$$

	κ ($\text{m}^2\,\text{s}^{-1}$)
Au	1.3×10^{-4}
Cu	1.1×10^{-4}
Fe	2.3×10^{-5}
air	1.9×10^{-5}
sandstone	1.1×10^{-6}
brick	0.5×10^{-6}
glass	3.4×10^{-7}
water	1.5×10^{-7}
pine	0.9×10^{-7}

Rounded to 10^{-6} square meters per second, this value is our canonical thermal diffusivity of a solid—for example, sandstone or brick. The table also shows a new phenomenon: For metals, κ is much larger than our canonical value. Although a small discrepancy could be explained by a few missing numerical factors, this significant discrepancy indicates a missing piece in our model.

Indeed, in metals, heat can also be carried by electron waves, not just phonons (lattice waves). Electron waves travel much faster and farther than phonons. Their speed, known as the Fermi velocity, is comparable to the orbital speed of an atom's outer-shell electron. As you found in Problem 5.36, for hydrogen this speed is αc, where α is the fine-structure constant ($\sim 10^{-2}$) and c is the speed of light. This speed, roughly 1000 kilometers per second, is much faster than any sound speed! As a result, the thermal diffusivity in metals is large. As you can see in the table, for a good conductor, such as copper or gold, $\kappa \sim 10^{-4}$ square meters per second.

7.3.3.1 Heating a skillet

To feel a thermal diffusivity, place a thin cast-iron skillet on a hot stove.

▷ *How long does it take for the top surface to feel hot?*

The hot stove supplies blobs of heat (of energy) that wander back and forth: The heat blobs perform a random walk. In a random walk, a particle with diffusion constant D wandering for a time t reaches a distance $z \sim \sqrt{Dt}$; we used this lumping model in Section 7.3.1 to estimate the diffusion time across a synaptic cleft. Here, the particle is a blob of heat, so the diffusion constant is κ. Thus, the hot front reaches a distance $z \sim \sqrt{\kappa t}$ into the skillet.

In this lumping picture, the temperature profile
is a rectangle with a moving right edge—representing the heat wave moving upward and into
the skillet. For my 1-centimeter-thick cast-iron
skillet, the hot front should reach the top of the
skillet in roughly 4 seconds:

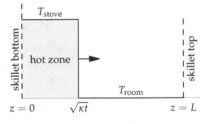

$$t \sim \frac{L^2}{\kappa_{Fe}} \approx \frac{(10^{-2}\,\text{m})^2}{2.3 \times 10^{-5}\,\text{m}^2\,\text{s}^{-1}} \approx 4\,\text{s}. \qquad (7.43)$$

Don't try the following experiment at home. But for the sake of lumping
and probabilistic reasoning, I set our flattest cast-iron skillet on a hot electric
stove while touching the top surface. After 2 seconds, my finger involuntar-
ily jumped off the skillet.

The discrepancy of a factor of 2 between the predicted
and actual times is not bad considering the simplicity, or crudity, of the lumping approximations that it
incorporates. However, there is a bigger discrepancy.
The model predicts that, for the first 4 seconds, the top
of the skillet remains at room temperature. One feels
nothing until, wham, the full temperature of the stove hits at 4 seconds.

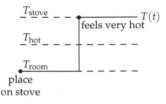

However, experience suggests that the skillet's temperature starts rising before the skillet becomes too
hot to touch, and then it monotonically approaches
the stove temperature—as sketched in the figure.

One reason for this discrepancy could be the skillet's
top surface. In our model, the skillet has only a bot-
tom surface and is infinitely thick. The top surface might alter the heat
flow. However, the infinite-slab assumption isn't the fundamental problem
(correcting the assumption turns out to speed up the heating process by a
factor of 2). Even if we fix it, the model would still make the bogus prediction of a sudden jump to the stove temperature. It's hard to believe after
the exhortations on the power of lumping, but we have lumped too much.

To improve the model, let's incorporate a more realistic temperature profile. Beyond the canonical lumping shape of a rectangle, the next simplest
shape is a triangle (the integral of a rectangle). We will therefore replace
the rectangular temperature profile with a triangle having the same area as
the rectangle. Because of the factor of $1/2$ in the area of a triangle, the hot
zone now extends to $2\sqrt{\kappa t}$ instead of to $\sqrt{\kappa t}$.

The area of the triangle is proportional to the heat transferred to the skillet. By matching the triangle's area to the rectangle's area, we preserve this integral quantity. When making lumping models, preserving integral quantities is usually more robust than is preserving differential quantities (such as slope).

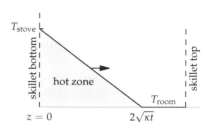

In the triangular-lumping model, you feel nothing until the tip of the triangle reaches the top of the skillet. Because the triangular hot front extends a factor of 2 farther than the rectangular hot front ($2\sqrt{\kappa t}$ versus $\sqrt{\kappa t}$), and because diffusion times are proportional to distance squared, the required time falls by a factor of 4. Thus, it falls from 4 seconds to 1 second. At 1 sec-

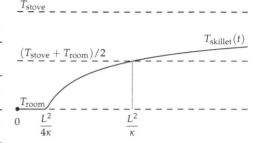

ond, the hot front arrives at the top of the skillet, which starts to feel warm. Then the temperature slowly increases toward the stove temperature. This next-simplest model makes quite realistic predictions!

Problem 7.19   Cooling the Moon
How long does it take for the Moon, with a radius of 1.7×10^6 meters to cool significantly by heat diffusion through rock? Given that the Moon is now cold, what do you conclude about the mechanism of cooling?

Problem 7.20   Diffusion with a bit of drift: Breaking the bank at Monte Carlo
You can play the card game blackjack such that your probability of winning a hand is $p = 0.51$ and of losing is $1 - p = 0.49$. You start with N betting units, the stake; and you bet 1 unit on each hand. The goal of this problem is to estimate the threshold N such that you are more likely to break the bank than to lose the stake.

Let x_n be your balance after the nth hand. Thus, $x_0 = N$. Losing your stake (the N units) corresponds to $x = 0$. To estimate N, extend the random-walk model to account for drift: that the probabilities of moving left and right are not equal.

a. What is the symbolic expression for breaking the bank?

b. Sketch your expected balance $\langle x \rangle$ versus the number of hands n (on linear axes).

c. Sketch, on the same axes, the dispersion x_{rms} versus n.

d. Explain graphically "a significant probability of breaking the bank."

e. Thus, estimate the required stake N.

7.3.3.2 Baking

A skillet on a stove gets heated from one side. An equally important kind of cooking, and one that helps us practice and extend our lumping model of random walks, is heating from two sides: baking. As an example, imagine baking a slab of fish that is $L = 1$ inch (2.5 centimeters) thick.

▷ *How long should it bake in the oven?*

A first, and quick, analysis predicts L^2/κ, the characteristic time for heat diffusion, where κ is the thermal diffusivity of water (organisms are mostly water). However, this simple model predicts an absurd time:

$$t \sim \frac{L^2}{\kappa_{\text{water}}} \approx \frac{(2.5\,\text{cm})^2}{1.5 \times 10^{-3}\,\text{cm}^2\,\text{s}^{-1}} \sim 70\,\text{minutes.} \tag{7.44}$$

After more than an hour in the oven, the fish will be so dry that it might catch fire, never mind not being edible. The model also ignores an important quantity: the oven temperature. Fixing this hole in the model will also improve the time estimate.

▷ *Why does the oven temperature matter?*

The inside of the fish must get cooked, which means that the proteins denature (lose their folded shape) and the fats and carbohydrates change their chemistry enough to become digestible. This process happens only if the food gets hot enough. Thus, a cold oven could not cook the fish, even after the fish reached oven temperature. What temperature is hot enough? From experience, a thin piece of meat on a hot skillet (at, say, 200 °C) cooks in less than a minute. At the other extreme, if the skillet is at 50 °C, unpleasantly hot to the touch but not much hotter than body temperature, the meat never cooks. A round, intermediate temperature of 100 °C, enough to boil water, should be enough to cook meat thoroughly.

If we set the oven to, say, 180 °C, the fish will then be cooked once its center reaches the midpoint of the room (20 °C) and oven temperatures, which is 100 °C. In this

> oven at 180 °C
> fish starts at 20 °C
> oven at 180 °C

improved model of cooking, the interior of the fish starts at $T_{\text{room}} \approx 20\,°C$, and the oven holds the top and bottom surfaces of the fish at $T_{\text{oven}} = 180\,°C$ (360 °F). With this model, and using triangular temperature profiles, we will estimate the time required for the center of the fish to reach the midpoint temperature of 100 °C.

Two triangular hot zones, one from each surface, move toward the center of the fish. Before the zones meet, the center of the fish, at $z = L/2$, is cold (room temperature). The fish, unless it is very fresh, is not ready to eat. But soon the triangles will meet.

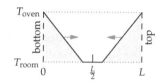

Each triangle extends a distance $2\sqrt{\kappa t}$. They first meet (at $x = L/2$) when $2\sqrt{\kappa t} = L/2$. Thus, $t_{\text{meet}} \sim L^2/16\kappa$. From that moment, the center warms up as the triangle fronts overlap ever more. The fish is cooked when the center reaches the mean of the room and oven temperatures (which is 100 °C).

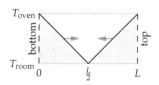

In dimensionless temperature units, where T_{room} corresponds to $\overline{T} = 0$ and T_{oven} to $\overline{T} = 1$, the cooking criterion is $\overline{T} = 1/2$. Each triangle wave therefore contributes $\overline{T} = 1/4$. The left triangle, representing the hot front invading from the bottom surface, then passes through the points $(z = 0, \overline{T} = 1)$ and $(z = L/2, \overline{T} = 1/4)$. It reaches the z axis ($\overline{T} = 0$) when $z = 2L/3$. The corresponding triangle-zone diffusion time is given by $2\sqrt{\kappa t} = 2L/3$, whose solution is

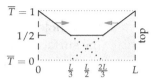

$$t_{\text{cooked}} \sim \frac{L^2}{9\kappa}. \tag{7.45}$$

For our 2.5-centimeter-thick fish, the time is roughly 7 minutes:

$$t \sim \frac{1}{9} \times \frac{(2.5\,\text{cm})^2}{1.5 \times 10^{-3}\,\text{cm}^2\,\text{s}^{-1}} \sim 7\,\text{minutes}. \tag{7.46}$$

This estimate is reasonable. My experience is that a fish fillet of this thickness requires about 10 minutes in a hot oven to cook all the way through (and further baking only dries it out).

Problem 7.21   Baking too long

In the model of two approaching triangle hot fronts, the central temperature starts to rise at $t \sim L^2/16\kappa$, and it reaches the halfway temperature at $t_{\text{cooked}} \sim L^2/9\kappa$. When does it reach the oven temperature? Sketch the central temperature versus time, labeling interesting values.

Problem 7.22   Cooking an egg

Baking a 6-kilogram turkey requires, from experience, 3 to 4 hours (you get to predict this time in Problem 7.34). Use proportional reasoning to estimate the time required to boil an egg.

7.3.4 Boundary layers

In cooking, the hot zone diffuses inward from the hot surface. For the most tangible example of this random walk, rub your finger on the blade of a (stationary!) window fan. Your finger comes away dusty and leaves a dust-free streak on the blade. But why is any dust on the blade at all? When the fan was turning, why didn't the air streaming by the blade blow off the dust?

The answer lies in the concept of the boundary layer. For the cooking examples (Sections 7.3.3.1 and 7.3.3.2), this layer is the expanding hot zone. It arises from the boundary constraint (the temperature), which diffuses into the skillet or the fish. For the fan blade, the analogous constraint is that, next to the blade, the fluid has zero velocity with respect to the blade. The condition, called the no-slip boundary condition, has the justification that the fluid molecules next the surface get caught by the inevitable roughness at the surface. (For a historical and philosophical discussion of the subtleties of this boundary condition, see Michael Day's article on "The no-slip condition of fluid dynamics" [8].)

Starting at the blade surface, a zero-speed or, equivalently, zero-momentum zone diffuses into the fluid—just as the stove temperature diffuses into the skillet or the oven temperature into the fish fillet. After a growth time t, the zero-momentum front has diffused a distance $\delta \sim \sqrt{\nu t}$, where ν is the diffusion constant for momentum (the kinematic viscosity). The distance δ is the boundary-layer thickness. Within the boundary layer, the fluid moves more slowly than the fluid in the free stream. Using a rectangular lumping picture, the fluid speed is zero within the layer and full speed outside it. Therefore, dust particles entirely in the layer remain on the blade.

To estimate the boundary-layer thickness, imagine a window fan with a blade width l and rotation speed v at the widest part of the blade. The growth time is $t \sim l/v$. For a generic window fan sweeping out a diameter of 0.5 meters, the blade width l may be roughly 0.15 meters. If the fan rotates at roughly 15 revolutions per second or $\omega \sim 100$ radians per second, the blade speed v is 10 meters per second:

$$v \sim \underbrace{0.1\,\text{m}}_{\text{arc radius}} \times \underbrace{100\,\text{rad s}^{-1}}_{\omega} = 10\,\text{m s}^{-1}. \tag{7.47}$$

At this speed, the growth time t is 0.015 seconds. In that time, the zero-momentum constraint diffuses roughly 0.5 millimeters from the fan blade:

$$\delta \sim \left(\underbrace{0.15\,\text{cm}^2\,\text{s}^{-1}}_{\nu_{\text{air}}} \times \underbrace{0.015\,\text{s}}_{t} \right)^{1/2} \approx 0.05\,\text{cm}. \tag{7.48}$$

Dust particles that are significantly smaller than 0.5 millimeters feel no air flow (in this simplest lumping picture) and thus remain on the fan blade until your finger rubs them off. For the same reason (boundary layers), simply rinsing dirty dishes in water will not remove the thin layer of food next to the surface. Proper cleaning requires scrubbing (a sponge) or soap (which gets underneath the food oils).

> **Problem 7.23   Reynolds number in the boundary layer**
> Based on the boundary-layer thickness $\delta \sim \sqrt{\nu t}$, estimate the Reynolds number in a boundary layer on an object of size L (a length) moving in a fluid at speed v.

> **Problem 7.24   Turbulence in the boundary layer**
> At Reynolds numbers comparable to 1000, flows usually become turbulent. Use Problem 7.23 to estimate the main-flow Reynolds number at which the boundary layer becomes turbulent. For a smooth ball similar in size to a golf ball, what flow speed is required? Suggest an explanation for the dimples on golf balls.

7.4 Transport by random walks

When a hot-oven temperature front diffuses into a slab of fish or the no-slip, zero-momentum front diffuses into a fluid, it comes with a heat or momentum flow. With our random-walk model, we can estimate the magnitude of these flows, and thereby understand the drag forces on fog droplets and bacteria (Problem 7.26) and why we feel cold on a winter day without thick clothing (Section 7.4.4).

7.4.1 Diffusion speed

The essential property of a random walk is that the distance traveled is not proportional to the time, as it would be in a regular walk, but rather to the square root of the time. This change in scaling exponent means that the speed at which heat, momentum, or particles diffuse depends on the diffusion distance. When the distance is L, the diffusion time is $t \sim L^2/D$, where D is the appropriate diffusion constant. Thus, the transport speed is comparable to D/L:

$$v \sim \frac{L}{t} \sim \frac{L}{L^2/D} = \frac{D}{L}. \tag{7.49}$$

This speed depends inversely on the distance. This scaling is consistent with our calculations showing that diffusion is a terribly slow means of transport over long distances (for example, for perfume molecules diffusing across a room) but fast over short distances (for example, for neurotransmitter molecules diffusing across a synaptic cleft).

When the diffusing quantity is momentum, the appropriate diffusion constant is ν, and the diffusion speed is ν/L. Thus, the Reynolds number, $v_{\text{flow}}L/\nu$, is the ratio $v_{\text{flow}}/v_{\text{diffusion}}$—for the same reason that it is the ratio of times $t_{\text{diffusion}}/t_{\text{flow}}$ (as you will find in Problem 7.32). Using the diffusion speed, we can estimate fluxes and flows.

7.4.2 Flux

Transport is measured by the flux:

$$\text{flux of stuff} = \frac{\text{amount of stuff}}{\text{area} \times \text{time}}. \tag{7.50}$$

As we found in Section 3.4.2, the flux is also given by

$$\text{flux of stuff} = \frac{\text{stuff}}{\text{volume}} \times \text{transport speed}. \tag{7.51}$$

When the stuff travels by diffusion, the transport speed is the diffusion speed D/L (from Section 7.4.1). The resulting flux is

$$\text{flux of stuff} \sim \frac{\text{stuff}}{\text{volume}} \times \frac{\text{diffusion constant}}{\text{distance}}. \tag{7.52}$$

In symbols,

$$F \sim n\frac{D}{L}, \tag{7.53}$$

where n is the concentration (stuff per volume), D is the appropriate diffusion constant (depending on what is diffusing), and L is the distance.

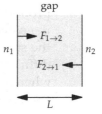

An important application is diffusion across a gap. The gap could be a synaptic cleft, with a gap width $L \sim 20$ nanometers and with different neurotransmitter concentrations on its two sides. Or it could be a shirt ($L \sim 2$ millimeters) with different temperatures—concentrations of energy—on the inside and outside. On one side, the density of stuff is n_1; on the other side it is n_2. Then there are two fluxes in the gap, left to right and right to left:

$$F_{1\to 2} \sim n_1 \frac{D}{L}$$
$$F_{2\to 1} \sim n_2 \frac{D}{L}. \tag{7.54}$$

The net flux F is their difference:

$$F \sim (n_2 - n_1)\frac{D}{L} = D\frac{n_2 - n_1}{L}. \tag{7.55}$$

The concentration difference $n_2 - n_1$ divided by the gap size L is an important abstraction: the concentration gradient. It measures how rapidly the concentration varies with distance. Using it, the flux becomes

flux = diffusion constant × concentration gradient. (7.56)

This result, called Fick's law, is exact (hence the equals sign instead of $\sim$). In calculus form, the concentration gradient is $\Delta n/\Delta x$, so

$$F = D\frac{\Delta n}{\Delta x}. \tag{7.57}$$

If the diffusing stuff is particles, then the appropriate diffusion constant is just D, and the result can be used as is. If the diffusing stuff is momentum, then the diffusion constant is the kinematic viscosity ν, and the flux is closely connected to a drag force (Problem 7.25).

Problem 7.25   Momentum flux produces drag

If the diffusing stuff is momentum, then the diffusion constant is the kinematic viscosity ν, and the concentration gradient is the gradient of momentum density. Thus, explain why Fick's law becomes

viscous stress = $\rho\nu$ × velocity gradient, (7.58)

where viscous stress is viscous force per area ($\rho\nu$ is the dynamic viscosity η).

Problem 7.26   Stokes drag

In this problem, you use momentum flux (Problem 7.25) to estimate the drag force on a sphere of radius r in a flow at low Reynolds number ($\mathrm{Re} \ll 1$). If $\mathrm{Re} \ll 1$, the boundary layer (Section 7.3.4)—the region over which the fluid velocity changes from zero to the free-stream velocity v—is comparable in thickness to r. Using that information, estimate the viscous drag force on the sphere.

If the diffusing stuff is heat (energy), the diffusion constant is the thermal diffusivity κ, and the concentration gradient is the gradient of energy density. Thus, heat flux is

heat flux = thermal diffusivity × energy-density gradient. (7.59)

To figure out the meaning of energy-density gradient, let's start with energy density itself, which is energy per volume. We usually measure it using temperature. Therefore, to make the eventual formula applicable, let's rewrite energy density in terms of temperature:

$$\frac{\text{energy}}{\text{volume}} = \frac{\text{energy}}{\text{volume} \times \text{temperature}} \times \text{temperature}. \tag{7.60}$$

The complicated quotient on the right splits into two simpler factors:

$$\frac{\text{energy}}{\text{volume} \times \text{temperature}} = \frac{\text{mass}}{\text{volume}} \times \frac{\text{energy}}{\text{mass} \times \text{temperature}}. \tag{7.61}$$

The first factor, mass per volume, is just the substance's density ρ. The second factor is called the specific heat c_p. The most familiar specific heat is water's: 1 calorie per gram per °C. That is, raising 1 gram of water by 1 °C requires 1 calorie ($\approx$ 4 joules).

Using these abstractions,

$$\frac{\text{energy}}{\text{volume} \times \text{temperature}} = \rho c_p, \tag{7.62}$$

To get energy density, or energy per volume, we multiply by temperature

$$\frac{\text{energy}}{\text{volume}} = \rho c_p T. \tag{7.63}$$

Now that we have energy density, we can find the energy-density gradient. Because ρ and c_p are constants (at least for small temperature and position changes), any gradient of energy density is due to the temperature gradient $\Delta T / \Delta x$:

$$\text{energy-density gradient} = \rho c_p \times \text{temperature gradient}. \tag{7.64}$$

Fick's law, when energy is the diffusing stuff, tells us that

$$\text{energy flux} = \text{thermal diffusivity} \times \text{energy-density gradient}, \tag{7.65}$$

so

$$\text{heat (energy) flux} = \underbrace{\kappa \times \rho c_p}_{\text{energy-density gradient}} \times \text{temperature gradient}. \tag{7.66}$$

The shaded combination $\rho c_p \kappa$ occurs in any heat flow driven by a temperature gradient. It is a powerful abstraction called the thermal conductivity K:

$$\underbrace{\text{thermal conductivity}}_{K} = \underbrace{\text{density}}_{\rho} \times \underbrace{\left(\begin{array}{c} \text{specific} \\ \text{heat} \end{array} \right)}_{c_p} \times \underbrace{\left(\begin{array}{c} \text{thermal} \\ \text{diffusivity} \end{array} \right)}_{\kappa}. \tag{7.67}$$

Thermal conductivity has dimensions of power per length per temperature, and is often quoted in units of watts per meter kelvin ($\text{W m}^{-1} \text{K}^{-1}$). Using K, Fick's law for energy flux in terms of temperature gradient $\Delta T / \Delta x$ becomes

$$F = K \frac{\Delta T}{\Delta x},$$
(7.68)

where ΔT is the temperature difference, and Δx is the gap size.

Now let's estimate a few important thermal conductivities in order to understand heat flow around us.

7.4.3 Thermal conductivity of air

To estimate our heat loss standing outside on a cold winter's day, we need to estimate the thermal conductivity of air.

▷ *Why do we estimate the thermal conductivity of air rather than of clothing?*

The purpose of clothing is to trap air so that heat flows via conduction—that is, by diffusion—rather than via the faster process of convection. (If the perfume molecules of Section 7.3.1 could be similarly limited to diffusion, the perfume aromas would travel very slowly.)

Because $K \equiv \rho c_p \kappa$, estimating K splits into three subproblems, one for each factor. The density of air ρ_{air} is just 1.2 kilograms per cubic meter (slightly more accurate than 1 kilogram per cubic meter). The thermal diffusivity κ_{air} is 1.5×10^{-5} square meters per second.

The specific heat c_p is not as familiar, but we can estimate it. As for water, it measures the thermal energy per mass per temperature:

$$c_p = \frac{\text{thermal energy}}{\text{mass} \times \text{temperature}}.$$
(7.69)

The thermal energy per particle is comparable to $k_B T$, where k_B is Boltzmann's constant, so the energy per temperature is comparable to k_B. Thus,

$$c_p \sim \frac{k_B}{\text{mass}}.$$
(7.70)

Our thermal energy, which is comparable to $k_B T$, is for one particle. The corresponding mass in the denominator is the mass of one particle, which is an air molecule.

To convert k_B and the mass to human-sized values, we multiply each by Avogadro's number N_A. Then we replace $k_B N_A$ with the universal gas constant R, and mass $\times N_A$ with the molar mass m_{molar}. The result is

$$c_p \sim \frac{k_B N_A}{\text{molecular mass} \times N_A} = \frac{R}{m_{molar}},$$
(7.71)

where m_{molar} is the molar mass of air.

This expression is correct except for a dimensionless prefactor. Air is mostly nitrogen, for which the prefactor is $7/2$. This magic number can be made slightly less magical as follows:

$$\frac{7}{2} = \frac{\text{spatial dimensions}}{2} + \frac{\text{rotational directions}}{2} + 1. \tag{7.72}$$

Each spatial dimension contributes one $1/2$ through a term in the translational kinetic energy of a nitrogen molecule; thus, the translational piece contributes $3/2$ to the prefactor. Each rotational direction contributes a $1/2$ through a term in the rotational energy of a nitrogen molecule. Because nitrogen is a linear molecule, there are only two rotational directions, so the rotational directions contribute $2/2$ to the prefactor. If we stop here, at $5/2$, we would have the prefactor to find c_v, the specific heat while holding the volume constant. The final term, $+1$, accounts for the energy required to expand a gas as it is heated and held at constant pressure. Therefore, c_p gets a prefactor of $7/2$. (For a more detailed discussion of the reasoning behind this calculation, see the classic *Gases, Liquids and Solids and Other States of Matter* [43, pp. 106].)

Air is mostly diatomic nitrogen, so m_{molar} is roughly 30 grams per mole. Then c_p is roughly 10^3 joules per kilogram kelvin:

$$c_p = \frac{7}{2} \times \frac{R}{m_{molar}} \approx \frac{7}{2} \times \underbrace{\frac{8\,J}{mol\,K}}_{R} \times \underbrace{\frac{1\,mol}{3 \times 10^{-2}\,kg}}_{m_{molar}^{-1}} \approx \frac{10^3\,J}{kg\,K}. \tag{7.73}$$

Putting together the three pieces,

$$K_{air} \approx \underbrace{1.2\,kg\,m^{-3}}_{\rho_{air}} \times \underbrace{10^3\,J\,kg^{-1}}_{c_p} \times \underbrace{1.5 \times 10^{-5}\,m^2\,s^{-1}}_{\kappa_{air}} \approx 0.02\,\frac{W}{m\,K}. \tag{7.74}$$

Before using the thermal conductivity, let's try out the specific heat of air on an old method of air conditioning. One summer I lived in a tiny Manhattan apartment (30 square meters). Summers are hot in New York City, and the beautiful people flee for the cooler beach areas—cooler thanks partly to the high specific heat of water (Problem 7.27). Because of global warming and the old electrical wiring in the apartment building, too old to handle an air-conditioning unit, the apartment reached 30 °C at night. A friend who grew up before air conditioning suggested taking a wet sheet and using a fan to blow air past it.

▶ *How much does this system cool the apartment?*

As heat from the room air evaporates the water from the sheet, the room air cools, as if the room were sweating. The energy required to evaporate the water is

$$E = m_{water}L_{vap}, \tag{7.75}$$

where m_{water} is the mass of water held by the wet sheet and L_{vap} is the heat of vaporization of water. This energy is the demand.

To lower the room's temperature by ΔT requires using up a thermal energy

$$E \sim \rho_{air}Vc_p\Delta T, \tag{7.76}$$

where V is the volume of the room and c_p is the specific heat of air. This energy is the supply.

Equating supply and demand gives an equation for ΔT:

$$\rho_{air}Vc_p\Delta T \sim m_{water}L_{vap}. \tag{7.77}$$

Its solution is

$$\Delta T \sim \frac{m_{water}L_{vap}}{\rho_{air}Vc_p}. \tag{7.78}$$

Now we estimate and plug in the needed values. A typical room is roughly 3 meters high, so the apartment's volume was roughly 100 cubic meters:

$$V \sim \underbrace{30\,m^2}_{area} \times \underbrace{3\,m}_{height} \approx 100\,m^3. \tag{7.79}$$

To estimate m_{water}, I imagined that the sheet was about as wet as it would be after coming out of the washing machine with its fast spin cycle. That mass feels like slightly more than 1 kilogram, so $m_{water} \sim 1$ kilogram. Finally, the heat of vaporization of water is about 2×10^6 joules per kilogram (as we estimated in Section 1.7.3 using a home experiment).

Putting in all the numbers, ΔT is about 20 °C (or 20 K):

$$\Delta T \sim \frac{\overbrace{1\,kg}^{m_{water}} \times \overbrace{2\times10^6\,J\,kg^{-1}}^{L_{vap}}}{\underbrace{1\,kg\,m^{-3}}_{\rho_{air}} \times \underbrace{100\,m^3}_{V} \times \underbrace{10^3\,J\,kg^{-1}\,K^{-1}}_{c_p}} = 20\,K. \tag{7.80}$$

This change would have turned the hot 30 °C room into a cold 10 °C room, if the cooling had been 100-percent efficient. Because some heat comes from the walls (and from the fan motor), ΔT will be less than 20 °C—perhaps 10 °C, leaving the room at a pleasant and sleepable temperature of 20 °C. This calculation shows not only that evaporative cooling is a reasonable method of air conditioning, but also that our estimate for the specific heat of air is reasonable.

> **Problem 7.27   Dimensionless specific heat of water**
> The specific heat per air molecule is $3.5k_B$, so the dimensionless specific heat of air, measured relative to k_B, is 3.5. What is the dimensionless specific heat of water?

7.4.4 Keeping warm on a cold day

Now we have assembled the pieces to understand why we dress warmly on a cold day. Our starting point is the heat flux:

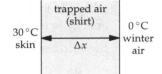

$$F = K \frac{\Delta T}{\Delta x},$$ (7.81)

where $\Delta T = T_2 - T_1$ is the temperature difference across a gap, Δx is the gap size, and K is the thermal conductivity of the gap material. Here, the gap material is air—the clothing serves to trap the air.

Let's say that the air outside is at $T_1 = 0$ °C and that skin is at $T_2 = 30$ °C (slightly lower than the internal body temperature of 37 °C). Then $\Delta T = 30$ K. Against the advice of your elders, you dress in a thin T-shirt—for decency, a very long one. A thin T-shirt has thickness Δx of roughly 2 millimeters. With these parameters, the heat flux through the shirt becomes 300 watts per square meter:

$$F \approx \underbrace{0.02 \, \frac{W}{m\,K}}_{K_{air}} \times \underbrace{\frac{\overbrace{30\,K}^{\Delta T}}{2 \times 10^{-3}\,m}}_{\Delta x} = 300\,W\,m^{-2}.$$ (7.82)

Flux is power per area, so the energy flow—the power—is the flux times a person's surface area. A person is roughly 2 meters tall and 0.5 meters wide, with a front and a back, so the surface area is about 2 square meters. Thus, the power (the energy outflow) is 600 watts.

▷ *Is this heat loss worrisome?*

Even though 600 might seem like a large number, we cannot therefore conclude that 600 watts is a large heat loss: As we learned in Chapter 5, a quantity with dimensions, such as heat flux, cannot be large or small on its own. It needs to be compared to a relevant quantity with the same dimensions. A relevant quantity is a normal human power output. When sitting around, a person produces 100 watts of heat; that's our basal metabolic rate. If 600 watts is escaping through your clothing, you are losing heat much faster than the basal metabolism is producing it. No wonder you feel so cold on a winter day wearing only a thin shirt and pants. Eventually, your core body temperature falls. Then essential chemical reactions in your body slow down, because the enzymes lose their shape optimized for body temperature and thereby become less efficient catalysts. Eventually you get hypothermia and, if it goes on too long, die.

One solution is to generate heat to make up the difference: by shivering or exercising. Cycling hard, which generates, say, 200 watts of mechanical power and another 600 watts of heat (thanks to the one-fourth metabolic efficiency), should be vigorous-enough exercise to keep you warm, even on a winter day in thin clothing.

Another simple solution is to dress warmly by putting on thick layers. Let's recalculate the power loss if you put on a jacket and thick pants, each 2 centimeters thick. We could redo the power calculation from scratch, but that approach is brute force. It is simpler to notice that the gap thickness Δx has increased by a factor of 10, yet nothing else changed. Because flux is inversely proportional to the gap size, the flux and the power drop by the same factor of 10. Therefore, wearing thick clothing reduces the energy outflow to a manageable 60 watts—comparable to the basal metabolism. As a result, your body heat can keep you warm. Indeed, when wearing thick clothing, only areas exposed directly to cold air, such as your hands and face, feel cold. Those regions are protected by only a thin layer of still air (the boundary layer analyzed in Section 7.3.4).

A thick gap means a small heat flux: When it is cold, bundle up!

Problem 7.28 Thermal conductivity of helium gas
Estimate the thermal conductivity of helium at standard temperature and pressure. The following fact will help you estimate the mean free path: The density of liquid helium is 125 grams per liter.

Problem 7.29 Comfortable outdoor temperature
You wear only that long thin T-shirt in which the winter temperature of 0 °C felt too cold. Estimate the outside temperature that would feel most comfortable.

7.4.5 Getting your clothes wet: More thermal conductivities

If your thick warm coat gets wet, you feel very cold. Let's use our knowledge of heat flow to explain why the coat becomes so useless. As we know from Section 7.4.4, you feel cold when the dimensionless ratio

$$\frac{\text{energy flow through your coat}}{\text{rate at which your body generates heat}} \tag{7.83}$$

is significantly larger than 1. The dry coat kept the energy flow (a power) comparable to the rate at which your body generated heat. Wetting the coat must increase the energy flow significantly. To see how, let's look again at the terms in the energy flow and apply proportional reasoning.

$$\text{energy flow} = \text{area} \times \underbrace{K\frac{\Delta T}{\Delta x}}_{\text{flux}}. \tag{7.84}$$

In comparing the wet to the dry coat, the area is unchanged. The temperature difference ΔT is also unchanged: It is still the 30 °C between skin temperature and winter-air temperature. The gap Δx, which is the thickness of the coat, is also unchanged.

The remaining possibility is that the thermal conductivity K of the gap material has increased significantly. If so, the thermal conductivity of water must be much higher than the thermal conductivity of air. Rather than studying water directly, let's first estimate the thermal conductivity of nonmetallic solids. (Metals have an even higher thermal conductivity, as we will discuss after studying water.) Using that estimate, we will estimate the thermal conductivity of water.

The problem is again comparative (proportional reasoning):

$$\frac{\text{thermal conductivity of nonmetallic solids}}{\text{thermal conductivity of air}}. \tag{7.85}$$

The ratio breaks into three ratios, corresponding to the three factors in the thermal conductivity (divide-and-conquer reasoning):

$$K = \underbrace{\text{density}}_{\rho} \times \underbrace{\text{specific heat}}_{c_p} \times \underbrace{\text{thermal diffusivity}}_{\kappa}. \tag{7.86}$$

Rather than using these factors directly, let's remix the first two (ρc_p) into a more insightful combination. The specific heat itself is

$$c_p = \frac{\text{energy}}{\text{mass} \times \Delta T}, \tag{7.87}$$

where ΔT is the temperature change.

Multiplying by the density ρ points us toward an interpretation of ρc_p:

$$\rho c_p = \underbrace{\frac{mass}{volume}}_{\rho} \times \underbrace{\frac{energy}{mass \times \Delta T}}_{c_p} = \frac{energy}{volume \times \Delta T}. \tag{7.88}$$

On the right side, the ratio of energy to volume can be subdivided as

$$\frac{energy}{volume} = \frac{energy}{mole} \times \left(\frac{volume}{mole}\right)^{-1}. \tag{7.89}$$

After all these reinterpretations, our remix of ρc_p becomes

$$\rho c_p = \frac{energy}{mole \times \Delta T} \times \left(\frac{volume}{mole}\right)^{-1}. \tag{7.90}$$

The first factor is the molar specific heat; it is usually denoted C_p (with a capital "C" to distinguish it from the usual, per-mass specific heat c_p). The volume per mole is also called the molar volume; it is usually denoted V_m. Therefore, $\rho c_p = C_p / V_m$. Then the thermal conductivity becomes

$$K = \rho c_p \kappa = \frac{C_p}{V_m} \kappa. \tag{7.91}$$

In words, the thermal conductivity K is

$$\frac{molar\ specific\ heat\ C_p}{molar\ volume\ V_m} \times thermal\ diffusivity\ \kappa. \tag{7.92}$$

This remix is more insightful than simply $\rho c_p \kappa$ because C_p varies less between substances than c_p does, and V_m varies less between substances than ρ does. The remixed form produces less quantity whiplash, in which the product swings wildly up and down as we include each factor. Another way to express the same advantage is that C_p and V_m are less correlated than are ρ and c_p; therefore, the remixed abstractions C_p and V_m offer more insight into the thermal properties of materials than do c_p and ρ.

Using the remixed form, let's apply divide-and-conquer reasoning and estimate the ratio corresponding to each factor.

1. *Ratio of molar specific heats C_p.* For air, C_p was $3.5R$. For most solids, whether metallic or nonmetallic, it is similar: $3R$ (where the 3 reflects the three spatial dimensions). Thus, this ratio is close to 1.

2. *Ratio of molar volumes V_m.* For any substance, 1 mole has a mass of A grams, where A is the dimensionless atomic mass (roughly, the number of protons and neutrons in the nucleus). Because a typical solid

density is A grams per 18 cubic centimeters (Section 6.4.1), the molar volume of the solid is 18 cubic centimeters per mole. In contrast, for air or any ideal gas (at standard temperature and pressure), 1 mole occupies 22 liters or 22 000 cubic centimeters. Thus, the ratio of molar volumes (solid to air) is 18/22 000 or roughly 10^{-3}.

3. *Ratio of thermal diffusivities κ.* From Section 7.3.3, this ratio is roughly 0.1:

$$\frac{\text{thermal diffusivity of nonmetallic solids}}{\text{thermal diffusivity of air}} \approx \frac{10^{-6}\,\text{m}^2\,\text{s}^{-1}}{10^{-5}\,\text{m}^2\,\text{s}^{-1}} \tag{7.93}$$
$$= 0.1.$$

The ratio of thermal conductivities is the product of the three ratios. As long as we remember that the molar volume appears in the denominator (the thermal conductivity is inversely proportional to the molar volume), we get a ratio of 100:

$$\frac{K_{\text{nonmetallic solid}}}{K_{\text{air}}} \sim \underbrace{1}_{C_\text{p}\,\text{ratio}} \times \underbrace{10^3}_{(V_\text{m}\,\text{ratio})^{-1}} \times \underbrace{10^{-1}}_{\kappa\,\text{ratio}} = 10^2. \tag{7.94}$$

Thus, in contrast to the thermal conductivity of 0.02 watts per meter kelvin for air, the typical (nonmetallic) solid has a thermal conductivity of about 2 watts per meter kelvin.

Now let's use proportional reasoning to compare this thermal conductivity to the thermal conductivity of water, which is the gap material in your wet coat. We'll estimate $K_{\text{water}}/K_{\text{nonmetallic solid}}$. The comparison has the same three ratios: molar specific heat, molar volume, and thermal diffusivity.

1. *Ratio of molar specific heats C_p.* We can find the specific heat of water from the definition of a calorie, as the energy required to raise 1 gram of water by 1 degree (celsius or kelvin). Thus, the regular specific heat is simple to state:

$$c_\text{p}^{\text{water}} = \frac{1\,\text{cal}}{\text{g}\,\text{K}}. \tag{7.95}$$

The resulting molar specific heat is 72 joules per mole kelvin:

$$C_\text{p}^{\text{water}} \approx \underbrace{\frac{1\,\text{cal}}{\text{g}\,\text{K}}}_{c_\text{p}} \times \underbrace{\frac{4\,\text{J}}{1\,\text{cal}}}_{1} \times \underbrace{\frac{18\,\text{g}}{\text{mol}}}_{m_{\text{molar}}} = \frac{72\,\text{J}}{\text{mol}\,\text{K}}. \tag{7.96}$$

The dimensionless specific heat $C_\text{p}^{\text{water}}/R$ is therefore approximately 9 (as you found in Problem 7.27):

$$\frac{C_{\text{p}}^{\text{water}}}{R} \approx \frac{72\,\text{J}\,\text{mol}^{-1}\,\text{K}^{-1}}{8\,\text{J}\,\text{mol}^{-1}\,\text{K}^{-1}} = 9. \tag{7.97}$$

For a nonmetallic solid, the dimensionless specific heat was only 3; it is a factor of 3 smaller than for water. Water stores heat very efficiently, which is why it is used as coolant fluid and why coastal weather is milder than inland weather. The ratio of molar specific heats is 3.

2. *Ratio of molar volumes V_m.* The molar volume of water, 18 cubic centimeters per mole, is identical to the canonical molar volume of a solid. Thus, the molar-volume ratio is 1.

3. *Ratio of thermal diffusivities κ.* As we found in Section 7.3.3 by comparing phonon mean free paths and propagation speeds, the thermal diffusivity κ_{water} is roughly a factor of 10 smaller than the thermal diffusivity of a typical solid (10^{-7} compared to 10^{-6} square meters per second). Thus, the thermal-diffusivity ratio contributes a factor of 0.1.

These three ratios result in the following comparison:

$$\frac{K_{\text{water}}}{K_{\text{nonmetallic solid}}} \approx \underbrace{3}_{C_{\text{p}} \text{ ratio}} \times \underbrace{1}_{(V_{\text{m}} \text{ ratio})^{-1}} \times \underbrace{0.1}_{\kappa \text{ ratio}} \approx 0.3. \tag{7.98}$$

In absolute terms,

$$K_{\text{water}} \approx 0.3 \times \underbrace{\frac{2\,\text{W}}{\text{m}\,\text{K}}}_{K_{\text{nonmetallic solid}}} = \frac{0.6\,\text{W}}{\text{m}\,\text{K}}. \tag{7.99}$$

This conductivity is a factor of 30 larger than K_{air}. As a result, wearing wet clothes on a cold day is so unpleasant and can even be dangerous. Thick clothing (a coat) allowed a comfortable 60 watts of heat flow—a factor of 10 lower than the T-shirt allowed. Wetting the thick coat increases the thermal conductivity by a factor of 30. The heat loss therefore increases by a factor of 30—making it higher even than the heat loss through the dry T-shirt. When you hike in the hills and mountains, bring waterproof clothing!

The table gives thermal conductivities for everyday substances (at room temperature). Our predictions for nonmetals match the data quite well. Having examined gases (in particular, air) and nonmetallic solids and liquids (water), let's turn to the remaining category of material. As the table shows, metals have an even higher thermal conductivity than a typical nonconducting solid. (For the unusually high thermal conductivity of diamond, try Problem 7.33.)

	$K\left(\frac{\text{W}}{\text{m K}}\right)$
diamond	2000
Cu	400
Ag	350
Al	240
cast iron	55
Hg	8.3
ice	2.2
sandstone	1.7
glass	1
asphalt	0.8
brick	0.8
concrete	0.6
water	0.6
soil (dry)	0.5
wood	0.15
He (gas)	0.14
methane (gas)	0.03
air	0.02

Metals, similarly, have a higher thermal diffusivity than most other substances. The reason, as we discussed in Section 7.3.3, is that, in a metal, heat is conducted not only by phonons but also by electrons. The electrons move much faster than the speed of sound, and the electron mean free path is much greater than the phonon mean free path.

Countering these increases, which increase the diffusivity, only a small fraction of the free electrons participate in heat conduction. However, this effect is not enough to overcome the greater speed and mean free path. Thus, metals will have a higher thermal conductivity than non-metallic liquids or solids. As a rule of thumb, the typical K_{metal} is 200 watts per meter kelvin: a factor of 100 higher than that of nonmetallic solid.

For this reason, a hot piece of metal, such as a seat-belt clip in a car outside on a hot day, feels much hotter than the plastic button on the same seat-belt clip, even though the plastic and metal are at the same temperature. The large heat flow from the metal into your finger pulls the surface temperature of your finger close to the temperature of the hot metal. Ouch!

Problem 7.30   Stone versus wood floors
Why, on a winter morning, do wood floors feel more comfortable than stone floors?

Problem 7.31   Mercury is special
Why does mercury (Hg) have such a low thermal conductivity for a metal?

7.5 Summary and further problems

In large, complex systems, the information is either overwhelming or not available. Then we have to reason with incomplete information. The tool for this purpose is probabilistic reasoning—in particular, Bayesian probability. Probabilistic reasoning helps us manage incomplete information. Using it, we can estimate the uncertainty in our divide-and-conquer estimates and understand the physics of random walks and thereby viscosity, boundary layers, and heat flow.

Problem 7.32   Reynolds number as a ratio of two times

For an object moving through a fluid, the Reynolds number is defined as vL/ν, where v is the object's speed, L is its size (a length), and ν is the fluid's kinematic viscosity. Show that the Reynolds number has the physical interpretation

$$\frac{\text{momentum-diffusion time over a distance comparable to the size } L}{\text{fluid-transport time over a distance comparable to the size } L}. \quad (7.100)$$

Problem 7.33   Diamond is special

Diamond has a high thermal conductivity, much higher even than many metals. The speed of sound in diamond is 12 kilometers per second, and diamond's specific heat c_p is 0.63 kilojoules per kilogram kelvin. Use those values to estimate the mean free path of phonons in diamond, as an absolute length and in units of typical interatomic spacings. How does the mean free path in diamond compare to a typical phonon mean free path of a few lattice spacings?

Problem 7.34   Baking in three dimensions

Extend the fish-cooking argument of Section 7.3.3.2 to three dimensions to predict the baking time of a 6-kilogram turkey (assumed to be a sphere). How well does the time agree with experience (for example, with the data given in Problem 7.22)?

Problem 7.35   Resistive networks to analyze random walks

Random walks are closely connected to infinite resistive networks (this connection is explored deeply in *Random Walks and Electric Networks* [11]). In particular, the probability of escape p_{esc}—the probability that an n-dimensional random walker escapes to infinity and never returns to the origin—is related to the resistance R to infinity of a n-dimensional electrical network of unit resistors: $p_{esc} = 1/2nR$. Use this connection, along with lumping arguments, to estimate R and thereby show that the two-dimensional random walk is recurrent ($p_{esc} = 0$) but that the three-dimensional walk is transient ($p_{esc} > 0$)—consistent with Pólya's theorem (Problem 7.17).

Problem 7.36   Turning differential equations into algebraic equations

The cold days of winter arrive, and the ice on a lake starts thickening as heat flows upward through the ice, turning ever more water into ice. Find the scaling exponent β in

$$\text{ice thickness} \propto (\text{time})^\beta. \quad (7.101)$$

cold air ($< 0\,^\circ$C)

heat

ice

water ($0\,^\circ$C)

Problem 7.37   Thermal and electrical conductivities

Among the metals, the better thermal conductors, such as copper and gold in comparison to aluminum, iron, or mercury, are also the better electrical conductors. (This connection is quantified in the Wiedemann—Franz law.) What is the reason for this connection?

Problem 7.38   Teacup spindown

You stir your afternoon tea to mix the milk (and sugar, if you have a sweet tooth). Once you remove the stirring spoon, the rotation begins to slow. In this problem you'll estimate the spindown time τ: the time for the angular velocity of the tea to fall by a significant fraction. To estimate τ, consider a lumped teacup: a cylinder with height l and diameter l, filled with liquid. Tea near the edge of the teacup—and near the base, but for simplicity neglect the effect of the base—is slowed by the presence of the edge (a result of the no-slip boundary condition).

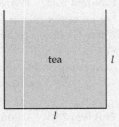

a. In terms of the viscous torque T, the initial angular velocity ω, and ρ and l, estimate the spindown time τ. Hint: Consider angular momentum, and drop all dimensionless constants, such as π and 2.

b. To estimate the viscous torque T, use the result of Problem 7.25:

$$\text{viscous force} = \rho v \times \text{velocity gradient} \times \text{surface area.} \qquad (7.102)$$

The velocity gradient is determined by the boundary-layer thickness δ. In terms of δ, estimate the velocity gradient near the edge and then the torque T.

c. Put your expression for T into your earlier estimate for τ, which should now contain only one quantity that you have not yet estimated (the boundary-layer thickness δ).

d. Estimate δ in terms of a growth time t, which is the time to rotate 1 radian. After 1 radian, the fluid is moving in a significantly different direction, so the momentum fluxes from different regions no longer add constructively to the growth of the boundary layer.

e. Put the preceding results together to estimate the spindown time τ: symbolically in terms of v, ρ, l, and ω; and then numerically.

f. Stir your tea and estimate τ experimentally, and compare with your prediction. Then enjoy a well-deserved cup.

8
Easy cases

A correct analysis works in all cases—including the simplest ones. This principle is the basis of our next tool for discarding complexity: the method of easy cases. We will meet the transferable ideas in an everyday example (Section 8.1.1). Then we will use them to simplify and understand complex phenomena, including black holes (Section 8.2.2.2), the temperature of the Sun (Section 8.3.2.3), and the diversity of water waves (Section 8.4.1).

8.1 Warming up

Let's start with an everyday example, so that we do not have to handle mathematical or physical complexity along with learning the new tool.

8.1.1 Everyday example of easy cases

One August, upon becoming eligible for one of the tax benefits at work, I chose to put $500 in an account for the calendar year's health costs (a feature of America's bureaucratic health system). The payroll office advised me that they would deduct $125 each month for the rest of the year—namely, for August through December. As August is the eighth month and December the twelfth month, the amount looked correct:

$$\frac{\$500}{\text{12th month} - \text{8th month}} = \frac{\$500}{4\ \text{months}} = \frac{\$125}{\text{month}}. \tag{8.1}$$

However, in January, the payroll office advised me that they should have deducted only $100 per month.

▶ *Which amount was correct?*

The simplest way to decide is the method of easy cases. Don't directly solve the hard problem of computing the correct deduction in general. Instead, imagine a simpler world in which I started deducting for the year in December. In this easy case, with only one month providing the year's deduction, the answer requires no calculation: Deduct $500 per month.

Then use the easy case and this result to check the proposed recipe, which predicted $125 per month when deductions started in August. In the easy case, when deductions start in December, the starting and ending months are both the twelfth month, so the recipe predicts an infinite deduction:

$$\frac{\$500}{\text{12th month} - \text{12th month}} = \frac{\$500}{0\ \text{months}} = \frac{\$\,\text{infinity}}{\text{month}}. \tag{8.2}$$

They do not pay me enough to survive that recipe. It needs an adjustment:

$$\frac{\$500}{\text{12th month} - \text{12th month} + 1\ \text{month}} = \frac{\$500}{1\ \text{month}} = \frac{\$500}{\text{month}}. \tag{8.3}$$

Applying this modified recipe to an August instead of a December start, the denominator is 5 months rather than 4 months, and the deduction is $100 per month. The revised advice from the payroll office was correct.

This analysis contains several features that we can abstract away and use to simplify difficult problems. First, the easy cases are specified by values of a dimensionless quantity. Here, it is the difference $m_2 - m_1$ between the first and last month numbers. Second, for particular values of the dimensionless quantity—for the easy cases—the problem has an obvious answer. Here, the easy case is $m_2 - m_1 = 0$. Third, understanding the easy cases transfers to the hard cases. Here, our understanding of the case $m_2 - m_1 = 0$ shows that the denominator should be $m_2 - m_1 + 1$ rather than $m_2 - m_1$.

8.1.2 Easy cases of the shared-birthday probability

The same easy-cases reasoning helps us check more abstruse formulas. As an example, cast your mind back to the birthday paradox of Section 4.4.

There, we used proportional reasoning to explain why we need only 23 people, rather than 183 people, in a room before two people are likely to share a birthday. Checking the prediction required the exact probability of a shared birthday:

$$p_{\text{shared}} = 1 - \left(1 - \frac{1}{365}\right)\left(1 - \frac{2}{365}\right)\cdots\left(1 - \frac{(n-1)}{365}\right). \tag{8.4}$$

The last part of Problem 4.23 asked why, with n people in the room, the last factor in this probability contains $(n-1)/365$ rather than $n/365$.

The simplest answer comes from the easy case $n = 1$. With one person in the room, the probability of sharing a birthday is zero. In this easy case, the prediction of the candidate formula with n in the numerator is easy to test. Its last factor would be $1 - 1/365$:

$$p_{\text{candidate with } n} = 1 - \left(1 - \frac{1}{365}\right). \tag{8.5}$$

This probability is $1/365$, which is incorrect. Thus, with n people in the room, the last factor in the probability of a shared birthday cannot contain $n/365$. In contrast, the candidate formula with $(n-1)/365$ in the last factor correctly predicts zero probability when $n = 1$:

$$p_{\text{candidate with } n-1} = 1 - \left(1 - \frac{0}{365}\right) = 0. \tag{8.6}$$

This candidate must be correct.

Problem 8.1   Sine or cosine by easy cases

For a mass sliding down an inclined plane, is its acceleration along the plane $g \sin\theta$ or $g\cos\theta$, where θ is the angle the plane makes with horizontal? Use an easy case to decide. Assume (the easy case of!) zero friction.

Problem 8.2   Friction by easy cases

For the mass of Problem 8.1, relax the assumption of zero friction. If the coefficient of sliding friction is μ, choose the block's correct acceleration along the plane: (a) $g(\sin\theta + \mu\cos\theta)$, (b) $g(1 + \mu)\sin\theta$, (c) $g(\sin\theta - \mu\cos\theta)$, or (d) $g(1 - \mu)\sin\theta$.

8.2 Two regimes

After warming up with the examples in Section 8.1, let's now look systematically at how to use the method of easy cases. The first step is to identify the dimensionless quantity. Once you know it—let's call it β—the behavior of the system almost always divides into three regimes: $\beta \ll 1$, $\beta \sim 1$ (or $\beta = 1$), and $\beta \gg 1$. In a common simplification, which is the subject of this section, one of the three regimes is impossible, so two regimes remain. (In Section 8.3, we will discuss what to do when all three regimes remain.) This simplification can happen because symmetry makes the bookend regimes $\beta \ll 1$ and $\beta \gg 1$ equivalent (Section 8.2.1) or because a geometric or physical constraint excludes one regime (Section 8.2.2).

8.2.1 Only two regimes because of symmetry

The bookend regimes $\beta \ll 1$ and $\beta \gg 1$ are often connected by a symmetry. Then the analysis at one bookend can be used as the analysis for the other bookend, and there are really two regimes: the symmetric bookends and the middle regime.

8.2.1.1 *Why multiplication is more important than addition*

As an example, here's an easy-cases explanation of why, when estimating, multiplication is a more important operation than addition. Let's say that we are estimating a cost, a force, or an energy consumption that splits into two pieces A and B to add together—for example, energy for heating and for transportation. Estimating $A + B$ seems to require addition.

The need disappears after we examine the easy-cases regimes. The regimes are determined by a dimensionless quantity. Because A and B have the same dimensions, their ratio A/B is dimensionless, and it categorizes the three easy-cases regimes.

Regime 1	*Regime 2*	*Regime 3*
$A \ll B$	$A \sim B$	$A \gg B$
$A/B \ll 1$	$A/B \sim 1$	$A/B \gg 1$

In the first regime, the sum $A + B$ is approximately B. In the symmetric third regime, the sum is approximately A. The common feature of the first and third regimes—their invariant—is that one contribution dominates the other. Only the second regime, where $A \sim B$, is different. To handle it, just make the lumping approximation that $A = B$; then $A + B \approx 2A$.

So, when estimating $A + B$, we do not need addition. In the first and third regimes, we just pick the larger contribution, A or B. In the second regime, we just multiply by 2.

8.2.1.2 *Area of an ellipse*

A mathematical example of this kind of symmetry happens when guessing the area of an ellipse. The dimensionless quantity is its aspect ratio a/b, where a is one-half of the horizontal diameter and b is one-half of the vertical diameter. Here are the regimes.

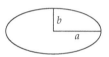

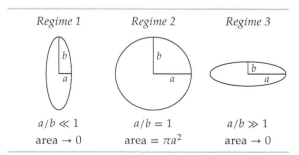

Regime 1	*Regime 2*	*Regime 3*
$a/b \ll 1$	$a/b = 1$	$a/b \gg 1$
area $\to 0$	area $= \pi a^2$	area $\to 0$

The first and third regimes are identical because of a symmetry: Interchanging a and b rotates the ellipse by 90° (and reflects it through its vertical axis) without changing its area. Therefore, any proposed formula for the area must work in the first regime, must satisfy the symmetry requirement that interchanging a and b has no effect on the area (thereby taking care of the third regime), and must work for a circle (the second regime).

Because of the symmetry requirement, simple but asymmetric modifications of the area of a circle—namely, πa^2 and πb^2—cannot be the area of an ellipse. A plausible, symmetric alternative is $\pi(a^2 + b^2)/2$. In the second regime, where $a = b$, it correctly predicts the area of a circle. However, it fails in the first regime: The area isn't zero even when $a = 0$.

▷ *Is there an alternative that passes all three tests?*

A successful alternative is πab. It predicts zero area when $a = 0$; it is symmetric; and it works when $a = b$. It is also the correct area.

8.2.1.3 *Atwood machine*

In the preceding examples, we had a ready-made dimensionless group and used it to categorize the three easy-cases regimes. The next example shows what you do when there is no group handy: Use dimensional analysis to make one. Then use the group to categorize the regimes and understand the system's behavior. Easy cases and dimensional analysis work together.

The example is the Atwood machine, a staple of the introductory mechanics curriculum. Two masses, m_1 and m_2, are connected by a massless, frictionless string and are free to move and down because of the pulley. This machine, invented by the Reverend George Atwood, reduces the effective gravitational acceleration by a known fraction and makes constant-acceleration motion easier to study. (A historical discussion of the Atwood machine is given by Thomas Greenslade in [27].) Today its principle is used in elevators: The elevator car is m_1, and the counterweight is m_2.

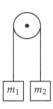

▷ *What is the acceleration of the masses?*

Our second step is to identify the easy-cases regimes, but our first step is to find the dimensionless group that categorizes them. Therefore, we make our list of relevant quantities and their dimensions. The goal is the acceleration of the masses. Because they are

a	LT^{-2}	acceleration
g	LT^{-2}	gravity
m_1	M	left block's mass
m_2	M	right block's mass

connected by a string, their accelerations have the same magnitude but opposite directions. Let a be the downward acceleration of the mass on the left—whether it is labeled m_1 or m_2 (planning for an application of symmetry). Because a is our goal quantity, it is the first quantity on the list. The motion is caused by gravity, so the list includes g. Finally, the masses affect the acceleration, so the list includes m_1 and m_2. And that's the whole list.

▷ *Doesn't the tension also affect the acceleration?*

The tension has an important effect: Without the tension, there would be no problem to solve, because each mass would accelerate downward at g. However, the tension is a consequence of m_1, m_2, and g—quantities already on the list. Thus, tension is redundant; adding it would only confuse the dimensional analysis.

This list contains only two independent dimensions: mass (M) and acceleration (LT^{-2}). Four quantities built from two independent dimensions produce two independent dimensionless groups. The most natural choices are the acceleration ratio a/g and the mass ratio m_1/m_2. Then the most general dimensionless statement is

$$\frac{a}{g} = f\left(\frac{m_1}{m_2}\right), \tag{8.7}$$

where f is a dimensionless function.

Although the next step is to use easy cases to guess the dimensionless function, first pause for a moment. The more we think now in order to choose a suitable representation, the less algebra we have to do later. Here, the ratio m_1/m_2 does not respect the symmetry of the problem. Interchanging the masses m_1 and m_2 turns m_1/m_2 into its reciprocal: It raises the ratio to the -1 power. Meanwhile, because the left mass is now m_2, which has the opposite acceleration to m_1, this symmetry operation changes the sign of the acceleration a (which measures the downward acceleration of the left mass): It multiplies the acceleration by -1. The function f will have to incorporate this change in the nature of the symmetry and will have to work hard to turn m_1/m_2 into an acceleration. Therefore, f will be complicated and hard to guess. (You can find it anyway in Problem 8.3.)

A more symmetric alternative to the ratio m_1/m_2 is the difference $m_1 - m_2$. Now interchanging m_1 and m_2 negates $m_1 - m_2$, just as it negates the acceleration. Unfortunately, $m_1 - m_2$ is not dimensionless! It has dimensions of mass. To make it dimensionless, we need to divide it by another mass, say m_1, to make $(m_1 - m_2)/m_1$. Unfortunately, that choice abandons our beloved symmetry. Dividing instead by the symmetric combination $m_1 + m_2$ solves all the problems. Let's call the result x:

$$x \equiv \frac{m_1 - m_2}{m_1 + m_2}. \tag{8.8}$$

This ratio is dimensionless and respects the problem's symmetry. With it as the dimensionless group, the most general dimensionless statement is

$$\frac{a}{g} = h\left(\frac{m_1 - m_2}{m_1 + m_2}\right), \tag{8.9}$$

where h is a dimensionless function (different from f). (This combination of m_1 and m_2 doesn't satisfy the definition of a group given in Section 5.1.1, as a dimensionless product of the quantities raised to various powers. However, it is worth extending the concept to include such combinations.)

To guess h, study the easy-cases regimes according to x. They are three: $x = -1$ (imagine that $m_1 = 0$), $x = 0$ (when $m_1 = m_2$), and $x = +1$ (imagine that $m_2 = 0$). Here you see how easy cases help us. The extremity of the first and third regimes and the simplicity of the second regime amplify our physical intuition and help our gut predict the behavior.

In the first regime, m_2 falls as if there were no mass on the other end of the string. Thus, m_1 rises with acceleration $+g$. Because a measures the downward acceleration of the left mass, $a = -g$. In the third, symmetric

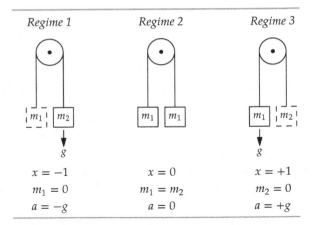

Regime 1	Regime 2	Regime 3
$x = -1$	$x = 0$	$x = +1$
$m_1 = 0$	$m_1 = m_2$	$m_2 = 0$
$a = -g$	$a = 0$	$a = +g$

regime, m_1 accelerates downward as if there were no mass on the other end, so $a = +g$. In the second regime, where $x = 0$ or $m_1 = m_2$, the masses are in equilibrium, and $a = 0$.

Here are the three easy cases plotted on a graph of the dimensionless acceleration a/g versus the dimensionless mass parameter x. Based on this graph, the simplest conjecture, an educated guess, is that the full graph of $h(x)$ is the straight line of unit slope passing through the three points. Thus, its equation is $a/g = x$. In terms of the masses, the equation is

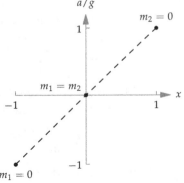

$$\frac{a}{g} = \frac{m_1 - m_2}{m_1 + m_2}. \qquad (8.10)$$

In Problem 8.4(d), you solve for the acceleration by using Newton's laws to find the tension in the string. Then you'll confirm our reasonable conjecture based on easy-cases reasoning.

Problem 8.3   Using the less symmetric dimensionless group

Find the dimensionless function f based on the simpler, but less symmetric, dimensionless group m_1/m_2:

$$\frac{a}{g} = f\left(\frac{m_1}{m_2}\right). \qquad (8.11)$$

Compare it to the dimensionless function that results from choosing the symmetric form $(m_1 - m_2)/(m_1 + m_2)$ as the independent variable.

Problem 8.4   Tension in the string in Atwood's machine
In this problem you use easy cases to guess the tension in the connecting string.

a. The tension T, like the acceleration, depends on m_1, m_2, and g. Explain why these four quantities produce two independent dimensionless groups.

b. Choose two suitable independent dimensionless groups so that you can write an equation for the tension in this form:

    group proportional to $T = f$(group without T).   (8.12)

c. Use easy cases to guess f; then sketch f.

d. Solve for T using freebody diagrams and Newton's laws. Then compare that result with your guess in part (c), and use T to find a, the downward acceleration of m_1.

8.2.2 Only two regimes because of a restriction

In the preceding examples, there were three regimes of a dimensionless quantity, but the two bookend regimes were connected by a symmetry operation and were therefore the same situation. Therefore, there were only two qualitatively different regimes. In the next group of examples, there are only two regimes for a different reason: A geometric or physical constraint excludes one bookend regime.

8.2.2.1 *Projectile range*

In Section 4.2.3, we used proportional reasoning to show that the range of a projectile should have the form $R \sim v^2/g$, where v is its launch velocity. However, we didn't determine how the launch angle θ affects the range. We can do so using easy cases, with help from dimensional analysis.

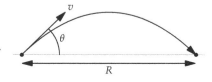

Now the problem contains four quantities: R, v, g, and θ. Four quantities containing two independent dimensions (for example, L and T) produce two independent dimensionless groups. A reasonable pair is the launch angle θ—which is already dimensionless—and, based

R	L	range
v	LT^{-1}	launch velocity
g	LT^{-2}	gravity
θ	1	launch angle

on the proportional-reasoning result, Rg/v^2. Then the most general dimensionless statement is

$$\frac{Rg}{v^2} = f(\theta),$$

(8.13)

where f is a dimensionless function. Solving for R,

$$R = \frac{v^2}{g} \times f(\theta).$$
(8.14)

Dimensional analysis does not tell us the form of f, but we can guess it by examining the easy cases. Easy-cases reasoning is a way of introducing physical knowledge.

Because angles are dimensionless, θ can categorize the easy-cases regimes. The natural easy cases of any quantity are its extremes: $\theta \ll 1$ (our usual first regime) and $\theta \gg 1$ (our usual third regime). The regime $\theta \ll 1$ includes the smallest launch angle $\theta = 0°$. Then the range is zero: The projectile starts at ground level, is launched horizontally, and hits the ground immediately. However, the opposite extreme $\theta \gg 1$ is not useful, because angles are periodic. Any angle beyond 2π is already handled by an angle less than 2π. Our usual third regime is therefore geometrically excluded.

Our usual second regime is $\theta \sim 1$. Here, the easy angle comparable to 1 (in radians) is $\theta = 90°$ ($\pi/2$ radians). This angle describes a vertical launch, which doesn't produce any range. Thus, R in this regime is also 0.

The two regimes are therefore $\theta = 0$ (an example of $\theta \ll 1$) and $\theta = \pi/2$ (an example of $\theta \sim 1$).

Regime 1	Regime 2
$\theta = 0$	$\theta = \pi/2$
$R = 0$	$R = 0$
$f(\theta) = 0$	$f(\theta) = 0$

▷ *What reasonable functions satisfy the two easy-cases constraints on $f(\theta)$?*

One such function is the product of two straight lines, each line ensuring that one of the constraints is satisfied:

$$f(\theta) = \theta\left(\frac{\pi}{2} - \theta\right).$$
(8.15)

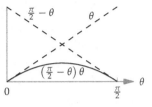

However, this functional form looks funny. The $\pi/2$ in the second factor appears by magic, with no obvious origin in the equations of free fall. Worse, the guess is not periodic. Increasing θ by 2π shouldn't change the range, but it changes this proposed f. Thus, we need to keep looking.

A similar but less magical form, also the product of two factors, is

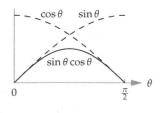

$$f(\theta) \sim \sin\theta\cos\theta. \tag{8.16}$$

The $\cos\theta$ factor ensures that $f(\pi/2) = 0$. The $\sin\theta$ factor ensures that $f(0) = 0$. This functional form is also periodic. And it has a physical justification: $\sin\theta$ is the trigonometric factor that converts the launch velocity v into its vertical component v_y; the $\cos\theta$ factor does the same for the horizontal component v_x. Because these velocities appeared in the proportional-reasoning analysis of Section 4.2.3, we already know that they are physically relevant, making this functional form more plausible than the preceding form.

The resulting range, including the effect of the angle, is then

$$R \sim \frac{v^2}{g}\sin\theta\cos\theta. \tag{8.17}$$

This form is correct, and the dimensionless prefactor turns out to be 2 (as you can show in Problem 8.5).

> **Problem 8.5   Finding the dimensionless constant**
> Extend the proportional-reasoning analysis of Section 4.2.3 to show that the missing dimensionless prefactor in the projectile range is 2.

8.2.2.2 *Light bending with large angles*

Our next example of a physical limit excluding a third regime also involves an angle: the bending of light by gravity.

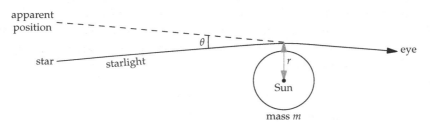

In Section 6.4.6, we used lumping to find that the gravitational field from a mass m bends light by an angle

$$\theta \sim \frac{Gm}{rc^2}, \tag{8.18}$$

where r is the distance of closest approach.

This angle is dimensionless. Thus, let's use it to categorize and to investigate the easy cases of light bending. The first regime is $\theta \ll 1$—for example, the Sun bending starlight by roughly 1 arcsecond. The second regime is $\theta \sim 1$. In this regime, a new physical phenomenon appears, and lumping will help us analyze it.

The lumping analysis is based on the American televi-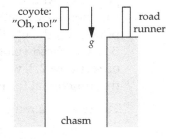
sion cartoon of the Road Runner pursued by the Wile
E. Coyote. The road runner runs across a chasm, followed by the coyote. They cruise as if gravity is gone,
until the road runner, safely across the gap, holds up
a sign with an arrow pointing downward explaining,
"Gravity this way." The coyote looks down, remembers
the existence of gravity, and falls into the chasm.

In the coyote model, light zooms past the mass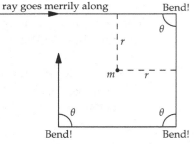
(say, a star) ignorant of gravity. After it has gone
too far—by a distance comparable to r—the star
holds up a sign saying, "You forgot about gravity! Please bend by θ now!" In this second regime,
$\theta \sim 1$. The following pictures and analysis are
clearest if $\theta = \pi/2$ (or 90°), so let's use 90° to represent the $\theta \sim 1$ regime.

The ray, on command, bends by 90°. It travels
along and gets reminded again. The distance of closest approach is still
r, so θ is still 90°. The beam traces out a square. Although the ray doesn't
actually follow this path with its sharp corners, the path illustrates the fundamental feature of the $\theta \sim 1$ regime: The ray is in orbit around the star.
When $\theta \sim 1$, light gets captured by the strong gravitational field. The third
regime, $\theta \gg 1$, doesn't introduce a new physical phenomenon—the light is
still captured by the gravitational field. In either regime, the mass is a black
hole.

Problem 8.6   Guessing a variance

The variance is a squared measure of the spread in a distribution:

$$\text{Var}\, x = \langle x^2 \rangle - \langle x \rangle^2. \tag{8.19}$$

where $\langle x^2 \rangle$ is the average or expected value of x^2 (the mean square) and $\langle x \rangle$ is the
average value of x (the mean).

a. Use an easy case to convince yourself that the mean square $\langle x^2 \rangle$ is never smaller than the squared mean $\langle x \rangle^2$.

b. The simplest possible distribution has only two possibilities, $x = 0$ and $x = 1$, with probabilities $1 - p$ and p, respectively. This distribution describes a coin toss where heads (represented by $x = 1$) comes up with probability p. Use easy cases to guess Var x.

Problem 8.7   Local black hole
What is roughly the largest radius that the Earth could have, with its current mass, and be a black hole?

8.3 Three regimes

Two regimes are easier than three. Therefore, we studied that case first to develop experience and identify the transferable ideas. Fortunately, even when a complex situation has three regimes, two simplifications are possible. First, the bookend regimes are often easier than the middle regime (Section 8.3.1). Then we study the bookends and, to predict the behavior in the middle, interpolate between the bookends. Alternatively, two effects compete and reach a draw in the middle regime (Section 8.3.2). This middle regime is then the regime found in nature.

8.3.1 Three regimes where the bookends are easier

In a common easy-cases situation, the dimensionless number categorizing the three regimes is a ratio between two physical effects. Then the two bookends are the easiest regimes to analyze—because at each bookend, one or the other effect vanishes. We'll practice this analysis in an example from introductory mechanics (Section 8.3.1.1); then we'll graduate to drag (Section 8.3.1.2).

8.3.1.1 Rolling down the plane

As our introductory example, let's revisit objects rolling without slipping down an inclined plane. The goal will be to predict an object's acceleration a along the plane. Dimensional analysis, from Problem 5.18, tells us that the most general dimensionless statement about a is

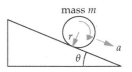

$$\frac{a}{g} = f\left(\theta, \frac{I}{mr^2}\right),$$

(8.20)

where f is a dimensionless function, θ is the incline angle, I is the object's moment of inertia, m is its mass, and r is its radius. The only quantities that affect the acceleration are the two dimensionless groups: the incline angle θ and the dimensionless mass distribution I/mr^2. The ratio I/mr^2, and therefore the acceleration, is invariant under changes to the object's mass or radius (for example, making a bigger ring or disc). In plainer language, which however doesn't connect as much to symmetry reasoning, simply changing the object's mass or radius without changing its shape does not change its acceleration.

▷ *Which shape rolls faster—a ring or a disc?*

In this comparison, the inclined plane and thus the incline angle θ remain fixed. However, the dimensionless group I/mr^2 changes and can categorize the easy-cases regimes. This group occurs repeatedly in the analysis, so we'll often symbolize it as β. By understanding the behavior in the regimes defined by β, we'll be able to predict the result of the ring–disc race.

Let's start with the first regime, $I/mr^2 = 0$. This regime is reached by making I zero. Then rolling, represented by I, becomes irrelevant. The object moves as if it were sliding down the plane without friction. Its acceleration is then $g \sin \theta$, and $a/g = \sin \theta$.

With this easy-cases result, we can simplify the unknown function f, which unfortunately is a function of two dimensionless groups. As a reasonable conjecture, the dependence on angle is probably still $\sin \theta$, even when I is not zero. Then the dimensionless statement simplifies to

$$\frac{a}{g} = h\left(\frac{I}{mr^2}\right) \sin \theta, \tag{8.21}$$

where h is a dimensionless function of only one group.

Let's guess h using easy cases of $\beta = I/mr^2$. The three regimes will be $\beta = 0$ (a special case of $\beta \ll 1$), $\beta \sim 1$, and $\beta \gg 1$. In the first regime, when $\beta = 0$, rolling is unimportant, as we just observed. Therefore, $a/g = \sin \theta$ and the dimensionless function $h(\beta)$ is just 1.

The middle regime $\beta \sim 1$ is difficult to analyze without a full calculation. Yet the goal of an easy-cases analysis is to make the situation simple enough that the behavior is easy to see and we can skip the full calculation. Even the simplest value within this regime, $\beta = 1$, is not any easier than other values. (It describes the ring, where all mass is distributed at the edge of the object, at a distance r from the center). The middle regime is, as promised, the difficult regime. Neither fish nor fowl, it mixes rolling and sliding in comparable amounts.

The solution is to study the third regime, where $\beta \gg 1$, and then to interpolate between the first and third regimes in order to predict the behavior in the middle regime. (We implicitly used this interpolation approach in Section 6.4.3, when we estimated the time and distance for a falling cone to reach its terminal speed.)

▷ *For an object rolling down the plane, how can β be increased beyond 1?*

There seems to be no way to increase β beyond 1: When $\beta = 1$, all mass is already distributed at the edge. Fortunately, the subtle point is that, in the dimensional analysis, r is the *rolling* radius, rather than the radius of the object, because the rolling radius determines the torque and the motion. The object's radius can be larger than the rolling radius. This possibility allows us to increase β as much as we want.

As an example, imagine a barbell used for weight lifting: A thin axle of radius r connects two large and massive ears of radii R. Place the axle on the inclined plane, with the ears beyond the plane. The rolling radius is the radius r of the axle. However, the moment of inertia I is mostly due to the large ears, because

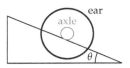

$$\text{moment of inertia} \propto \left(\text{distance from the axis of rotation}\right)^2. \tag{8.22}$$

Thus, $I \sim mR^2$, where m is the combined mass of the two ears, which is almost the same as the mass of the whole object. By choosing $R \gg r$, we can make $\beta \gg 1$:

$$\beta \equiv \frac{I}{mr^2} \sim \left(\frac{R}{r}\right)^2 \gg 1. \tag{8.23}$$

▷ *How quickly does this large-β object accelerate down the plane?*

This extreme regime is easier to think about than the intermediate regime $\beta = 1$. The extremity of the $\beta \gg 1$ regime amplifies our intuition and shouts loudly enough for our gut to hear. Then our gut's answer is loud enough for us to hear: The barbell, rolling on its small axle, as long as it doesn't slip, will *creep* down the plane (the italics represent my gut's shouting). Going fast would, in spinning up the ears, demand far more energy than gravity can provide. Thus, as the ears get ever larger and β goes to infinity, the dimensionless rolling factor $h(\beta)$ goes to zero.

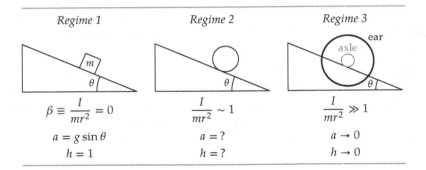

Regime 1	Regime 2	Regime 3

$$\beta \equiv \frac{I}{mr^2} = 0$$ $$\frac{I}{mr^2} \sim 1$$ $$\frac{I}{mr^2} \gg 1$$

$$a = g \sin \theta \qquad\qquad a = ? \qquad\qquad a \to 0$$

$$h = 1 \qquad\qquad\qquad h = ? \qquad\qquad h \to 0$$

A simple guess that accounts for the two extreme regimes is

$$h(\beta) = \frac{1}{1 + \beta}. \tag{8.24}$$

In terms of the original quantities, the acceleration would be

$$a = \frac{g \sin \theta}{1 + I/mr^2}. \tag{8.25}$$

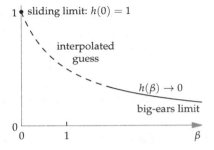

This educated guess turns out to be correct (Problem 8.8). Now we can answer our original question about which shape rolls faster. The farther outward the mass is distributed, the greater β is, so $\beta_{\mathrm{ring}} > \beta_{\mathrm{disc}}$. Because a decreases as β increases, the disc rolls faster than the ring.

Problem 8.8   Exact solution to rolling down the plane
Use conservation of energy to find the acceleration of the object along the plane, and confirm the guess based on the easy-cases regimes.

Problem 8.9   A pendulum with chewing gum
A pendulum with a disc as the pendulum bob is oscillating with period T. You then stick a small piece of chewing gum onto the center of the disc. How does the gum affect the pendulum's period: an increase, no effect, or a decrease?

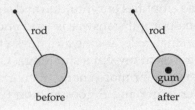

8.3.1.2 *Drag using easy cases*

Drag, like any phenomenon related to fluids, is a hard problem. In particular, there is no way to calculate the drag coefficient c_d as a function of Reynolds number Re—even for simple shapes such as a sphere or a cylinder. However, dimensional analysis (Section 5.3.2) has told us that

$$\underbrace{\text{drag coefficient}}_{c_d} = f(\underbrace{\text{Reynolds number}}_{\text{Re}}). \tag{8.26}$$

However, dimensional analysis could not tell us the function f. That's no slur on dimensional analysis; finding the whole function is beyond our present-day understanding of mathematics. However, we can understand much about f in two easy cases: the bookend regimes Re $\gg$ 1 and Re $\ll$ 1. In the difficult middle regime Re $\sim$ 1, the function f interpolates between its behavior in the two extremes.

Low Reynolds number. Flows at low Reynolds number, although not as frequent in everyday experience as flows at high Reynolds number, include a fog droplet falling in air (Problem 8.13), a bacterium swimming in water (discussed in the classic paper "Life at low Reynolds number" [39]), and ions conducting electricity in seawater (Problem 8.10). Our goal here is to find the drag coefficient in the regime Re $\ll$ 1.

The Reynolds number (based on radius) is vr/ν, where v is the speed, r is the object's radius, and ν is the kinematic viscosity of the fluid. Therefore, to shrink Re, either make the object small, make the object's speed low, or use a fluid with high viscosity. The method does not matter, as long as Re is small: The drag coefficient is determined not by any of the individual parameters r, v, or ν, but rather by the abstraction Re. We'll choose the means that leads to the most transparent physical reasoning: making the viscosity huge. Imagine, for example, a tiny bead oozing through a jar of cold honey. (The tininess of the bead further reduces Re.)

In this extremely viscous flow, the drag force, as you might expect, is produced directly by viscous forces. As you reasoned in Problem 7.25, viscous forces, which are proportional to the kinematic viscosity ν, are given by

$$F_{\text{viscous}} \sim \rho\nu \times \text{velocity gradient} \times \text{area}. \tag{8.27}$$

The area is the surface area of the object. The velocity gradient is the rate at which velocity changes with distance; it is analogous to the concentration gradient $\Delta n/\Delta x$ in the Fick's-law analysis of Section 7.4.2 or to the temperature gradient $\Delta T/\Delta x$ in the heat-flow analysis of Section 7.4.4.

Because the drag force is due to viscous forces directly, it should be proportional to ν. This constraint determines the form of the function f and therefore the drag force. Let's write out the dimensional-analysis result connecting the drag coefficient and the Reynolds number:

$$c_{\mathrm{d}} \equiv \frac{F_{\mathrm{d}}}{\frac{1}{2}\rho A_{\mathrm{cs}}v^2} = f\underbrace{\left(\frac{vr}{\nu}\right)}_{\mathrm{Re}}, \tag{8.28}$$

where F_{d} is the drag force and A_{cs} is the object's cross-sectional area. In terms of the radius, $A_{\mathrm{cs}} \sim r^2$, so

$$\underbrace{\frac{F_{\mathrm{d}}}{\rho r^2 v^2}}_{\sim c_{\mathrm{d}}} \sim f\underbrace{\left(\frac{vr}{\nu}\right)}_{\mathrm{Re}}. \tag{8.29}$$

The viscosity ν appears only in the Reynolds number, where it appears in the denominator. To make F_{d} proportional to ν requires making the drag coefficient proportional to $1/\mathrm{Re}$. Equivalently, the function f, when $\mathrm{Re} \ll 1$, is given by $f(x) \sim 1/x$. Then

$$\frac{F_{\mathrm{d}}}{\rho r^2 v^2} \sim \frac{\nu}{vr}. \tag{8.30}$$

For the drag force itself, the consequence is (as you found in Problem 7.26)

$$F_{\mathrm{d}} \sim \rho r^2 v^2 \frac{\nu}{vr} = \rho \nu v r. \tag{8.31}$$

This result is valid for almost all shapes; only the dimensionless prefactor changes. For a sphere, the British mathematician George Stokes showed that the missing dimensionless prefactor is 6π:

$$F_{\mathrm{d}} = 6\pi\rho\nu vr. \tag{8.32}$$

Accordingly, this result is called Stokes drag.

High Reynolds number. Because the Reynolds number rv/ν contains three quantities, the limit of high Reynolds number can also be reached in three ways, all of which produce the same behavior. We'll choose the method that is easiest to feel, which is to shrink the viscosity ν to 0. In this limit, called form drag, viscosity disappears from the problem and the drag force should not depend on viscosity. This reasoning contains several untruths, but they are subtle and the conclusion is mostly correct. (Clarifying the subtleties required centuries of progress in mathematics, culminating in the theory of singular perturbations and boundary layers, the subject of Section 7.3.4 and discussed much more in *Physical Fluid Dynamics* [46].)

Without viscosity and thus without the Reynolds number to depend on, the dimensionless function f must be a constant! Its value depends on the shape of the object and typically ranges from 0.5 to 2. Because the drag coefficient is $F_d / \frac{1}{2}\rho v^2 A_{cs}$, the drag force is

	c_d
car	0.4
sphere	0.5
cylinder	1.0
flat plate	2.0

$$F_d \sim \rho v^2 A_{cs}. \tag{8.33}$$

This result agrees with our analysis in Section 3.5.1 based on energy conservation. In that analysis, we had implicitly assumed a high Reynolds number without knowing it.

Problem 8.10   Conductivity of seawater

Estimate the conductivity of seawater by assuming that the current-carrying ions (Na^+ or Cl^-) travel with a shell one layer thick of water molecules around them.

Problem 8.11   Drag coefficient resulting from Stokes drag

At low Reynolds number, the drag force on a sphere is $F = 6\pi\rho\nu vr$ (Stokes drag). Find the drag coefficient c_d as a function of Reynolds number, basing the Reynolds number on the sphere's diameter rather than its radius.

Problem 8.12   Choosing the high- or low-Reynolds-number regime

To estimate the terminal speed of a raindrop (Problem 3.37), you implicitly used the high-Reynolds-number regime. Now that you know another regime, how can you choose which regime to use? Without knowing which regime, you cannot find the terminal speed. Without the terminal speed, how do you find Reynolds number, which you need to choose the regime? The answer is to choose a regime and see whether the result is self-consistent. If it is not, choose the other regime.

As practice with this reasoning, assume that the Reynolds number for the falling raindrop is small and therefore that the drag force F_d is given by Stokes drag:

$$F_d \sim \rho_{air}\nu vr. \tag{8.34}$$

Estimate the resulting terminal speed v for the raindrop (diameter of about 0.5 centimeters). Then check the assumption that $Re \ll 1$.

Problem 8.13   Terminal speed of fog droplets

a. Estimate the terminal speed of fog droplets ($r \sim 10\,\mu m$). In estimating the drag force, use either the limit of low or high Reynolds numbers—whichever limit you guess is more likely to be valid. (Problem 8.12 introduces this reasoning.)

b. Use the speed to estimate the Reynolds number and check whether you used the correct limit for the drag force. If not, try the other limit!

c. Fog is a low-lying cloud. How long does a fog droplet require to fall 1 kilometer (a typical cloud height)? What is the everyday effect of this settling time?

Interpolation. We now know the drag force in the two extreme regimes: viscous drag (low Reynolds number) and form drag (high Reynolds number). Interpolating between these regimes is easiest in dimensionless form—that is, in terms of the drag coefficient rather than the drag force.

When Re ≫ 1, the drag coefficient c_d is roughly 0.5 (for a sphere). On log–log axes, the Re ≫ 1 behavior is a straight line with zero slope. In the Re ≪ 1 regime, you should have found in Problem 8.11 that the drag coefficient (for a sphere) is 24/Re, where the Reynolds number is based on the diameter rather than the radius: The 1/Re

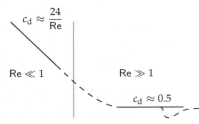

scaling happens because, as we just deduced, $f(\text{Re}) \sim 1/\text{Re}$ to make the drag force proportional to viscosity ν; the dimensionless prefactor of 24 is what you had to compute. On log–log axes, the Re ≪ 1 behavior is also a straight line but with a −1 slope.

The dashed line interpolates between the extremes. The final twist is that at Re ∼ 3×10^5, the boundary layer becomes turbulent (Problem 7.23), and the drag coefficient drops significantly. With that caveat, we have explained the main features of drag, a ubiquitous force in the living world.

Problem 8.14   Q as an easy-cases parameter

A damped spring–mass system has a dimensionless measure of damping known as the quality factor Q, which you studied in Problem 5.53. Choose $Q < 0.5$, $Q = 0.5$, or $Q > 0.5$ to describe each of the three regimes: (a) underdamped, (b) critically damped, and (c) overdamped.

Problem 8.15   Floating on water

Some insects can float on water thanks to the surface tension of water. Extend the analysis of this effect from Problem 5.52 by estimating the dimensionless ratio

$$R \equiv \frac{\text{force due to surface tension}}{\text{weight of the bug}} \tag{8.35}$$

as a function of the bug's size l (as a length). Interpret the regimes $R \ll 1$ and $R \gg 1$, and find the critical bug size l_0 for floating on water.

Problem 8.16   Energy loss in highway versus city driving

Explain why the following dimensionless ratio measures the importance of drag:

$$\frac{\text{mass of fluid swept out}}{\text{mass of object}}. \tag{8.36}$$

An equivalent computation, except for a dimensionless factor of order unity, is

$$\frac{\text{fluid density}}{\text{object density}} \times \frac{\text{distance traveled}}{\text{size of object}},$$ (8.37)

where the size of the object is its linear dimension in the direction of travel. When this ratio is comparable to unity, drag significantly affects the trajectory.

Apply either form of the ratio to a car during city driving, and find the distance d at which the ratio becomes significant (say, roughly 1). How does the distance compare with the distance between stop signs or traffic lights on city streets? What therefore is the main mechanism of energy loss in city driving? How does this analysis change for highway driving?

Problem 8.17   Gain of an *RC* circuit

The low-pass *RC* circuit of Section 2.4.4 is amenable to easy-cases analysis. With the abstraction of capacitive impedance (Section 2.4.4), explain the unity gain at low frequency and why the gain goes to zero at high frequency. What is the dimensionless way to say "high frequency"?

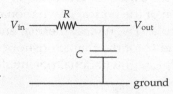

Problem 8.18   Bode magnitude sketch for an *RC* circuit

A Bode magnitude plot is a log–log plot of |gain| versus frequency. In a lumped Bode sketch, which often gives the most insight into the behavior of a system, the segments of the plot are straight lines. Make the Bode magnitude sketch for the low-pass *RC* circuit of Problem 8.17, labeling the slopes and intersection points.

8.3.2   Three regimes where two effects compete

In the final group of three-regime examples, the three regimes are again based on the relative size of two physical effects. However, in contrast to the examples in Section 8.3.1, where we chose a regime—for example, by choosing a flow speed and thus the Reynolds number—here nature chooses. Nature chooses the middle regime, where the two competing physical effects reach a draw. The method of easy cases shows how that choice is made.

8.3.2.1   *Height of the atmosphere*

Competition between physical effects is the theme of the wonderful *Gases, Liquids and Solids and Other States of Matter* [43, pp. xiii], where we learn how "the three primary states of matter are the result of competition between thermal energy and intermolecular forces." The following example, estimating the height of our atmosphere, illustrates an analogous competition: between thermal energy and gravity. Here is how these two physical effects determine the height of the atmosphere.

1. *Thermal energy.* Thermal energy gives the molecules speed. Unfettered by gravity, the air molecules and the atmosphere would disperse into space. Thermal energy is therefore responsible for increasing the atmosphere's height.

2. *Gravity.* Gravity pulls air molecules downward. If gravity is the only effect, meaning that thermal motion is absent (the atmosphere is at absolute zero), then gravity pulls the molecules all the way to the ground. Gravity therefore decreases the atmosphere's height.

A dimensionless measure of the relative size of the two effects is a comparison of the typical gravitational energy with the thermal energy of one molecule. The typical thermal energy is k_BT. With m for the molecule's mass and H for the characteristic or typical or scale height of the atmosphere, the typical gravitational potential energy is mgH. The energy ratio is then

$$\beta \equiv \frac{mgH}{k_BT}. \tag{8.38}$$

This ratio categorizes the three easy-cases regimes: $\beta \ll 1$, $\beta \sim 1$, and $\beta \gg 1$.

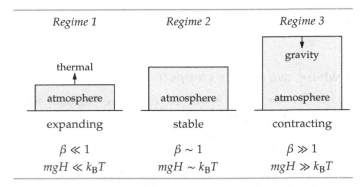

Regime 1	*Regime 2*	*Regime 3*
expanding	stable	contracting
$\beta \ll 1$	$\beta \sim 1$	$\beta \gg 1$
$mgH \ll k_BT$	$mgH \sim k_BT$	$mgH \gg k_BT$

In the first regime, the atmosphere is expanding: The molecules have so much thermal energy that they escape farther from Earth. In the third regime, the atmosphere is contracting: The molecules do not have enough thermal energy to resist gravity as it pulls them back to Earth. The happy medium, where the atmosphere is stable, is the middle regime.

Therefore, the regime is chosen not by us but by nature. The height of the atmosphere is determined by the requirement that the two effects compete and reach a draw—when the two effects are comparable in strength. In that regime, $mgH \sim k_BT$, so the atmosphere's scale height is

$$H \sim \frac{k_{\mathrm{B}}T}{mg}. \tag{8.39}$$

For the Earth's atmosphere, where $T \approx 300\,\mathrm{K}$ and the molecular mass m is approximately the mass of a nitrogen molecule, this height is roughly 8 kilometers—as we predicted in Section 5.4.1 using dimensional analysis. The easy-cases reasoning complements the dimensional analysis by providing a physical model.

As a bonus, this general model of competition explains why we guess that an unknown dimensionless number is comparable to 1—for example, when we estimated an atomic blast energy in Section 5.2.2. Often, the dimensionless number represents a ratio between two physical effects. Then "comparable to 1" means "the effects have reached a draw." By guessing that unknown dimensionless numbers were comparable to 1, we foreshadowed the easy-cases reasoning that we use for these examples of competition.

8.3.2.2 *Hydrogen's binding energy*

Our next example, before we turn to the giant scales of stars, is our old friend the hydrogen atom. In Section 5.5.1, we predicted its size, the Bohr radius a_0, using dimensional analysis. The prediction was correct, but, as with the dimensional analysis of the atmosphere height, dimensional analysis didn't give us a physical model. Easy-cases reasoning will help us make our tacit physical knowledge about hydrogen explicit and thereby build a physical model.

As we found in Section 6.5.2 using lumping, two physical effects compete:

1. *Electrostatics.* The electrostatic attraction between the proton and electron shrinks the atom.

2. *Quantum mechanics.* Via the uncertainty principle, quantum mechanics gives an electron confined to a small region a high momentum uncertainty and, therefore, a high kinetic energy. With this confinement energy, analogous to the thermal energy of an air molecule, the electron can resist the inward pull of electrostatics. Therefore, quantum mechanics expands the atom.

A dimensionless ratio that measures the relative size of the two effects is the energy ratio

$$\beta \equiv \frac{|\text{electrostatic potential energy}|}{\text{confinement energy}}, \tag{8.40}$$

where the absolute-value bars say that, although the electrostatic potential energy is negative, we care only about its magnitude.

In a hydrogen atom of radius r, the electrostatic energy is $e^2/4\pi\epsilon_0 r$. The confinement energy—which we estimated in Section 6.5.2 by using lumping—is comparable to $\hbar^2/m_e r^2$. Therefore,

$$\beta \sim \frac{e^2/4\pi\epsilon_0 r}{\hbar^2/m_e r^2}. \tag{8.41}$$

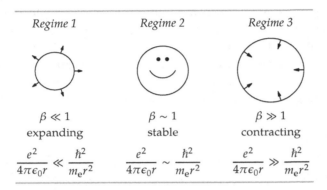

Regime 1	Regime 2	Regime 3
$\beta \ll 1$	$\beta \sim 1$	$\beta \gg 1$
expanding	stable	contracting
$\dfrac{e^2}{4\pi\epsilon_0 r} \ll \dfrac{\hbar^2}{m_e r^2}$	$\dfrac{e^2}{4\pi\epsilon_0 r} \sim \dfrac{\hbar^2}{m_e r^2}$	$\dfrac{e^2}{4\pi\epsilon_0 r} \gg \dfrac{\hbar^2}{m_e r^2}$

In the first regime, $\beta \ll 1$, quantum mechanics is much stronger than electrostatics, and the huge kinetic energy from quantum mechanics (the confinement energy) expands the atom. In the third regime, $\beta \gg 1$, electrostatics, now much stronger than quantum mechanics, contracts the atom.

The radius of hydrogen, like the height of the atmosphere, is determined by the requirement that two physical effects compete and reach a draw. This truce happens in the middle regime. There, $e^2/4\pi\epsilon_0 r \sim \hbar^2/m_e r^2$, so

$$r \sim \frac{\hbar^2}{m_e(e^2/4\pi\epsilon_0)}, \tag{8.42}$$

which is the Bohr radius a_0 that we found using dimensional analysis (Section 5.5.1) and lumping (Section 6.5.2).

An alternative approach, which also uses easy cases but gives different insights, is to base the dimensionless ratio on the size r directly. Making r dimensionless requires another size. Fortunately, we have one, because the two energies, electrostatic and confinement, have different scaling exponents. The electrostatic energy is proportional to r^{-1}. The confinement energy is proportional to r^{-2}.

Regime 1	*Regime 2*	*Regime 3*
$\beta \ll 1$	$\beta \sim 1$	$\beta \gg 1$
expanding	stable	contracting
$r \ll r_0$	$r \sim r_0$	$r \gg r_0$

On log–log axes, they are straight lines but with different slopes: -2 for the confinement line and -1 for the electrostatic-energy line. Therefore, they intersect. The size at which they intersect is a special size. Let's call it r_0. Then our dimensionless ratio is $\beta \equiv r/r_0$, and the three regimes are shown in the table.

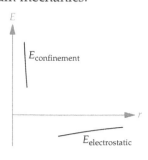

The stable middle regime is the one chosen by nature. Then r, the radius of hydrogen, is comparable to our special radius r_0. The special radius, and therefore r, is determined by equating electrostatic and confinement energies, which is how we found the Bohr radius a_0 in Section 6.5.2. Thus, the radius of hydrogen is the Bohr radius and results from competition between electrostatics and quantum mechanics.

Our next project is to use the easy-cases regimes of size to understand thermal expansion—why substances expand upon heating. The first step is to sketch the total energy E, which is the sum of confinement and electrostatic energies.

This sketch is easiest in the bookend regimes. In the first regime, where $r \ll r_0$, the atom is too small, and quantum mechanics is stronger: Its kinetic-energy contribution is the dominant term in the energy, because its $1/r^2$ scaling overwhelms the $1/r$ scaling of the potential energy. In the third regime, where $r \gg r_0$, the atom is too large, and electrostatics is the dominant term in the energy: Its $1/r$ scaling goes to zero more slowly than the $1/r^2$ scaling of the kinetic energy. Therefore, in the extremes (the bookends), the total energy has two different scalings:

$$E \propto \begin{cases} 1/r^2 & \text{(regime 1: } r \ll r_0) \\ -1/r & \text{(regime 3: } r \gg r_0) \end{cases}$$

(8.43)

To complete the sketch, we interpolate between the bookend regimes.

The slopes in the two regimes have opposite signs, so there
is no smooth way to connect the two extreme segments
without introducing a point where the slope is zero. This
point, where the energy is a minimum, is nature's choice
for the ground state of hydrogen.

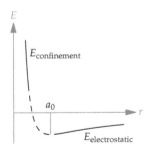

With this sketch, we can explain thermal expansion. Thermal expansion first appeared in Problem 5.38, where you
used dimensional analysis to estimate a typical thermal-expansion coefficient. However, the problem glided over the
fundamental question: the coefficient's sign. Knowing nothing about thermal expansion, we might guess that positive and negative coefficients are
equally likely. Equivalently, substances are equally likely to expand as to
contract when heated: Some substances might expand, and others might
contract. However, this symmetry reasoning fails empirically: Almost all
substances expand rather than contract when heated. The symmetry between positive and negative thermal-expansion coefficients, or between expansion and contraction, must get broken somewhere.

The curve of energy versus size for hydrogen shows where the symmetry
gets broken. In its generic shape, the curve applies to all bonds, whether
intra-atomic (such as the electron–proton bond in hydrogen), interatomic
(such as hydrogen–oxygen bonds in a water molecule), or even intermolecular (such as the hydrogen bonds between water molecules). Bonds result
from a competition between (1) attraction, which is important at long distances and which looks like the electrostatic piece of the generic curve, and
(2) repulsion, which is important at short distances and which looks like
the confinement piece of the generic curve. (Even the gravitational bond
between the Sun and a planet, represented by the curve that you drew in
Problem 5.55c, has the same shape.)

Therefore, the curve of bond energy versus separation cannot be symmetric.
At the $r \ll r_0$ end, which is the first regime, r has a minimum possible value,
namely zero. However, at the $r \gg r_0$ end, which is the third regime, r is
unbounded. Thus, the two bookend regimes are not symmetric, and the
bond-energy curve skews toward larger r.

To see how this asymmetry leads to thermal expansion, look at how thermal
energy affects the average bond length. First, think of the bond as a spring
(a model that we will discuss further in Section 9.1). As the bond vibrates
around its equilibrium length r_0, thermal energy, which is a kinetic energy,

converts into and out of potential energy in the bond. The minimum bond length r_{min} and the maximum bond length r_{max} are determined by where the vibration speed is zero—where the bond has slurped up all the kinetic energy (the thermal energy) and turned it into potential energy.

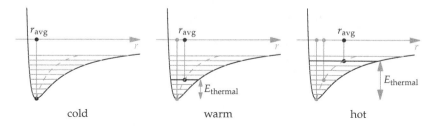

cold   warm   hot

Graphically, we draw a horizontal line at a height E_{thermal} above the minimum energy. This line intersects the potential-energy curve twice, at $r = r_{\mathrm{min}}$ and at $r = r_{\mathrm{max}}$. Because the potential-energy curve skews to the right, toward $r \gg r_0$, the average bond length $r_{\mathrm{avg}} = (r_{\mathrm{min}} + r_{\mathrm{max}})/2$ is larger than r_0. As E_{thermal} grows, the skew affects the average more, and the difference between r_{avg} and r_0 grows: Adding thermal energy increases the bond length. Therefore, substances expand when heated.

8.3.2.3 *Core temperature of the Sun*

As our final example, we'll estimate the temperature of the core of the Sun. The Sun, like the atmosphere, is a result of competition between thermal motion and gravity: Thermal motion expands the Sun; gravity compresses the Sun. A dimensionless measure of the two effects' relative strength is

$$\beta \equiv \frac{\text{thermal energy}}{|\,\text{gravitational potential energy}\,|}. \tag{8.44}$$

The numerator, the thermal energy for one particle, is just $k_{\mathrm{B}}T$. For the denominator, we need to know the mass of a particle. The Sun itself is made of hydrogen, but in the core of the Sun, the thermal energy is more than enough to strip an electron from the proton (as you verify in Problem 8.19). Each proton gets gravitational potential energy from the rest of the Sun, which has mass M_{Sun}. The rest of the Sun, if lumped into a massive particle, would be at a distance comparable to R_{Sun}, where R_{Sun} is the Sun's radius. Therefore, the typical gravitational potential energy (per particle) is

$$E_{\mathrm{gravitational}} \sim \frac{GM_{\mathrm{Sun}}m_{\mathrm{p}}}{R_{\mathrm{Sun}}}, \tag{8.45}$$

where the $\sim$ contains the minus sign in the potential energy. The ratio of energies is therefore

$$\beta \sim \frac{k_\mathrm{B}T}{GM_\mathrm{Sun}m_\mathrm{p}/R_\mathrm{Sun}},\tag{8.46}$$

and it produces the following three regimes.

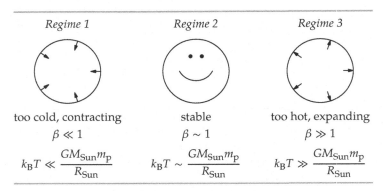

Regime 1	*Regime 2*	*Regime 3*
too cold, contracting	stable	too hot, expanding
$\beta \ll 1$	$\beta \sim 1$	$\beta \gg 1$
$k_\mathrm{B}T \ll \dfrac{GM_\mathrm{Sun}m_\mathrm{p}}{R_\mathrm{Sun}}$	$k_\mathrm{B}T \sim \dfrac{GM_\mathrm{Sun}m_\mathrm{p}}{R_\mathrm{Sun}}$	$k_\mathrm{B}T \gg \dfrac{GM_\mathrm{Sun}m_\mathrm{p}}{R_\mathrm{Sun}}$

In the first regime, the Sun is too cold for its size, so gravity wins the competition and compresses the sun. This contraction shows the difference between the thermal–gravitational competition in the Sun and in the Earth's atmosphere (Section 8.3.2.1). The temperature of the atmosphere is determined by the blackbody temperature of the Earth, roughly 300 K (Problem 5.43). In our analysis of the height of the atmosphere, we therefore held the temperature fixed and let only the height vary until it produced the gravitational energy to match the fixed thermal energy. In the Sun, however, the temperature is determined by the speed of the fusion reactions, which depends on the temperature and the density: Higher density means more frequent collisions that might result in fusion, and higher temperature (faster thermal motion) means that each collision has a higher chance of resulting in fusion. Thus, as the Sun contracts, the density increases, as does the temperature and the reaction rate. The contraction stops when the temperature is high enough for thermal motion to balance gravity.

(There is a caveat. If a star contracts very quickly, its core can heat up faster than the negative-feedback process can oppose the change. Then the core ignites like a giant hydrogen bomb and the star becomes a supernova.)

In the third regime, the Sun is too hot for its size, so thermal motion wins the competition and expands the Sun. As the Sun expands, the reaction rate, the temperature, and the thermal energy fall—until thermal motion again balances gravity. The result is the middle regime. In the middle regime, the Sun has just the right temperature—determined by the condition $k_B T \sim GM_{Sun}m_p/R_{Sun}$, so

$$T \sim \frac{GM_{Sun}m_p}{k_B R_{Sun}}. \tag{8.47}$$

The proton mass m_p in the numerator and Boltzmann's constant k_B in the denominator are easier to handle if we multiply each constant by Avogadro's number N_A. The product $N_A k_B$ is the universal gas constant R. The product $m_p N_A$ is the molar mass of protons, which is approximately the molar mass of hydrogen: 1 gram per mole. Then

$$T \sim \frac{\overbrace{6.7\times10^{-11}\,\mathrm{kg^{-1}\,m^3\,s^{-2}}}^{G} \times \overbrace{2\times10^{30}\,\mathrm{kg}}^{M_{Sun}} \times \overbrace{10^{-3}\,\mathrm{kg\,mol^{-1}}}^{N_A m_p}}{\underbrace{8\,\mathrm{J\,mol^{-1}\,K^{-1}}}_{N_A k_B} \times \underbrace{0.7\times10^{9}\,\mathrm{m}}_{R_{Sun}}}. \tag{8.48}$$

To evaluate T, start with the most important piece.

1. *Units.* The inverse moles in the numerator and denominator cancel. The numerator then contributes $\mathrm{kg\,m^3\,s^{-2}}$, which is $\mathrm{J\,m}$. The denominator contributes $\mathrm{J\,K^{-1}\,m}$. Therefore, the $\mathrm{J\,m}$ in the numerator and denominator cancel, and the inverse kelvins in the denominator become the kelvins in the Sun's core temperature.

2. *Powers of ten.* The numerator contributes 16 powers of ten; the denominator contributes 9. The result is 7 powers of ten. Thus, the temperature will be comparable to 10^7 K.

3. *Numerical prefactor.* The 6.7 in the numerator is mostly canceled out by the 8×0.7 in the denominator, leaving only the 2 in the numerator. Thus, the temperature is roughly 2×10^7 K.

No one has measured the internal temperature directly, but the current best estimate for the core temperature is 1.5×10^7 K. Our easy-cases model of competition between gravity and thermal motion is surprisingly accurate.

Problem 8.19 Thermal versus electronic energy
Compare the thermal energy (per particle) in the core of the Sun to the binding energy of hydrogen. Were we justified in assuming that the electrons get stripped from the protons?

8.4 Two dimensionless quantities

In the examples of Section 8.3, one dimensionless quantity categorized the easy-cases regimes. Whether the quantity was small, comparable to 1, or large, the regimes could all be arranged on one axis. To extend that idea, we'll use easy cases to categorize the world using two dimensionless quantities and therefore two axes. For practice, we'll organize the world of waves onto a two-dimensional map (Section 8.4.1). Then we'll organize the four fundamental branches of physics (Section 8.4.2).

8.4.1 The two-dimensional world of water waves

Water waves come in several varieties. Perhaps the most evident regime represents waves near a beach. These waves, the subject of Problem 5.15, have a wavelength much larger than the depth of the water. You can make them at home: Fill a bathtub or baking dish with water, disturb the water, and create waves sloshing back and forth. Their speed v is given by $v^2 = gh$, where h is the depth of the water.

Before these shallow-water waves reached the beach, they traveled on the open ocean, where their wavelength was much smaller than the depth. Their speed—technically, the phase velocity—is given by $v^2 = g\lambdabar$ (Problem 5.11), where $\lambdabar$ is the reduced wavelength $\lambda/2\pi$.

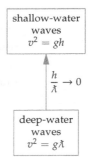

These two regimes—deep and shallow water—are distinguished by the dimensionless ratio $h/\lambdabar$. (You can also use h/λ, but the mathematical descriptions turn out to be simpler using $h/\lambdabar$.) Therefore, the two regimes sit on a dimensionless axis that measures depth. Because the axis measures depth, let's orient the axis vertically and place deep water at the bottom.

Another familiar kind of wave is produced by dropping a pebble in a pond. Small ripples zoom outward from the point of impact. These waves have a small wavelength, much smaller than the depth of the pond. Therefore, $h/\lambdabar$ is large—as it is for deep-water waves.

However, these ripples are different from the deep-water waves on the open ocean. Ocean waves are driven by the water's weight—that is, by gravity. In contrast, ripples are driven by the water's surface tension—the same effect that allows small bugs to walk on water (see Problem 8.15). In order to distinguish ripples from deep-water gravity waves, our second axis should measure the relative importance of gravity and surface tension.

This axis is therefore labeled by the ratio of gravity to surface tension. This ratio is a dimensionless group, so we can find it using dimensional analysis. The ratio will depend on two characteristics of the water: its density ρ (to make the weight) and its surface tension γ. It will depend on gravity g (to make the weight). And it will depend on the (reduced) wavelength $\lambdabar$. These four quantities, built from three dimensions, form one independent dimensionless group. A useful choice is $\rho g \lambdabar^2 / \gamma$.

At long wavelengths (large $\lambdabar$), in dense fluids (large ρ), or in strong gravity (large g), this dimensionless ratio is large, indicating that gravity drives the waves. For short wavelengths (ripples) or, equivalently, high surface tension, this ratio is small, indicating that surface tension drives the waves. Here are the three regimes categorized using both groups.

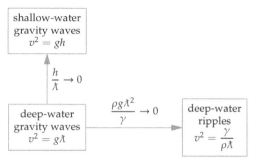

The empty corner stares at us, asking to be filled. It incorporates both limits:

$$\frac{h}{\lambdabar} \to 0 \quad \text{(shallow water)} \qquad \text{and} \qquad \frac{\rho g \lambdabar^2}{\gamma} \to 0 \quad \text{(ripples).} \qquad (8.49)$$

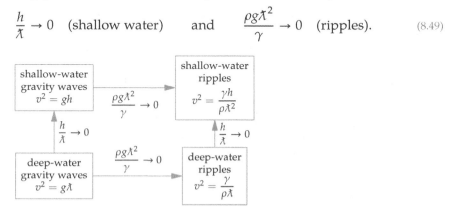

These waves are therefore ripples on shallow water. (They are hard to make, because ripples are tiny, smaller than a few millimeters, yet the water depth must be even smaller.)

The four corners are the bookend regimes on two axes. If each axis had three regimes, two axes could have made three squared or nine regimes. The missing five middle regimes can be constructed by interpolating between the corners. The easy-cases map shows how: First fill in the four regions between the corners, and then fill in the center. That analysis, which involves addition and a tanh function, gives the following map.

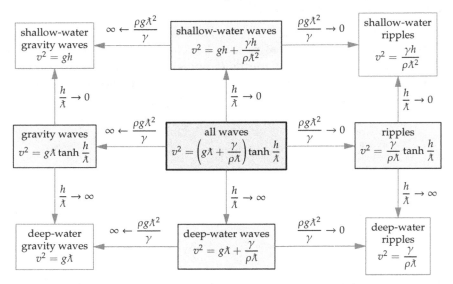

On this map, the middle regimes are not our usual middle regimes. Our usual middle regimes represented a particular regime (which was usually of the form $\beta \sim 1$). However, on this map, the middle regimes represent the general solution $\beta =$ anything. As an example, look at the bottom, deep-water row of three regimes. The bookend regimes, deep-water gravity waves and deep-water ripples, are easy cases of the middle, deep-water regime. For fun, check the other limiting cases, including that the central regime—which covers waves driven by any mixture of gravity and surface tension and traveling on any depth of water—turns into the other eight regimes in the appropriate limits.

8.4.2 The two-dimensional world of physics

Now we'll use the same method to organize the four fundamental branches of physics: classical (Newtonian) mechanics, quantum mechanics, special relativity, and quantum electrodynamics.

As a first step, which will determine the first axis, let's compare classical mechanics and special relativity. Special relativity is Einstein's theory of motion. It unifies classical mechanics and classical electrodynamics (the theory of radiation), giving a special role to the speed of light c. This role is popularized in the T-shirt showing Einstein in a policeman's hat. He holds out his arm and palm to make a "Stop" signal and warns: "186 000 miles per second. It's not just a good idea. It's the law!" The speed of light is the universe's speed limit, and special relativity obeys it. In contrast, classical mechanics knows no speed limit. Classical mechanics and special relativity therefore sit on an axis connected by the speed of light. In the limit $c \to \infty$, special relativity turns into classical mechanics.

For the second axis, we compare one of the two remaining branches of physics—either quantum mechanics or quantum electrodynamics—with either classical mechanics or special relativity. Because quantum electrodynamics, if only from its name, looks frightening, let's select quantum mechanics. We've seen its effect several times: Quantum mechanics contributes a new constant of nature $\hbar$. This constant appears in the Heisenberg uncertainty principle $\Delta p \Delta x \sim \hbar$, where Δp and Δx are a particle's momentum and position uncertainties, respectively. The Heisenberg uncertainty principle restricts how small we can make these uncertainties, and therefore how accurately we can determine the position and momentum.

However, if $\hbar$ were zero, then the uncertainty principle would not restrict anything. We could exactly determine the position and momentum of a particle simultaneously, as we expect in classical mechanics. Classical mechanics is the $\hbar \to 0$ limit of quantum mechanics. Therefore, classical and quantum mechanics are connected on a second, $\hbar$ axis. The map including quantum mechanics therefore has two dimensions.

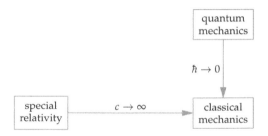

In this two-dimensional map of physics, one corner sits empty. Furthermore, one branch of physics—quantum electrodynamics—hasn't been considered. We need only a bit of courage to set quantum electrodynamics in the empty corner. Quantum mechanics must be the $c \to \infty$ limit of quantum electrodynamics. And it is. Quantum electrodynamics is the result of marrying special relativity ($c < \infty$) and quantum mechanics ($\hbar > 0$). Thus, in the $\hbar \to 0$ limit, quantum electrodynamics turns into special relativity.

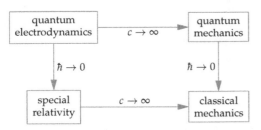

▷   *What happened to the bookend regimes?*

The bookend regimes are here implicitly, because this example introduced a new feature: These axes are not labeled using dimensionless quantities! Only for a dimensionless quantity can we sensibly distinguish the three regimes $\ll 1$, ~ 1, and $\gg 1$. Because c and $\hbar$ have dimensions, the only valid comparisons are with zero or with infinity (which are zero and infinity in any system of units). Thus, there are only two regimes on each axis. The special-relativity–classical-mechanics axis compares c with infinity; the quantum-mechanics–classical-mechanics axis compares $\hbar$ with zero.

8.5 Summary and further problems

When the going gets tough, the tough lower their standards. In this chapter, you learned how to do that by studying the easy cases of a problem. This tool is based on the idea that a correct solution works in all cases, including the easy cases. Therefore, look at the easy cases first. Often, we can completely solve a problem simply by understanding the easy cases.

Problem 8.20   Easy cases for the period of a pendulum
Does the period of a pendulum increase, decrease, or remain constant as the amplitude is increased? Decide by selecting an amplitude for which you can easily predict the period.

Problem 8.21   Pyramid volume

Use easy cases to find the dimensionless prefactor in the
volume of a pyramid with height h and a square ($b \times b$)
base:

$$V = \text{dimensionless prefactor} \times b^2 h. \qquad (8.50)$$

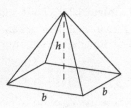

In particular, choose the easy case of a pyramid that, with
a few more copies of itself, can be assembled into a cube.

Problem 8.22   Power means

The arithmetic and geometric means are easy cases of a higher-level abstraction:
the power mean. The kth power mean of two positive numbers a and b is defined
by

$$M_k(a,b) \equiv \left(\frac{a^k + b^k}{2} \right)^{1/k}. \qquad (8.51)$$

You raise the numbers to the kth power, take the (regular) mean, and then undo
the exponentiation by taking the kth root.

a. What is k for an arithmetic mean?

b. What is k for the rms (root mean square)?

c. The harmonic mean of a and b is sometimes written as $2(a \parallel b)$, where $\parallel$ denotes
    the parallel combination of a and b (introduced in Section 2.4.3). What is k for
    the harmonic mean?

d. (Surprising!) What is k for the geometric mean?

Problem 8.23   Easy case of the compound pendulum

For the compound pendulum of Problem 5.25, what easy case produces an ordinary, noncompound pendulum? Check that, in this limit, your formula for the
period from Problem 5.25 behaves correctly.

Problem 8.24   Means in an elliptical orbit

A planetary orbit (an ellipse) has two important
radii: $r_{\min}$ and $r_{\max}$. The other lengths in the ellipse are power means of these radii (see Prob-
lem 8.22 about power means).

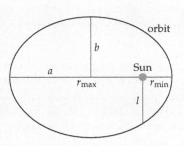

Match the three power means—arithmetic, geometric, and harmonic—to the three lengths: the
semimajor axis a (which is related to the orbital
period), the semiminor axis b, and the semilatus
rectum l (which is related to the orbital angular
momentum).

Hint: The power-mean theorem says that $M_m(a,b) < M_n(a,b)$ if and only if $m < n$.

Problem 8.25   Four regimes in orbital motion

Once in a while, there are *four* interesting easy-cases regimes. An example is orbits.
The dimensionless parameter β that characterizes the type of an orbit is

$$\beta \equiv \frac{\text{kinetic energy}}{\left|\text{gravitational potential energy}\right|}, \tag{8.52}$$

where the absolute value handles a possibly negative potential energy.

A related dimensionless parameter is the orbit eccentricity ϵ. In terms of the eccentricity, a planet's orbit in polar coordinates is

$$r(\theta) = \frac{l}{1 + \epsilon \cos \theta}, \tag{8.53}$$

where the Sun is at the origin, and l is the length scale of the orbit (l is diagrammed
in Problem 8.24). Sketch and classify the four orbit shapes according to their values
of β and ϵ (giving a point value or a range, as appropriate): (a) circle, (b) ellipse,
(c) parabola, and (d) hyperbola.

Problem 8.26   Superfluid helium

Helium, when cold, turns into a liquid. When very cold, the liquid turns into a
superfluid—a quantum liquid. Here is a dimensionless ratio determining how
quantum the liquid is

$$\beta \equiv \frac{\text{quantum uncertainty in the position of a helium atom}}{\text{separation between atoms}}. \tag{8.54}$$

a. Estimate β in terms of the quantum constant $\hbar$, helium's density ρ (as a liquid),
the thermal energy $k_B T$, and the atomic mass m_{He}.

b. In the $\beta \sim 1$ regime, helium becomes a quantum liquid (a superfluid). Thus,
estimate the superfluid transition temperature.

Problem 8.27   Adiabatic atmosphere

The simplest model of the atmosphere is isothermal: The atmosphere has one temperature throughout it. A better approximation, the adiabatic atmosphere, relaxes
this assumption and incorporates the adiabatic gas law:

$$pV^\gamma \propto 1, \tag{8.55}$$

where p is atmospheric pressure, V is the volume of a parcel of air, and $\gamma \equiv c_p/c_v$
is the ratio of the two specific heats in the gas. (For dry air, $\gamma = 1.4$.) Imagine an
air parcel rising up a mountain. As the parcel rises into air with a lower pressure,
it expands, and its volume and temperature change according to a combination of
the adiabatic and ideal-gas laws.

a. What easy case of γ reproduces the isothermal atmosphere?

b. For $\gamma = 1.4$ (dry air), will air temperature decrease with, increase with, or be
independent of height?

Problem 8.28   Loan payments

A fixed-term, fixed-interest-rate loan has four important parameters: the principal P (the amount borrowed), the interest rate r, the repayment interval τ, and the number of payments n. The loan is repaid in n equal payments over the loan term $n\tau$. Each payment consists of a principal and an interest portion. The interest portion is the interest accumulated on the principal outstanding during the term; the principal portion reduces the outstanding principal.

The dimensionless quantity determining the type of loan is $\beta \equiv n\tau r$.

a. Estimate the payment (the amount per term) in the easy case $\beta = 0$, in terms of P, n, and τ. (The term $n\tau$ and the repayment interval τ don't vary that much—τ is usually 1 month and $n\tau$ is somewhere between 3 to 30 years—so $\beta \ll 1$ is usually reached by lowering the interest rate r.)

b. Estimate the payment in the slightly harder case where $\beta \ll 1$ (which includes the $\beta = 0$ case). In this regime, the loan is called an installment loan.

c. Estimate the payment in the easy case $\beta \gg 1$. In this regime, the loan is called an annuity. (This regime is usually reached by increasing n.)

Problem 8.29   Heavy nuclei

In this problem, you study the innermost electron in an atom such as uranium that has many protons, and analyze a surprising physical consequence of its binding energy. Imagine a nucleus with Z protons around which orbits one electron. Let $E(Z)$ be the binding energy (the hydrogen binding energy is the case $Z = 1$).

a. Show that the ratio $E(Z)/E(1)$ is Z^2.

b. In Problem 5.36, you showed that $E(1)$ is the kinetic energy of an electron moving with speed αc where α is the fine-structure constant (roughly 10^{-2}). How fast does the innermost electron move around a heavy nucleus with charge Z?

c. When that speed is comparable to the speed of light, the electron has a kinetic energy comparable to its (relativistic) rest energy. One consequence of such a high kinetic energy is that the electron has enough kinetic energy to produce a positron (an anti-electron) out of nowhere; this process is called pair creation. That positron leaves the nucleus, turning a proton into a neutron as it exits: The atomic number Z decreases by one. The nucleus is unstable! Relativity therefore places an upper limit to Z. Estimate this maximum Z and compare it with the Z for the heaviest stable nucleus (uranium).

Problem 8.30   Minimum wave speed

For deep-water waves, estimate the minimum wave speed in terms of ρ, g, and γ (the surface tension). Test your prediction in two different ways. (1) Drop a pebble into water, and observe how fast the slowest ripples move outward. (2) Move a toothpick through a pan of water, and look for the fastest speed at which the toothpick generates no waves.

Problem 8.31   Surface tension and the size of raindrops

A liquid's surface tension, usually denoted γ, is the energy required to create new surface divided by the area of the new surface. For a falling raindrop, surface tension and drag compete: the drag force flattens the raindrop, and surface tension keeps it spherical. If the drop gets too flat, it can lower its surface energy by breaking into smaller and more spherical droplets. The fluid dynamics is complicated, but we don't need to know it. Instead, use competition reasoning (easy cases) to estimate the maximum size of raindrops.

Problem 8.32   Waves driven by surface tension

Imagine a wave with reduced wavelength λ on the surface of a fluid.

a. Show that the dimensionless ratio

$$R \equiv \frac{\text{potential energy due to gravity}}{\text{potential energy due to surface tension}} \qquad (8.56)$$

is, after ignoring dimensionless constants, the dimensionless group $\rho g \lambda^2 / \gamma$ that we used in Section 8.4.1 to distinguish waves driven by gravity from waves driven by surface tension.

b. For water, estimate the critical wavelength λ at which $R \sim 1$.

Problem 8.33   Including buoyancy

The terminal speed v of a raindrop with radius r can be written in the following dimensionless form:

$$\frac{v^2}{rg} = f\left(\frac{\rho_{\text{water}}}{\rho_{\text{air}}}\right). \qquad (8.57)$$

In this problem, you use easy cases of $x \equiv \rho_{\text{water}}/\rho_{\text{air}}$ to guess how buoyancy affects this result. (Imagine that you may vary the density of air or water as needed.) In dimensional analysis, including the buoyant force requires including ρ_{air}, g, and r in order to compute the weight of the displaced fluid (which is the buoyant force)—but those variables are already included in the dimensional analysis. Therefore, including buoyancy doesn't require a new dimensionless group. So it must change the form of the dimensionless function f.

a. Before you account for buoyancy: What is the dimensionless function $f(x)$? Assume spherical raindrops and that $c_{\text{d}} \approx 0.5$.

b. What would be the effect of buoyancy if ρ_{water} were equal to ρ_{air}? This thought experiment is the easy case $x = 1$. Therefore, find $f(1)$.

c. Guess the general form of f with buoyancy, and thereby find v including the effect of buoyancy.

d. Explain physically the difference between v without and with buoyancy. Hint: How does buoyancy affect g?

9

Spring models

Our final tool for mastering complexity is making spring models. The essential characteristics of an ideal spring, the transferable abstractions, are that it produces a restoring force proportional to the displacement from equilibrium and stores an energy proportional to the displacement squared. These seemingly specific requirements are met far more widely than we might expect. Spring models thereby connect chemical bonds (Section 9.1), xylophone notes (Section 9.2.3), gravitational radiation (Section 9.3.3), and the colors of the sky and sunsets (Section 9.4).

9.1 Bond springs

A ubiquitous spring is the bond between the electron and proton in hydrogen—the bond that is our model for all chemical bonds. In Section 9.1.1, we'll build a spring model of hydrogen, giving us a physical model for the Young's modulus (Section 9.1.2) and for the speed of sound (Section 9.1.3).

9.1.1 Finding the spring

In Section 8.3.2.2, we saw how hydrogen is a competition between electrostatics and quantum mechanics. When the electron–proton separation x is

much smaller than the Bohr radius a_0, quantum mechanics wins, and the
net force on the electron is outward (positive). When the separation is much
larger than the Bohr radius, electrostatics wins, and the net force is inward
(negative). Between these two extremes, when the separation is the Bohr
radius (when $x = a_0$), the force crosses the zero-force line (the x axis).

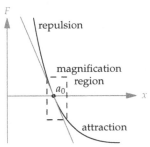

This equilibrium point is an easy case that helps us describe the bond force simply. We magnify the force curve
at the equilibrium point. The curve now looks straight,
because any curve looks straight at a large-enough magnification. Equivalently, any curve can be approximated
locally by its tangent line—which is an example of lumping shapes and graphs (Section 6.4) and is where spring
models discard actual information and complexity.

Physically, the straight-line approximation means that,
as long as the bond distance differs from a_0 by only a small amount Δx, the
force is linearly proportional to the deviation Δx. Furthermore, the force
curve has a negative slope: A negative deviation produces a positive force,
and vice versa. The force therefore opposes the deviation and is a restoring force. A linear restoring force is the force from an ideal spring, so the
electron–proton bond is an ideal spring! It has an equilibrium length a_0 at
which $F = 0$ and spring constant k, where $-k$ is the slope of the force curve.

$$F = -k\Delta x. \tag{9.1}$$

To make this small Δx stretch, the required energy ΔE is

$$\Delta E \sim \text{force} \times \Delta x. \tag{9.2}$$

Because the force varies from 0 to $k\Delta x$, the typical or characteristic force is
comparable to $k\Delta x$. Then

$$\Delta E \sim k\Delta x \cdot \Delta x = k(\Delta x)^2. \tag{9.3}$$

The scaling exponent connecting ΔE and Δx is 2. This quadratic dependence on the displacement is the energy signature of a spring. For small displacements around the equilibrium point (the minimum), the energy-ver-
sus-displacement curve is a parabola. (This analysis is a physical version of
a Taylor-series approximation.)

Because almost any energy curve has a minimum, almost every system contains a spring. For the bond spring, the energy relation $\Delta E \sim k(\Delta x)^2$ gives
us an estimate of the spring constant k. Pretend that the energy curve is exactly a parabola, even for large displacements, and increase Δx to the bond

size a (for hydrogen, a is the Bohr radius a_0). That stretch requires an energy $\Delta E \sim ka^2$. It is a characteristic energy of the bond, so it must be comparable to the bond energy E_0. Then $E_0 \sim ka^2$, and

$$k \sim \frac{E_0}{a^2}. \tag{9.4}$$

This relation is an estimate for the spring constant in terms of quantities that we know in other ways: E_0 from the heat of vaporization and a from the density and atomic mass. The estimate is often off by a factor of 3 or 10, because of the inaccuracy in extending the parabolic, ideal-spring approximation to displacements comparable to the bond length. But it gives us an order of magnitude that will be useful in subsequent estimates.

9.1.2 **Young's modulus**

From the spring constant of one bond, we could find the spring constant of a block of material. But as we discussed in Section 5.5.4, a better measure is the Young's modulus Y: It is an intensive quantity, so not dependent on the block's dimensions. The Young's modulus is measured by stretching a block of material with a force F at each end.

$$Y \equiv \frac{\text{stress}}{\text{strain}}. \tag{9.5}$$

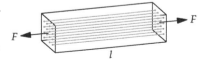

The stress is F/A, where A is the block's cross-sectional area. Estimating the strain requires more steps. Fortunately, the estimate breaks into a tree. To grow it, imagine the block as a bundle of fibers, each a chain of springs (bonds) and masses (atoms). Because strain is the *fractional* length change, the strain in the block is the strain in each fiber and the strain in each spring of each fiber:

$$\text{strain} = \frac{\text{spring extension}}{\text{bond length } a}. \tag{9.6}$$

That's the root of the tree. Here are the internal nodes:

$$\text{spring extension} = \frac{\text{force/fiber}}{\text{spring constant } k}; \tag{9.7}$$

$$\frac{\text{force}}{\text{fiber}} = \frac{F}{N_{\text{fibers}}}. \tag{9.8}$$

The number of fibers will be

$$N_{\text{fibers}} = \frac{\text{cross-sectional area } A}{\text{per-fiber cross-sectional area } a^2}. \qquad (9.9)$$

Here is how the leaf values propagate to the root.

1. The force per fiber becomes Fa^2/A.

2. The per-spring extension becomes Fa^2/kA.

3. The strain becomes Fa/kA.

Finally, the Young's modulus becomes k/a:

$$Y \equiv \frac{\text{stress}}{\text{strain}} = \frac{F/A}{Fa/kA} = \frac{k}{a}. \qquad (9.10)$$

Because $k \sim E_0/a^2$ (Section 9.1.1), $Y \sim E_0/a^3$, which confirms with a spring model our dimensional-analysis prediction in Section 5.5.4.

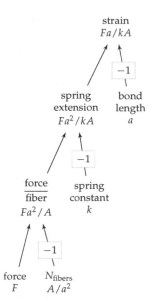

9.1.3 Sound speed in solids and liquids

The spring model of solids and even liquids also gives a physical model for the speed of sound. Start with a fiber of atoms and bonds:

(m)WWWWWWW(m)WWWWWWW(m) $\cdots$

$|\!\!\leftarrow\!\!-\!\!-\!\! a_{\text{bond}} \!\!-\!\!-\!\!\rightarrow\!\!|$

The sound speed is the speed at which a vibration signal travels along the fiber. Let's study the simplest lumped signal: In one instant, the first mass moves to the right by a distance x, the signal amplitude.

$(\ )$ (m)WWWWWW(m)WWWWWWW(m) $\cdots$

$|\!\!\leftarrow\!\! x \!\!\rightarrow\!\!|$

The compressed bond pushes the second atom to the right. When the second atom has moved to the right by the signal amplitude x, the signal has traveled one bond length. The sound speed is the propagation distance a_{bond} divided by this one-bond propagation time.

$(\ )$ (m)WWWWWWW(m)WWWWWW(m) $\cdots$

$|\!\!\leftarrow\!\! x \!\!\rightarrow\!\!|$ $|\!\!\leftarrow\!\! x \!\!\rightarrow\!\!|$

▷ *What is the propagation time? That is, roughly how much time does the second mass require to move to the right by the signal amplitude x?*

The second mass moves because of the spring force from the first spring. At $t = 0$, the spring force is kx. As the mass moves, the spring loses compression, and the force on the second mass falls. So an exact calculation of the propagation time requires solving a differential equation. But lumping will turn the calculation into algebra: Just replace the changing spring force with its typical or characteristic value, which is comparable to kx.

The force produces a typical acceleration $a \sim kx/m$. After a time t, the mass will have a velocity comparable to at and therefore will have moved a distance comparable to at^2. This force acts for long enough to move the mass by a distance x, so $at^2 \sim x$. Using $a \sim kx/m$ gives $kxt^2/m \sim x$. The amplitude x cancels, as it always does in ideal-spring motion. (Thus, the sound speed does not depend on loudness.) The propagation time is then

$$t \sim \sqrt{\frac{m}{k}}. \tag{9.11}$$

This characteristic time is just the reciprocal of the natural frequency $\omega_0 = \sqrt{k/m}$. In this time, the signal travels a distance a_{bond}, so

$$c_{\text{s}} \sim \frac{\text{distance traveled}}{\text{propagation time}} \sim \frac{a_{\text{bond}}}{\sqrt{m/k}} = \sqrt{\frac{ka_{\text{bond}}^2}{m}}. \tag{9.12}$$

To make this expression more meaningful, let's convert the numerator and denominator, which are in terms of microscopic (atomic) quantities, into macroscopic quantities. To do so, divide by $a_{\text{bond}}^3/a_{\text{bond}}^3$ in the square root:

$$c_{\text{s}} \sim \sqrt{\frac{ka_{\text{bond}}^2/a_{\text{bond}}^3}{m/a_{\text{bond}}^3}} = \sqrt{\frac{k/a_{\text{bond}}}{m/a_{\text{bond}}^3}}. \tag{9.13}$$

The numerator k/a_{bond} is, as we saw in Section 9.1.2, the Young's modulus Y. The denominator is the mass per molecular volume, so it is the substance's density ρ. Thus, $c_{\text{s}} \sim \sqrt{Y/\rho}$. Our physical, spring model therefore confirms our estimate for c_{s} in Section 5.5.4 based on dimensions analysis.

9.2 Energy reasoning

The analysis of sound propagation in Section 9.1.3 required estimating the spring forces. Often, however, the forces or their effects are harder to track than are the energies. Then, as you'll see in the next examples, we track the energy and look for the energy signature of a spring: the quadratic dependence of energy on displacement.

9.2.1 Oscillation frequency of a spring–mass system

To illustrate the energy method, let's practice on the most familiar spring system by finding its natural (angular) oscillation frequency ω_0. The method requires finding its kinetic and potential energies. Because these energies vary in complicated ways, we use typical or characteristic energies.

In terms of the amplitude A_0, the typical potential energy is comparable to kA_0^2. The typical kinetic energy is comparable to mv^2, where v is the typical speed of the oscillating mass. This speed is comparable to $A_0\omega_0$, because the mass travels a distance comparable to A_0 in the characteristic time $1/\omega_0$ (which corresponds to 1 radian, or approximately one-sixth of an oscillation period). Thus, the typical kinetic energy is comparable to $mA_0^2\omega_0^2$.

In spring motion, kinetic and potential energy interconvert, so the ratio

$$\frac{\text{typical potential energy}}{\text{typical kinetic energy}} \tag{9.14}$$

should be comparable to 1. This bold conclusion is not limited to spring motion. For example, for gravitational orbits, the ratio, defined carefully using the time-averaged energies, is -2. More generally, the virial theorem says that, with a potential $V \propto r^n$, the energy ratio will be $2/n$.

Equating the typical energies gives an equation for ω_0:

$$\underbrace{kA_0^2}_{E_{\text{potential}}} \sim \underbrace{mA_0^2\omega_0^2}_{E_{\text{kinetic}}}. \tag{9.15}$$

The amplitude A_0 divides out—another illustration that a spring's period is independent of amplitude—giving $\omega_0 \sim \sqrt{k/m}$. Because the energy ratio is 1 (due to the virial theorem), the missing dimensionless prefactor is 1.

9.2.2 Vibrations of a piano string

From springs to strings: A piano string is a steel wire stretched close to its breaking point—the high tension makes the string's resistance to bending less important and the sound cleaner (as you investigated in Problem 9.17). When you push a piano key, a hammer bangs on the string and sets it into vibration—whose frequency we'll estimate with a spring model.

For a physical model, start with an unstretched piano string of length L. It is a bundle of springs and masses, so it acts like one large spring. Now stretch the string by putting it under tension T. Then hammer it.

The hammer gives the atoms vertical velocities. Eventually the kinetic energy turns into potential energy, and the string gets a sinusoidal shape with a wavelength $\lambda = 2L$ and a small amplitude y_0.

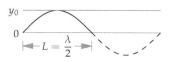

As we did for the simple spring–mass system (Section 9.2.1), we'll find the typical potential energy in the string and the typical kinetic energy in the motion of the string. The potential energy comes from the tension force: The curved string is longer than the equilibrium string, so the tension force has done work on the string by stretching it; the string stores that work as its potential energy. The work done is the force times the distance, so

$$E_{\text{potential}} \sim T \times \text{extra length.} \qquad (9.16)$$

To estimate the extra length, lump a piece of the curved string as the hypotenuse of a right triangle with base $\lambdabar$, where $\lambdabar \equiv \lambda/2\pi$. This base represents 1 radian of the sine-wave shape. In 1 radian, a sine wave attains almost its full height ($\sin 1 \approx 0.84$), so the height of the triangle is comparable to the amplitude y_0. The triangle then has slope $\tan\theta \approx y_0/\lambdabar$. Because $y_0 \ll \lambdabar$, the opening angle θ is small; thus, $\tan\theta \approx \theta$ and $\theta \approx y_0/\lambdabar$.

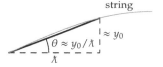

Using our lumping triangle, we'll find the *fractional* change in length between the hypotenuse and the base. Fractional changes, being dimensionless, require less algebra and are more widely applicable than absolute changes. To find the fractional change, rescale the triangle so that the base has unit length; then it has height θ and hypotenuse $\sqrt{1 + \theta^2}$. Because θ is small, the square root is, by the binomial theorem, approximately $1 + \theta^2/2$. Thus, the fractional increase in length is comparable to θ^2, which is $y_0^2/\lambdabar^2$.

The fractional increase applies to the whole string, whose length was L, so

$$\text{extra length} \sim L\left(\frac{y_0}{\lambdabar}\right)^2. \qquad (9.17)$$

The work done in making this much stretch, which is the potential energy in the string, is $T \times \text{extra length}$, where T is the tension, so

$$E_{\text{potential}} \sim TL\left(\frac{y_0}{\lambdabar}\right)^2. \qquad (9.18)$$

As befits a spring, even a giant one composed of individual bond springs, the potential energy is proportional to the square of the amplitude y_0.

Now let's estimate the kinetic energy in the motion of the string. As the string vibrates at the so-far-unknown angular frequency ω, the pieces of the string move up and down with a typical speed ωy_0. Thus,

$$E_{\text{kinetic}} \sim \text{mass} \times (\text{typical speed})^2 \sim \underbrace{\rho b^2 L}_{m} \times \underbrace{\omega^2 y_0^2}_{\sim v^2}, \tag{9.19}$$

where ρ is the string's density and b is its diameter. The kinetic energy is also proportional to the squared amplitude y_0^2—the other energy signature of a spring. Equating the energies gives the equation for ω:

$$\underbrace{\rho b^2 \cancel{L} \omega^2 \cancel{y_0^2}}_{\sim E_{\text{kinetic}}} \sim \underbrace{T \cancel{L} \left(\frac{\cancel{y_0}}{\lambdabar} \right)^2}_{\sim E_{\text{potential}}}. \tag{9.20}$$

The length L and the squared amplitude y_0^2 cancel, leaving

$$\omega = \frac{1}{\lambdabar} \sqrt{\frac{T}{\rho b^2}}. \tag{9.21}$$

Despite the extensive use of lumping, this result turns out to be exact—as do many energy-based spring analyses. The circular frequency $f = \omega/2\pi$ has the same structure:

$$f = \frac{1}{\lambda} \sqrt{\frac{T}{\rho b^2}}. \tag{9.22}$$

The wave propagation speed is $f\lambda$ (or $\omega\lambdabar$), which is just $\sqrt{T/\rho b^2}$. Let's check that this speed makes sense. As a first step, it can be rewritten as

$$v = \sqrt{\frac{T/b^2}{\rho}}. \tag{9.23}$$

Because the numerator T/b^2 is the pressure (force per area) applied to the ends of the string to put it under the tension T, the speed of these transverse waves is $\sqrt{\text{applied pressure}/\rho}$. (They are called transverse waves because the direction of vibration is perpendicular, or transverse, to the direction of travel.) This speed is analogous to the speed of sound that we found in Section 5.5.4, $\sqrt{\text{pressure}/\rho}$, where the pressure was the gas pressure or the elastic modulus. So our speed makes good sense.

▷ *How long is the middle-C string on a piano?*

We can find the length from the propagation speed and the frequency. The frequency of middle C is roughly 250 hertz. The propagation speed is

$$v = \sqrt{\frac{\sigma}{\rho}}, \tag{9.24}$$

where σ is the stress T/b^2. This stress is also ϵY, where ϵ is the strain (the fractional change in length). Using $\sigma = \epsilon Y$ and $c_s = \sqrt{Y/\rho}$,

$$v = \sqrt{\frac{\epsilon Y}{\rho}} = \sqrt{\epsilon}\sqrt{\frac{Y}{\rho}} = \sqrt{\epsilon}\, c_s. \tag{9.25}$$

Thus, in terms of the Mach number M, which measures speeds relative to the full wave speed, transverse waves have a Mach number of $\sqrt{\epsilon}$.

In steel, $c_s \approx 5$ kilometers per second. For piano wire, which is made of high-strength carbon steel, the yield strain is roughly 0.01. However, the string is not stretched quite so far. To provide a margin of safety, the strain ϵ is kept to roughly 3×10^{-3}. Then the transverse-wave speed is

$$v \approx \underbrace{0.06}_{\sqrt{\epsilon}} \times \underbrace{5 \times 10^3\,\text{m s}^{-1}}_{c_s} = 300\,\text{m s}^{-1}. \tag{9.26}$$

At $f \approx 250$ hertz, the wavelength is roughly 1.2 meters:

$$\lambda = \frac{v}{f} \approx \frac{300\,\text{m s}^{-1}}{250\,\text{Hz}} = 1.2\,\text{m}. \tag{9.27}$$

The wavelength of this lowest, fundamental frequency is twice the string's length, so the string should be 0.6 meters long. To check, I looked into our piano: 0.6 meters is almost exactly the length of the strings set into motion when I play middle C.

9.2.3 Musical notes from bending beams

Another musical device that we can model is a wooden or metal slat in a marimba or xylophone. Using spring models, proportional reasoning, and dimensional analysis, we'll find how the frequency of the slat's musical note depends on its dimensions.

Thus, imagine a thin block of wood of length l, width w, and thickness h. It is supported at the two dots (or held at one of them) and tapped in the center. As it vibrates, its shape varies from bent to straight and back to bent. Here are the shapes shown in side view.

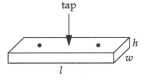

▷ *How does the slat's width w affect the frequency?*

The answer comes from that cheapest kind of experiment, a thought experiment. Using a type of argument
developed by Galileo in his study of free fall, lay two
identical slats of width w side by side. Tapping both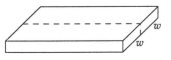
slats simultaneously produces the same motion as does tapping a slat of
width $2w$ (the result of gluing the two slats along the long, thin edge). So
the width cannot affect the frequency.

▷ *How does the slat's thickness h affect the frequency?*

The slat, made of atoms connected by bond springs, acts
like a giant spring–mass system. When the slat is straight
(the equilibrium position), it has zero potential energy. The
energy increases upon bending the slat, which stretches or
compresses the bond springs. Thus, the slat resists bending. As befits a
giant spring, its resistance to bending can be represented by a stiffness or
spring constant k. (Mechanical engineers define a related quantity called
the bending stiffness or the flexural rigidity, which has dimensions of energy times length. Our stiffness is an actual spring constant, with dimensions of force per length.)

Then, as befits a giant spring–mass system, the slat has a vibration frequency comparable to $\sqrt{k/m}$, where m is its mass. Deciding how the thickness affects the frequency has split into two smaller problems: how the
thickness affects the mass and how the thickness affects the stiffness. The
first decision is not difficult: The mass is proportional to the thickness.

▷ *How does the stiffness k depend on thickness?*

To answer this proportional-reasoning question, we'll perform the thought
experiment of bending each slat by the same vertical deflection y.

The stiffness k is proportional to the force F required to bend the beam
($F = -ky$). However, we won't try to understand k by finding how the
thickness affects F itself. Force is a vector, so finding the required force requires carefully bookkeeping many little forces and their directions to know

which contributions cancel. Instead, we'll find how the thickness affects the stored energy (the potential energy). As a positive scalar quantity, potential energy has no direction or even sign and is therefore easier to bookkeep.

Because a slat is a big spring, the energy required to produce the vertical deflection y is given by

$$E \sim ky^2. \tag{9.28}$$

Because y is the same for the slats (an easy condition to enforce in a thought experiment), the energy relation becomes the proportionality $E \propto k$. To find how k depends on thickness, let's redraw the bent slats with a dotted line showing the neutral line (the line without compression or extension):

Above the neutral line, the bond springs along the length of the block get extended; below the neutral line, they get compressed. The compression or extension Δl determines the energy stored in each bond spring. Then the stored energy in the whole slat is

$$E \sim E_{\text{typical spring}} \times N_{\text{springs}}. \tag{9.29}$$

Because $E \propto k$,

$$k \propto E_{\text{typical spring}} \times N_{\text{springs}}. \tag{9.30}$$

To find how $E_{\text{typical spring}}$ depends on the slat's thickness h, break the energy into factors (divide and conquer):

$$E_{\text{typical spring}} \sim k_{\text{bond}} \times (\Delta l)^2_{\text{typical spring}}. \tag{9.31}$$

Because the bonds in the two blocks are the same (the blocks differ only in thickness), this relation becomes the proportionality

$$E_{\text{typical spring}} \propto (\Delta l)^2_{\text{typical spring}}. \tag{9.32}$$

Therefore, the stiffness is

$$k \propto (\Delta l)^2_{\text{typical spring}} \times N_{\text{springs}}. \tag{9.33}$$

To find how $(\Delta l)_{\text{typical spring}}$ depends on h, compare typical bond springs in the thick and thin blocks—for example, a spring halfway from the neutral line to the top surface. Because the thick block is twice as thick as the thin block, this spring is twice as far from the neutral line in absolute distance.

The extension is proportional to the distance from the neutral line—as you can guess by observing that it is the simplest scaling relationship that predicts zero extension at the neutral line (or try Problem 9.1). In symbols,

$$(\Delta l)_{\text{typical spring}} \propto h. \tag{9.34}$$

Furthermore, N_{springs} is also proportional to h. Therefore, $k \propto h^3$:

$$k \propto (\Delta l)^2_{\text{typical spring}} \times N_{\text{springs}} \propto h^3. \tag{9.35}$$

Doubling the thickness multiplies the stiffness by eight! The vibration frequency of a slat, considered as a giant spring, is

$$\omega \sim \sqrt{\frac{k}{m}}. \tag{9.36}$$

The mass is proportional to h, so $\omega \propto h$:

$$\omega \propto \sqrt{\frac{h^3}{h}} = h. \tag{9.37}$$

Doubling the thickness should double the frequency. To test this prediction, I tapped two pine slats having these dimensions:

$$\underbrace{30\,\text{cm}}_{l} \times \underbrace{5\,\text{cm}}_{w} \times \begin{cases} 1\,\text{cm} & (h \text{ for the thin slat}); \\ 2\,\text{cm} & (h \text{ for the thick slat}). \end{cases} \tag{9.38}$$

To measure the frequencies, I matched each slat's note to a note on a piano. The thin slat sounded like C one octave above middle C. The thick slat sounded like A in the octave above the thin block. The interval between the two notes is almost an octave or a factor of 2 in frequency.

Now let's extend the analysis to the xylophone slats, which vary not in thickness but in length. This extension bring us to the third scaling question.

▷ *How does the slat's length l affect the frequency?*

We already found the scaling relation between ω (frequency) and w (width), namely $\omega \propto w^0$; and between ω and h (thickness), namely $\omega \propto h$. By adding the constraints of dimensional analysis to these scaling relations, we can find the scaling relation between ω and l.

The quantities relevant to the frequency ω are the speed of sound c_s and two of the three dimensions: thickness h and length l. The third dimension, the width, is not on the list, because the frequency, we already found, is independent of the width. (Instead of the speed of sound, the list could include the Young's modulus Y and the density ρ. As the only two variables containing mass, Y and c_s would end up combining anyway to make c_s.)

These four quantities, made from two dimensions, make two independent dimensionless groups. The first group should be proportional to the goal ω. Because ω and c_s are the only quantities containing time, each as T^{-1}, the group has to contain ω/c_s. Making ω/c_s dimensionless requires multiplying by a length. Either of the two lengths works. Let's choose l. (For the alternative, try Problem 9.2). Then the group is $\omega l/c_s$.

ω	T^{-1}	frequency
c_s	LT^{-1}	sound speed
h	L	thickness
l	L	length

The other group should not contain the goal ω. Then the only choices are powers of the aspect ratio h/l. If we use h/l itself, the most general dimensionless statement is

$$\frac{\omega l}{c_s} = f\left(\frac{h}{l}\right). \tag{9.39}$$

The thickness scaling relation, $\omega \propto h$, determines the form of f, giving

$$\frac{\omega l}{c_s} \sim \frac{h}{l}. \tag{9.40}$$

Solving for the frequency,

$$\omega \sim \frac{c_s h}{l^2}. \tag{9.41}$$

As a scaling relation, $\omega \propto l^{-2}$. Let's check the scaling exponent against experimental data. When my older daughter was small, she got a toy xylophone from her uncle. Its (metal) slats have the tabulated dimensions and frequencies. The lower and higher C notes (C and C′) are a factor of 2 apart in frequency. If the scaling relation is correct, C should come from the longer slat, and the ratio of slat lengths should be $\sqrt{2}$. Indeed, the measured length ratio is almost exactly $\sqrt{2} \approx 1.414$:

$$\frac{12.2\,\text{cm}}{8.6\,\text{cm}} \approx 1.419. \tag{9.42}$$

	l (cm)	f (Hz)
C	12.2	261.6
D	11.5	293.6
E	10.9	329.6
F	10.6	349.2
G	10.0	392.0
A	9.4	440.0
B	8.9	493.8
C′	8.6	523.2

Problem 9.1   Spring extension versus distance from the neutral line

Consider a bent slat as an arc of a circle, and thereby explain why the spring extension is proportional to the distance from the neutral line.

Problem 9.2   Alternative dimensionless group

Repeat the dimensional analysis for the dependence of the oscillation frequency on slat length but using $\omega h/c_s$ and h/l as the two independent dimensionless groups. Do you still conclude that $\omega \propto l^{-2}$?

Solving the beam differential equation gives the vibration frequency as

$$f \approx \frac{3.56}{\sqrt{12}} \times c_s \frac{h}{l^2}. \tag{9.43}$$

The dimensionless prefactor $3.56/\sqrt{12}$ is almost exactly 1. This example is one of the rare cases where using circular frequency (f) rather than angular frequency (ω) makes the prefactor closer to 1.

For my block of pine, a light wood, $\rho \approx 0.5\rho_{\text{water}}$ and $Y \approx 10^{10}$ pascals, so

$$c_s = \sqrt{\frac{Y}{\rho}} \sim \sqrt{\frac{10^{10}\,\text{Pa}}{0.5 \times 10^3\,\text{kg m}^{-3}}} \approx 4.5\,\text{km s}^{-1}. \tag{9.44}$$

For the thin block, $h = 1$ centimeter and $l = 30$ centimeters, so

$$f \approx c_s \frac{h}{l^2} \approx 4.5 \times 10^3\,\text{m s}^{-1} \times \frac{10^{-2}\,\text{m}}{10^{-1}\,\text{m}^2} \sim 450\,\text{Hz}. \tag{9.45}$$

This estimate is reasonably accurate. The thin block's note was approximately one octave above middle C, with a frequency of roughly 520 hertz.

Problem 9.3   Graphing the data on frequency versus length

Check the scaling $\omega \propto l^{-2}$ by plotting the xylophone data for frequency versus length on log–log axes. What slope should the graph have?

Problem 9.4   Finding the stiffness, then the frequency

Use dimensional analysis to write the most general dimensionless statement connecting stiffness k to a slat's Young's modulus Y, width w, length l, and thickness h. What is the scaling exponent q in $k \propto w^q$? Use that scaling relation and $k \propto h^3$ to find the missing exponents in $k \sim Y^p w^q l^r h^3$.

Problem 9.5   Xylophone notes

If you double the width, thickness, and length of a xylophone slat, what happens to the frequency of the note that it makes?

Problem 9.6   Location of the node

Here's how you can predict the location of the node (where to hold the wood block) using conservation. Because the bar vibrates freely without an external force, the center of mass (the dot) stays fixed. Approximate the bar's shape as a shallow parabola, find the center of mass (CM), and therefore find the node locations (as a fraction of the bar's length). My daughter's longest xylophone slat is 12.2 centimeters long with holes 2.7 centimeters from the ends. Is that fraction consistent with your prediction?

9.3 Generating sound, light, and gravitational radiation

The sound generated by the vibrating wood blocks is an example of the most pervasive spring: radiation. It comes in three varieties. Electromagnetic radiation (or, more informally, light) is produced by an accelerating charge. Sound (acoustic radiation) can be produced simply by a changing but nonmoving charge (such as an expanding or contracting speaker membrane). Therefore, sound is simpler than light—which in turn is simpler than gravitational radiation. Do the easy cases first: We'll first apply spring models to sound (Section 9.3.1). By adding the complexity of motion, we'll extend the analysis to light (Section 9.3.2). Then we'll be ready for the complexity of gravitational radiation (Section 9.3.3).

9.3.1 Acoustic radiation from a charge monopole

When we think of radiation, we think first of electromagnetic radiation, which we see (pun intended) everywhere. To analyze sound radiation while benefiting from what we know about electromagnetic radiation, we'll find an analogy between electromagnetic and acoustic radiation—starting at the source of radiation, namely a single charge (a monopole).

The search for the acoustic analog of charge is aided by scaling relations. An electric charge q produces a disturbance, the electric field E. Their connection is $E \propto q$. Because the symbols E and q amplify the mental connection to electromagnetism, let's write the relation between E and q in words. Words promote a broader, more abstract view not limited to electromagnetism:

$$\text{field} \propto \text{charge.} \tag{9.46}$$

Another transferable scaling comes from the energy density $\mathcal{E}$ (energy per volume) in the field. For an electric field, $\mathcal{E} \propto E^2$. In words,

$$\text{energy density} \propto \text{field}^2. \tag{9.47}$$

Sound waves move fluid, and motion means kinetic energy. Because the kinetic-energy density is proportional to the fluid velocity squared, the acoustic field could be the fluid velocity v itself. Then our first scaling relation, that field $\propto$ charge, becomes

$$v \propto \text{charge.} \tag{9.48}$$

Thus, an acoustic charge moves fluid and with a speed proportional to the charge. In contrast to electromagnetism, acoustic charge cannot measure a fixed amount of stuff. For example, it cannot measure simply the volume of a speaker. A fixed volume would produce no motion and no acoustic field.

Instead, acoustic charge must measure a *change* in the source. As an example of such a change, imagine an expanding speaker. As it expands, it pushes fluid outward. The faster it expands, the faster the fluid moves. To identify the kind of change to measure, let's work backward from the electric field E of a point electric charge to the velocity field v of a point acoustic charge. The electric field points outward with magnitude

$$E = \frac{q}{4\pi\epsilon_0 r^2}.$$
(9.49)

Furthermore, the electrostatic ϵ_0 appears in the energy density

$$\mathcal{E} = \frac{1}{2}\epsilon_0 E^2,$$
(9.50)

whose acoustic counterpart is the kinetic-energy density

$$\mathcal{E} = \frac{1}{2}\rho v^2.$$
(9.51)

Because E and v are analogous, the electrostatic ϵ_0 corresponds, in acoustics, to the fluid density ρ. Therefore, in the electric field, let's replace ϵ_0 by ρ, E by v, and q by acoustic charge to get

$$v = \frac{\text{acoustic charge}}{4\pi\rho r^2},$$
(9.52)

or

$$\text{acoustic charge} = \rho v \times 4\pi r^2.$$
(9.53)

Here, r is the distance from the charge, and v is the fluid's speed outward (just as the electric field points outward). Then each factor in the acoustic charge has a meaning, as does the product. The factor ρv is the mass flux:

$$\underbrace{\text{flux}}_{\rho v} = \underbrace{\text{density of stuff}}_{\rho} \times \underbrace{\text{speed}}_{v}.$$
(9.54)

The factor $4\pi r^2$ is the surface area of a sphere of radius r. Thus, the acoustic charge $\rho v \times 4\pi r^2$ measures the rate at which mass flows out of this sphere.

The acoustic charge itself, at the center of this sphere, must displace mass at this rate. So the acoustic analog of charge is a mass source. The source could be an expanding speaker that directly forces fluid outward. Alternatively, it could be a hose supplying new fluid that forces the old fluid outward. For the rate, a convenient notation is $\dot{M}$: The dot represents the time derivative, turning mass into a mass rate—which is the charge strength.

	acoustics	electrostatics
field	fluid velocity v	electric field E
source strength (charge)	$\dot{M}$	q
field from a point source	$\dfrac{\dot{M}}{4\pi\rho}\dfrac{1}{r^2}$	$\dfrac{q}{4\pi\epsilon_0}\dfrac{1}{r^2}$

We have found an acoustic field v proportional to r^{-2}. But, as we found in Section 5.4.3, the signature of radiation is that the field is proportional to r^{-1}. So, we have constructed the acoustic analog of a static electric field and charge, but we have not yet constructed a radiating acoustic system.

Producing radiation requires change—for example, due to a speaker. As a model of a speaker, a small pulsating sphere grows and shrinks in response to the music that it broadcasts. Maybe you put the sphere in a fancy box and slap a brand name on it, but growing and shrinking is still its fundamental operating principle and how it makes sound. A simple model of this change is spring motion—a sinusoidal oscillation in the charge:

$$\dot{M} = \dot{M}_0 \cos\omega t. \tag{9.55}$$

At $t = 0$, when $\cos\omega t = 1$, the speaker is expanding at its maximum rate, displacing mass at a rate $\dot{M}_0$. At $t = \pi/\omega$, the speaker is contracting at its maximum rate. Here is one cycle of its oscillation.

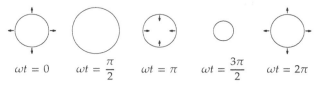

$\omega t = 0 \qquad \omega t = \dfrac{\pi}{2} \qquad \omega t = \pi \qquad \omega t = \dfrac{3\pi}{2} \qquad \omega t = 2\pi$

▷ *How much power does this changing acoustic charge radiate?*

This analysis requires easy cases and lumping. Easy cases will help us find the velocity field; because the charge and field are changing, lumping will help us find the resulting energy flow. The easiest case is near the charge: News about the changes requires no time to arrive, so the fluid responds to a changing $\dot{M}$ instantaneously. In this region, we know v:

$$v = \frac{\dot{M}}{4\pi\rho r^2} \tag{9.56}$$

Far from the charge, however, the fluid cannot know right away that $\dot{M}$ has changed. But what does "far" mean? As we learned in Chapter 8, easy cases are defined by a dimensionless parameter, so the distance from the source is not enough information by itself to decide between near and far.

The decision needs a comparison length. This length is based on how the news is transmitted: It travels as a sound wave, so it has speed c_s. Because the change happens at a rate ω (the angular frequency of the charge oscillation), the characteristic timescale of the changes is $\tau = 1/\omega$. In this time, the charge changes significantly, and the news has traveled a distance $c_s \tau$ or $\sim c_s/\omega$. This distance is the reduced wavelength $\lambdā$ of the sound wave produced by the speaker ($\lambdā \equiv \lambda/2\pi$).

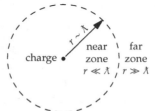

Therefore, "near the charge" (in the near field or zone) means $r \ll \lambdā$. "Far from the charge" (in the far field or the radiation field or zone) means $r \gg \lambdā$. For example, for middle C ($f \approx 250$ hertz, and $\lambda \approx 1.3$ meters), near and far are measured relative to 20 centimeters.

With $\lambdā$ as the comparison length, the dimensionless ratio determining whether r is small or large is $r/\lambdā$, which is $r\omega/c_s$. At the boundary between the near and far zones, $r \sim \lambdā$.

In the lumping model, the velocity field in the near zone follows the changes in $\dot{M}$ instantly, with energy flowing outward and inward in rhythm with the speaker's motion. At the zone boundary, at $r \sim \lambdā$, the velocity field changes its character. It becomes a signal describing those changes, and this signal, a sound wave, travels outward at the speed of sound c_s.

To estimate the power carried by this signal—which is the power radiated by the source—start with the power per area, which is energy flux. At the zone boundary, $r \sim \lambdā$, so

$$\text{energy flux} = \underbrace{\text{energy density (at } r \sim \lambdā)}_{\rho v^2/2} \times \underbrace{\text{propagation speed.}}_{c_s} \qquad (9.57)$$

To estimate the energy density at $r \sim \lambdā$, return to the lumping approximation—that the velocity field tracks the changes in $\dot{M}$ throughout the near zone—and gather enough courage to extend the assumption. Assume that the instantaneous tracking happens all the way out to the zone boundary—that is, it applies not just when $r \ll \lambdā$ but even when $r \sim \lambdā$ (where the field abruptly changes its character and becomes a radiation field).

In this approximation, the velocity field at $r \sim \lambdā$ is

$$v \sim \frac{\dot{M}}{4\pi\rho\lambdabar^2},$$ (9.58)

so the energy density $\mathcal{E} \sim \rho v^2/2$ becomes

$$\mathcal{E} \sim \frac{1}{2}\rho\left(\frac{\dot{M}}{4\pi\rho\lambdabar^2}\right)^2.$$ (9.59)

The radiated power P is the energy flux times the surface area of the sphere enclosing the near zone:

$$P \sim \underbrace{4\pi\lambdabar^2}_{\text{surface area}} \times \underbrace{\frac{1}{2}\rho\left(\frac{\dot{M}}{4\pi\rho\lambdabar^2}\right)^2}_{\text{energy density } \rho v^2/2} \times \underbrace{c_{\text{s}}.}_{\text{propagation speed}}$$ (9.60)

Using $\lambdabar = c_{\text{s}}/\omega$, the power radiated by our acoustic monopole becomes

$$P_{\text{monopole}} = \frac{1}{8\pi}\frac{\dot{M}^2\omega^2}{\rho c_{\text{s}}}.$$ (9.61)

Despite the absurd number of lumping approximations, this result is exact! Let's use it to estimate the acoustic power output of a tiny speaker.

▷ *How much power is radiated by an R = 1 centimeter speaker whose radius varies by ±1 millimeter at f = 1 kilohertz (roughly two octaves above middle C)?*

This calculation becomes slightly simpler if we replace $\dot{M}$ by $\rho\dot{V}$, where $\dot{V}$ is the rate of volume change. (In acoustics, $\dot{V}$ is often called the source strength Q—for example, in the classic work *The Physics of Musical Instruments* [15, p. 172]. However, for the sake of the analogy with electromagnetism, it is more consistent to make the source strength $\dot{M}$ rather than $\dot{V}$.) In terms of $\dot{V}$,

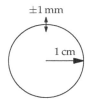

$$P_{\text{monopole}} = \frac{1}{8\pi}\frac{\rho\dot{V}^2\omega^2}{c_{\text{s}}}.$$ (9.62)

Here, $\dot{V} = \dot{V}_0\cos\omega t$, where $\dot{V}_0$ is the amplitude of the oscillations in $\dot{V}$. Therefore, the power P_{monopole} is also oscillating. By symmetry, the average value of $\cos^2\omega t$ is 1/2 (Problem 3.38), so the time-averaged power is one-half the maximum power:

$$P_{\text{avg}} = \frac{1}{16\pi}\frac{\rho\dot{V}_0^2\omega^2}{c_{\text{s}}}.$$ (9.63)

To find the amplitude $\dot{V}_0$, write $\dot{V}$ in terms of the speaker dimensions:

$$\dot{V} = \underbrace{4\pi R^2}_{\text{surface area}} \times v_{\text{surface}}, \tag{9.64}$$

where v_{surface} is the outward speed of the speaker membrane. Because the surface is oscillating like a mass on a spring with amplitude A_0, its maximum velocity is $A_0\omega$ and its time variation is

$$v_{\text{surface}}(t) = A_0\omega\cos\omega t. \tag{9.65}$$

Then

$$\dot{V} = 4\pi R^2 A_0\omega\cos\omega t. \tag{9.66}$$

The corresponding amplitude is everything except the $\cos\omega t$:

$$\dot{V}_0 = 4\pi R^2 A_0\omega. \tag{9.67}$$

The radiated power is then

$$P_{\text{avg}} = \frac{1}{16\pi}\frac{\rho\,\overbrace{(4\pi R^2 A_0\omega)^2}^{\dot{V}_0^2}\,\omega^2}{c_{\text{s}}} = \frac{\pi\rho R^4 A_0^2\omega^4}{c_{\text{s}}}. \tag{9.68}$$

By multiplying P_{avg} by $c_{\text{s}}^3/c_{\text{s}}^3$, the power can be written in terms of the dimensionless ratio $R\omega/c_{\text{s}}$, which is also $R/\lambdabar$:

$$P = \pi\rho c_{\text{s}}^3 A_0^2\left(\frac{R\omega}{c_{\text{s}}}\right)^4 = \pi\rho c_{\text{s}}^3 A_0^2\left(\frac{R}{\lambdabar}\right)^4. \tag{9.69}$$

Physically, $R/\lambdabar$ is the dimensionless speaker size (measured relative to $\lambdabar$). The scaling exponent of 4 tells us that the radiated power depends strongly on the dimensionless size of the speaker. As a result, big speakers (large R) are loud; and long wavelengths (low frequencies) require big speakers.

For this speaker, the radius is R is 1 centimeter, the surface-oscillation amplitude A_0 is 1 millimeter, and f is 1 kilohertz. So the wavelength of the sound is roughly 30 centimeters ($\lambda = c_{\text{s}}/f$), and $\lambdabar$ is roughly 5 centimeters. Then the dimensionless speaker size $R/\lambdabar$ is approximately 0.2, and

$$P_{\text{avg}} \approx \underbrace{\frac{3}{\pi}}_{} \times \underbrace{1\,\text{kg}\,\text{m}^{-3}}_{\rho} \times \underbrace{\left(3\times10^2\,\text{m}\,\text{s}^{-1}\right)^3}_{c_{\text{s}}^3} \times \underbrace{\left(10^{-3}\,\text{m}\right)^2}_{A_0^2} \times \underbrace{0.2^4}_{(R/\lambdabar)^4}. \tag{9.70}$$

To evaluate this power mentally, divide and conquer as usual:

1. *Units.* They are watts:

$$\text{kg}\,\text{m}^{-3}\times\text{m}^3\,\text{s}^{-3}\times\text{m}^2 = \text{kg}\,\text{m}^2\,\text{s}^{-3} = \text{W}. \tag{9.71}$$

2. *Powers of ten.* They contribute 10^{-4}:

$$\underbrace{10^6}_{\text{from } c_s^3} \times \underbrace{10^{-6}}_{\text{from } A_0^2} \times \underbrace{10^{-4}}_{\text{from } (R/\lambdabar)^4} = 10^{-4}. \tag{9.72}$$

So far, the power is 10^{-4} watts.

3. *Remaining numerical factors.* They are

$$\underbrace{3}_{\text{from } \pi} \times \underbrace{3^3}_{\text{from } c_s^3} \times \underbrace{2^4}_{\text{from } (R/\lambdabar)^4}. \tag{9.73}$$

Because $2^4 = 16$ and 3×3^3 is $\left(3^2\right)^2$ or roughly 10^2, the remaining numerical factors contribute roughly 1600. Let's round it to 2000.

Then the power is roughly 2000×10^{-4} or 0.2 watts.

▷ *Does that power represent a loud or a soft sound?*

It depends on how close you stand to the speaker. If you are 1 meter away, the 0.2 watts are spread over a sphere of area $4\pi \times (1 \text{ meter})^2$, or roughly 10 square meters. Then the power flux is roughly 0.02 watts per square meter. In decibels, which is the more familiar measure of loudness (introduced in Problem 3.10), this power flux corresponds to just over 100 decibels, which is very loud, almost enough to cause pain.

▷ *What radius fluctuations would produce a barely audible, 0-decibel flux?*

Shrinking the flux to 0 decibels, which is a drop of 100 decibels, is a drop of a factor of 10^{10} in energy flux and power. Because the energy flux is proportional to A_0^2, A_0 must fall by a factor of 10^5: from 10^{-3} meters to 10^{-8} meters. Thus, 10-nanometer fluctuations in a small speaker's radius are (barely) enough to produce an audible sound. The human ear has an amazing dynamic range and sensitivity.

9.3.2 Electromagnetic radiation from a dipole

In the spirit of laziness, let's transfer, by analogy, the radiated power from acoustic to electromagnetic radiation. In Section 9.3.1, we developed the analogy between acoustics and electrostatics. Using it, the acoustic radiated power,

$$P_{\text{monopole}} = \frac{1}{8\pi} \frac{\dot{M}^2 \omega^2}{\rho c_s}, \tag{9.74}$$

acoustics	electrostatics
ρ	ϵ_0
v	E
c_s	c
$\dot{M}$	q

implies an electromagnetic radiated power of $q^2 \omega^2 / 8\pi\epsilon_0 c$.

Alas, this conjecture has three problems. First, if it represents the power radiated by an oscillating charge—moving, say, on a spring oscillating with frequency ω—then its acceleration is $\propto \omega^2$, so the radiated power is proportional to the acceleration. However, we learned from dimensional analysis (Section 5.4.3) that the power had to be proportional to the acceleration squared. Second, the power should depend on the amplitude of the motion, which is a length, yet the proposed power contains no such length.

These two problems are symptoms of the third problem, that transferring the acoustic analysis to electromagnetism has made an illegal situation. A single changing electromagnetic charge $q(t)$ violates charge conservation: If $q(t)$ is increasing, from where would the new charge come?

This question suggests the valid physical model, that the new charge comes from a nearby charge. As a model of this flow, imagine a pair of opposite, nearby charges. As the positive charge flows to the negative charge, the two charges swap places; and vice versa. This model is an oscillating dipole. Here is a full cycle of its oscillation.

$$\oplus \qquad\qquad \ominus \qquad\qquad \oplus$$

$$\ominus \qquad\qquad \oplus \qquad\qquad \ominus$$

$$\omega t = 0 \qquad\qquad \omega t = \pi \qquad\qquad \omega t = 2\pi$$

If the charges are $\pm q$ and their separation is l, then ql is called their dipole moment d. Here, the time-varying dipole moment is $d(t) = d_0 \cos \omega t$, where d_0 is the amplitude of the oscillations in the dipole moment. To estimate the power radiated, we'll reuse the structure from acoustics,

$$P \sim \underbrace{4\pi r^2}_{\text{surface area}} \times \underbrace{\frac{1}{2}\epsilon_0 E^2}_{\text{energy density}} \times \underbrace{c,}_{\text{propagation speed}} \qquad (9.75)$$

and evaluate it at the near-field–far-field boundary ($r \sim \lambdabar$). The only change from acoustics is that the electric field E is not the field from a single point charge (a monopole source) but rather from two opposite charges (a dipole source). Let's evaluate their field at the position of a test charge at $r \sim \lambdabar$.

The two charges (the monopoles) contribute slightly different electric fields $\mathbf{E}_+$ and $\mathbf{E}_-$. Because these fields are vectors, adding them correctly requires tracking their individual components. Let's therefore make the lumping approximation that we can add the vectors using only their magnitudes:

$$E_{\text{dipole}} \approx E_+ - E_-. \qquad (9.76)$$

This approximation would be exact if the vectors lay along the same line—which they would if the dipole were an ideal dipole, with zero separation ($l = 0$). By making this approximation even for this nonideal dipole, we will obtain an important and transferable insight about the field of a dipole.

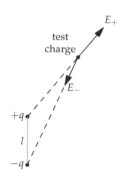

Because the distances from the test charge to the monopoles are almost identical, the two fields E_+ and E_- have almost the same magnitude. Therefore, the difference $E_+ - E_-$ is almost zero. (The key word is almost. If the difference were exactly zero, there would be no light and no radiation.) Let's approximate the difference using a further lumping approximation.

The dipole field is the difference $\Delta E = E(r_+) - E(r_-)$, where $E(r)$ is the monopole field. The difference is approximately

$$\underbrace{\Delta E}_{\text{rise}} \approx \underbrace{E'(r)}_{\text{slope}} \times \underbrace{\Delta r}_{\text{run}} , \qquad (9.77)$$

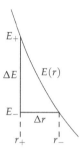

where $\Delta r = r_- - r_+$. (This formula ignores a minus sign, but we are interested only in the magnitude of the field, so the sign doesn't matter.) In Leibniz's notation, the slope $E'(r)$ is also dE/dr. Using the lumping approximation of Section 6.3.4 (which I remember as $d \sim d$), the ds cancel and dE/dr, and therefore $E'(r)$, is roughly E/r.

The trickiest factor is Δr, which is $r_- - r_+$. It depends on the charge separation l and on the position of the test charge relative to the dipole's orientation. When the test charge is directly above the dipole (at the north pole), Δr is just the charge separation l. When the test charge is at the equator, $\Delta r = 0$. With our lumping approximation, Δr is comparable to l.

Then the difference ΔE, which is the dipole field, becomes

$$E_{\text{dipole}} \sim E_{\text{monopole}} \times \frac{l}{r}. \qquad (9.78)$$

The $1/r$ factor comes from differentiating the monopole field. The factor of l, the dipole size, turns the derivative of the field back into a field and makes the overall operation dimensionless: Making a dipole from two monopoles dimensionlessly differentiates the monopole field.

Because the energy density $\mathcal{E}$ in the electric field is proportional to E^2, and the power radiated is proportional to the energy density, the power is

$$P_{\text{dipole}} \sim P_{\text{monopole}} \left(\frac{l}{r} \right)^2 . \qquad (9.79)$$

▷ *In evaluating the radiated power, what r should we use?*

The power radiated is determined by the energy density at the boundary between the near- and far-field regions: at $r \sim c/\omega$. With that substitution,

$$P_{\text{dipole}} \sim \underbrace{\frac{1}{8\pi}\frac{q^2\omega^2}{\epsilon_0 c}}_{P_{\text{monopole}}} \times \left(\frac{l}{c/\omega}\right)^2 = \frac{1}{8\pi}\frac{q^2 l^2 \omega^4}{\epsilon_0 c^3}. \tag{9.80}$$

With a $1/6\pi$ instead of $1/8\pi$, this result is exact. Furthermore, if the source is a single accelerating charge, instead of a charge flow, then its acceleration is comparable to $l\omega^2$, and $P_{\text{dipole}} \sim q^2 a^2/\epsilon_0 c^3$, which is now consistent with what we derived in Section 5.4.3 using dimensional analysis.

In terms of the dipole moment $d = ql$,

$$P_{\text{dipole}} = \frac{1}{6\pi}\frac{\omega^4 d^2}{\epsilon_0 c^3}. \tag{9.81}$$

Dipole radiation is the strongest kind of electromagnetic radiation. In Section 9.4, we'll use the dipole power to explain why the sky is blue and a sunset red.

Problem 9.7   Lifetime of hydrogen if it could radiate

Assuming that the ground state of hydrogen could radiate as an oscillating dipole (because of the orbiting electron), estimate the time τ required for it to radiate its binding energy E_0. The ground state of hydrogen is protected by quantum mechanics—there is no lower-energy state to go to—but many of hydrogen's higher-energy states have a lifetime comparable to τ.

9.3.3 Gravitational radiation from a quadrupole

Having started with acoustics and practiced with electromagnetics, we can extend our analysis of radiation to gravitational waves—without solving the equations of general relativity. In acoustics, radiation could be produced by a monopole (a point charge). In electromagnetics, radiation could be produced by a dipole but not by a monopole. Building a dipole requires charges of two signs. Because the gravitational equivalent of charge is mass, which comes in one sign, there is no way to make a gravitational dipole.

Therefore, gravitational radiation requires a quadrupole. A quadrupole is to a dipole what a dipole is to a monopole. It is two nearby dipoles with opposite strengths—so that their fields *almost* cancel.

An example is an oblate sphere. Relative to a sphere, it is fat at the equator (represented by the + signs) and thin at the poles (represented by the − signs). One plus–minus pair forms one dipole. The other plus–minus pair forms the second dipole. They have the same magnitude but point in opposite directions. As the sphere shifts to pro-lateness (tall and thin), the signs of the charges flip, as do the directions of the dipoles.

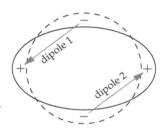

Just as the dipole field is the dimensionless derivative of the monopole field (Section 9.3.2), the quadrupole field is the dimensionless derivative of the dipole field. Thus, if the pulsating object has size l, so that the two dipoles are separated by a distance comparable to l, then at the boundary between the near and far fields (at $r \sim c/\omega$), the fields are related by

$$E_{\text{quadrupole}} \sim E_{\text{dipole}} \times \frac{l}{c/\omega} = E_{\text{dipole}} \times \frac{\omega l}{c}. \tag{9.82}$$

The radiated powers are related by the square of the extra factor:

$$P_{\text{quadrupole}} \sim P_{\text{dipole}} \times \left(\frac{\omega l}{c} \right)^2. \tag{9.83}$$

That dimensionless ratio in parenthesis has a physical interpretation. Its numerator ωl is the characteristic speed of the objects making the field. Its denominator c is the wave speed. Their ratio is the Mach number M, so $\omega l / c$ is the characteristic Mach number of the sources. In terms of M,

$$P_{\text{quadrupole}} \sim P_{\text{dipole}} \times \text{M}^2. \tag{9.84}$$

For the radiated power from an electromagnetic dipole, we found an analogous relation:

$$P_{\text{dipole}} \sim P_{\text{monopole}} \times \text{M}^2. \tag{9.85}$$

In general,

$$P_{2^m\text{-pole}} \sim P_{\text{monopole}} \times \text{M}^{2m}, \tag{9.86}$$

where a 2^0-pole is a monopole, a 2^1-pole is a dipole, and so on.

With the analogy between electrostatics and gravity (from Section 2.4.2), we can convert the power radiated by an electromagnetic dipole to the power radiated by a gravitational dipole, if it existed. Then we just adjust for the difference between a dipole and a quadrupole. From the analogy, the electrostatic field E is analogous to the gravitational field g, and electrostatic charge q is analogous to mass m. Finally, because $g = Gm/r^2$ is analogous to $E = q/4\pi\epsilon_0 r^2$, the electrostatic combination $1/4\pi\epsilon_0$ is analogous to Newton's constant G.

For electromagnetic dipole radiation, the power radiated is

$$P_{\text{dipole}} = \frac{1}{6\pi\epsilon_0} \frac{q^2 l^2 \omega^4}{c^3}. \tag{9.87}$$

Replacing $1/4\pi\epsilon_0$ with G and q with m, but leaving c alone because gravitational waves also travel at the speed of light (a limitation set by relativity), the power radiated by a gravitational dipole, if it existed, would be

$$P_{\text{dipole}} \sim \frac{Gm^2 l^2 \omega^4}{c^3}. \tag{9.88}$$

Changing from dipole to quadrupole radiation adds a factor of $\omega l/c$ to the field and $(\omega l/c)^2$ to the power, so

$$P_{\text{quadrupole}} \sim \frac{Gm^2 l^4 \omega^6}{c^5}. \tag{9.89}$$

Let's use this formula to estimate the gravitational power radiated by the Earth–Sun system. We'll divide the system into two sources. One source is the Earth as it rotates around the system's center of mass (CM). The other source is the Sun as it rotates around the system's center of mass.

$$(9.90)$$

Which source generates more gravitational radiation?

The two sources share the constants of nature G and c. Because they orbit around the center of mass like a spinning dumbbell, they also share the angular velocity ω. Therefore, the power simplifies to a proportionality without G, c, or ω:

$$P_{\text{quadrupole}} \propto m^2 l^4, \tag{9.91}$$

where m is the mass of the object (either the Earth or Sun) and l is its distance from the center of mass. Furthermore, ml is shared, because the center of mass is defined as the point that makes ml the same for both objects.

$$m_{\text{Earth}} \times l_{\text{Earth–CM distance}} = M_{\text{Sun}} \times l_{\text{Sun–CM distance}}. \tag{9.92}$$

By factoring out two powers of ml and discarding them, the proportionality further simplifies from $P_{quadrupole} \propto m^2 l^4$ to $P_{quadrupole} \propto l^2$. The Earth, with the longer lever arm l, generates most of the gravitational wave energy. Equivalently, we can factor out four powers of ml and get $P_{quadrupole} \propto m^{-2}$; the Earth, with the smaller mass, still wins. So,

$$P_{quadrupole} \sim \frac{Gm_{Earth}^2 l^4 \omega^6}{c^5}. \qquad (9.93)$$

As the last simplification before evaluating the power, let's eliminate the angular frequency ω. For motion in a circle of radius l, the centripetal acceleration is v^2/l (as we found in Section 5.1.1). In terms of the angular velocity, this acceleration is $\omega^2 l$. It is produced by the gravitational force

$$F \approx \frac{GM_{Sun}m_{Earth}}{l^2} \qquad (9.94)$$

(approximately, because l is slightly smaller than the Earth–Sun distance). The resulting centripetal acceleration is F/m_{Earth} or GM_{Sun}/l^2, so

$$\omega^2 l = \frac{GM_{Sun}}{l^2}. \qquad (9.95)$$

Using this relation to replace $(\omega^2 l)^3$ in $P_{quadrupole}$ with $(GM_{Sun}/l^2)^3$ gives

$$P_{quadrupole} \sim \frac{G^4 m_{Earth}^2 M_{Sun}^3}{l^5 c^5}. \qquad (9.96)$$

The power based on a long and difficult general-relativity calculation is almost the same:

$$P_{quadrupole} \approx \frac{32}{5} \frac{G^4}{l^5 c^5} (m_{Earth} M_{Sun})^2 (m_{Earth} + M_{Sun}). \qquad (9.97)$$

With the approximation that $m_{Earth} + M_{Sun} \approx M_{Sun}$, the only difference between our estimate and the exact power is the dimensionless prefactor of $32/5$. Including that factor and approximating $m_{Earth} + M_{Sun}$ by M_{Sun},

$$P_{quadrupole} \approx \frac{32}{5} \frac{G^4 m_{Earth}^2 M_{Sun}^3}{l^5 c^5}. \qquad (9.98)$$

To avoid exponent whiplash and promote formula hygiene, let's rewrite the power using a dimensionless ratio, by pairing another velocity with the c in the denominator. It would also be helpful to get rid of G, which seems like a random, meaningless value. To fulfill both wishes, we again equate the two ways of finding the Earth's centripetal acceleration, as the acceleration produced by the Sun's gravity and as the circular acceleration v^2/l:

$$\frac{GM_{\text{Sun}}}{l^2} = \frac{v^2}{l}.$$

(9.99)

Therefore, $GM_{\text{Sun}}/l = v^2$ and

$$\frac{G^4 M_{\text{Sun}}^4}{l^4} = v^8.$$

(9.100)

That substitution gives

$$P_{\text{quadrupole}} \approx \frac{32}{5} \frac{m_{\text{Earth}}^2}{M_{\text{Sun}}} \frac{v^8}{lc^5}.$$

(9.101)

The ratio v^5/c^5 is M^5, where M is the Mach number of the Earth (its orbital velocity compared to the speed of light). Of the three remaining powers of v, one power combines with l in the denominator to give back the angular velocity $\omega = v/l$. The remaining two powers of v combine with one power of m_{Earth} to give (except for a factor of 2) the orbital kinetic energy of the Earth. The remaining masses make the Earth–Sun mass ratio $m_{\text{Earth}}/M_{\text{Sun}}$. Therefore, in a more meaningful form, the power is

$$P_{\text{quadrupole}} \approx \frac{32}{5} \frac{m_{\text{Earth}}}{M_{\text{Sun}}} \mathsf{M}^5 \times m_{\text{Earth}} v^2 \, \omega.$$

(9.102)

In this processed form, the dimensions are more obviously correct than they were in the unprocessed form. The factors before the × sign are all dimensionless. The factor of $m_{\text{Earth}} v^2$ is an energy. And the factor of ω converts energy into energy per time—which is power.

Now that we have reorganized the formula into meaningful chunks, we are ready to evaluate its factors.

1. The mass ratio is 3×10^{-6}:

$$\frac{m_{\text{Earth}}}{M_{\text{Sun}}} \approx \frac{6 \times 10^{24}\,\text{kg}}{2 \times 10^{30}\,\text{kg}} = 3 \times 10^{-6}.$$

(9.103)

2. The Mach number $\mathsf{M} \equiv v/c$ turns out to have a compact value. The Earth's orbital velocity v is 30 kilometers per second (Problem 6.5):

$$v = \frac{\text{circumference}}{\text{orbital period}} \approx \frac{2\pi \times 1.5 \times 10^{11}\,\text{m}}{\pi \times 10^7\,\text{s}} = 3 \times 10^4\,\text{m s}^{-1},$$

(9.104)

which uses the estimate from Section 6.2.2 of the number of seconds in a year. The corresponding Mach number is 10^{-4}:

$$\mathsf{M} \equiv \frac{v}{c} = \frac{3 \times 10^4\,\text{m s}^{-1}}{3 \times 10^8\,\text{m s}^{-1}} = 10^{-4}.$$

(9.105)

3. For the factor $m_{Earth}v^2$, we know m_{Earth} and have just evaluated v. The result is 6×10^{33} joules:

$$\underbrace{6 \times 10^{24}\,\text{kg}}_{m_{Earth}} \times \underbrace{10^9\,\text{m}^2\,\text{s}^{-2}}_{v^2} = 6 \times 10^{33}\,\text{J}. \tag{9.106}$$

4. The final factor is the Earth's orbital angular velocity ω. Because the orbital period is 1 year, $\omega = 2\pi/1$ year, or

$$\omega \approx \frac{2\pi}{\pi \times 10^7\,\text{s}} = 2 \times 10^{-7}\,\text{s}^{-1}. \tag{9.107}$$

With these values,

$$P_{quadrupole} \approx \frac{32}{5} \times \underbrace{3 \times 10^{-6}}_{m_{Earth}/M_{Sun}} \times \underbrace{10^{-20}}_{M^5} \times \underbrace{6 \times 10^{33}\,\text{J}}_{m_{Earth}v^2} \times \underbrace{2 \times 10^{-7}\,\text{s}^{-1}}_{\omega}. \tag{9.108}$$

The resulting radiated power is about 200 watts. At that rate, the Earth's orbit will not soon collapse due to gravitational radiation (Problem 9.8).

Quadrupole radiation depends strongly on the Mach number v_{source}/c, and the Earth's Mach number is tiny. However, when a star gets captured by a black hole, the orbital speed can be a large fraction of c. Then the Mach number is close to 1 and the radiated power can be enormous—perhaps large enough for us to detect on distant Earth.

Problem 9.8   Energy loss through gravitational radiation
In terms of a gravitational system's Mach number and mass ratio, how many orbital periods are required for the system to lose a significant fraction of its kinetic energy? Estimate this number for the Earth–Sun system.

Problem 9.9   Gravitational radiation from the Earth–Moon system
Estimate the gravitational power radiated by the Earth–Moon system. Use the results of Problem 9.8 to estimate how many orbital periods are required for the system to lose a significant fraction of its kinetic energy.

9.4 Effect of radiation: Blue skies and red sunsets

We have developed almost all the pieces and tools to understand our final two phenomena: blue skies (Section 9.4.1) and red sunsets (Section 9.4.2). The only missing piece is the amplitude of a spring–mass system when it is driven by an oscillating force. We'll build that piece where it is first needed, and then put all the pieces together.

9.4.1 Skies are blue

The light that we see from a clear sky is dipole radiation scattered from air molecules (on the Moon, with no atmosphere, there are no blue skies). Plenty of sunlight does not get affected by air molecules, but unless we look at the sun, which is almost always very hazardous, the direct sunlight does not reach our eye. (The exception is just at sunset, and in Section 9.4.2 we'll use the exception to help explain why sunsets are red.)

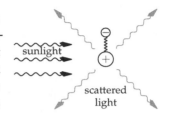

The analysis makes the most sense as a causal sequence from the sunlight to this scattered radiation. Sunlight is an oscillating electric field. The electric field exerts a force on the charged particles in an air molecule (say, in N_2). The charged particles, the electrons and protons, form spring–mass systems, and the electric force accelerates the masses. Accelerating charges radiate, which is the radiation that reaches our eye. By quantifying the steps in the sequence from sunlight to scattered radiation, we'll see why the scattered radiation looks blue.

1. *Electric field of the sunbeam.* Sunlight contains many colors of light, each simple to describe as an electric field (divide and conquer!):

   $$E(\omega) = E_0(\omega) \cos \omega t, \qquad\qquad (9.109)$$

   where ω is the angular frequency corresponding to that color. For example, for red light, ω is 3×10^{15} radians per second. The amplitude $E_0(\omega)$ depends on the intensity of the color in sunlight, so $E(\omega)$ is a distribution over ω, and it and the amplitude $E_0(\omega)$ have dimensions of field per frequency. However, as a simple lumping approximation, think of seven separate fields, one for each color in the color mnemonic Roy G. Biv: red, orange, yellow, green, blue, indigo, and violet.

   For the color of the sky, what matters is the relative distribution of colors in the sunbeam and how that relative distribution is different in scattered light. Therefore, let's do our calculations in terms of an unknown E_0 and, in the spirit of proportional reasoning, determine the dependence of E_0 on ω in the scattered light.

2. *Force on the charged particles.* The electric field produces a force on the electrons and protons in an air molecule. Using e for the electron charge and ignoring signs,

   $$F = eE = eE_0 \cos \omega t, \qquad\qquad (9.110)$$

3. *Amplitude of the motion of the charged particles.* Because protons are much heavier than an electrons, the electrons move faster and farther than the protons and produce most of the scattered radiation. Therefore, let's assume that the protons are fixed and analyze the motion of an electron.

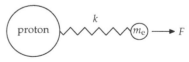

The electron, which is connected to its proton by a spring, is driven by the oscillating force F. Fortunately, we do not need to solve for the motion in general, because we can use easy cases based on the ratio between the driving frequency ω and the system's natural frequency ω_0 (which is $\sqrt{k/m_e}$, where k is the bond's spring constant). The three regimes are then (1) $\omega \ll \omega_0$, (2) $\omega = \omega_0$, and (3) $\omega \gg \omega_0$.

To decide which regime is relevant, let's make a rough estimate of how the two frequencies compare. For air molecules, the natural frequency ω_0 corresponds to ultraviolet radiation—the radiation required to break the strong triple bond in N_2. The driving frequency ω corresponds to one of the colors in visible light (sunlight), so the electron's motion is in the first, low-frequency regime $\omega \ll \omega_0$. (For the analysis of the other regimes, try Problems 9.12 and 9.15.)

The low-frequency regime is easiest to study in the $\omega = 0$ extreme. It represents a constant force $F = eE_0$ pulling on the electron and stretching the electron–proton bond. When there is change, *make* what does not change! The bond stretches until the spring force balances the stretching force eE_0. The forces balance when the stretch is $x = F/k$ or eE_0/k.

Because $\omega = 0$, the force has been constant since forever, so the bond stretched to its extended length back in the mists of time. When ω is nonzero, but still much smaller than ω_0, the bond still behaves approximately as it did when $\omega = 0$: It stretches so that the spring force balances the slowly oscillating force F. Therefore, the stretch, which is the displacement of the electron, is

$$x(t) \approx \frac{eE_0}{k} \cos \omega t. \qquad (9.111)$$

4. *Acceleration of the electron.* We can find the acceleration using dimensional analysis. In driven spring motion, the important length is the displacement x, and the important time is $1/\omega$. Therefore, the acceleration, which has dimensions of LT^{-2}, must be comparable to $x\omega^2$. Because we used the angular frequency (ω rather than f), the dimensionless prefactor turns out to be 1. Then the electron's acceleration is

$$a(t) = x(t)\omega^2 \approx \frac{eE_0\omega^2}{k}\cos\omega t. \tag{9.112}$$

Because the spring constant k is related to the electron mass m_e and to the natural frequency ω_0 by $k = m_e\omega_0^2$,

$$a(t) \approx \frac{eE_0}{m_e}\frac{\omega^2}{\omega_0^2}\cos\omega t. \tag{9.113}$$

5. *Power radiated by the accelerating electron.* We found in Sections 5.4.3 and 9.3.2 that the power radiated by an accelerating charge e is

$$P_{\text{dipole}} = \frac{e^2}{6\pi\epsilon_0}\frac{a^2}{c^3}. \tag{9.114}$$

For the squared acceleration a^2, we'll use the time average to find the average radiated power. Because the time average of $\cos^2\omega t$ is $1/2$,

$$\langle a^2 \rangle = \frac{1}{2}\frac{e^2E_0^2}{m_e^2}\frac{\omega^4}{\omega_0^4}. \tag{9.115}$$

Then

$$P_{\text{dipole}} = \frac{1}{12\pi\epsilon_0}\frac{1}{c^3}\frac{e^4E_0^2}{m_e^2}\left(\frac{\omega}{\omega_0}\right)^4. \tag{9.116}$$

This mouthful is useful in explaining the color of sunsets (Section 9.4.2), so it has been worthwhile carrying lots of baggage on the trip from sunlight to scattered light. But to explain the color of the daytime sky, we can simplify the power using proportional reasoning. Because ϵ_0, c, ω_0, m_e, and e are independent of the driving frequency ω (which represents the color),

$$P_{\text{dipole}} \propto E_0^2\omega^4. \tag{9.117}$$

The factor of E_0^2 is proportional to the energy density in the incoming sunlight (at frequency ω), which itself is proportional to the incoming power in the sunlight (also at the frequency ω). Thus,

$$P_{\text{dipole}} \propto P_{\text{sunlight}}\omega^4. \tag{9.118}$$

Let's review how the four powers of ω got here. For low frequencies—and visible light is a low frequency compared to the natural electronic-vibration frequency of an air molecule—the amplitude of spring motion is independent of the driving frequency. The acceleration is then proportional to ω^2. And the radiated power is proportional to the square of the acceleration, so it is proportional to ω^4.

Therefore, the air molecule acts like a filter that takes in (some of the) incoming sunlight and produces scattered light, altering the distribution of colors—similar to how a circuit changes the amplitude of each incoming frequency. However, unlike the low-pass *RC* circuit of Section 2.4.4, which preserves low frequencies and attenuates high frequencies, the air molecule amplifies the high frequencies.

Here are the sunlight and the scattered spectra based on the ω^4 filter. Each spectrum shows, by the area of each band, the relative intensities of the various colors. (The unlabeled color band between red and yellow is orange.)

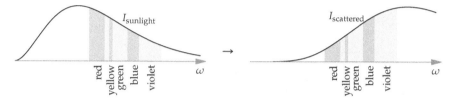

Sunlight looks white. In the scattered light, the high-frequency colors such as blue and violet are much more prominent than they are in sunlight. For example, because $\omega_{\text{blue}}/\omega_{\text{red}} \approx 1.5$ and $1.5^4 \approx 5$, the blue part of the sunlight is amplified by a factor of 5 compared to the red part. As a result, the scattered light—what comes to us from the sky—looks blue!

9.4.2 Sunsets are red

As sunlight passes through the atmosphere, ever more of its energy gets taken in by air molecules and then scattered (reradiated) in all directions. As we found in Section 9.4.1, this process happens more rapidly at higher (bluer) frequencies, due to the factor of ω^4 in the radiated power. Therefore, the sunbeam becomes redder as it travels through the atmosphere. If the beam travels far enough in the atmosphere, sunsets should look red. To estimate the necessary travel distance, we can adapt the analysis of Section 9.4.1. Then we will use geometry to estimate the actual travel distance.

▷ *How far must sunlight travel in the atmosphere before the beam looks red?*

To estimate this length, let's estimate the rate at which energy gets scattered out of the beam. The beam carries an energy flux

$$F = \frac{1}{2}\epsilon_0 E_0^2 c, \tag{9.119}$$

which has the usual form of energy density times propagation speed. (The electric and magnetic fields each contribute one-half of this flux.) To measure the effect of one scattering electron on the beam, let's compute one electron's radiated power divided by the flux:

$$\frac{P_{\mathrm{dipole}}}{F} = \underbrace{\frac{1}{12\pi\epsilon_0}\frac{1}{c^3}\frac{e^4 E_0^2}{m_e^2}\left(\frac{\omega}{\omega_0}\right)^4}_{P_{\mathrm{dipole}}} \Big/ \underbrace{\frac{1}{2}\epsilon_0 E_0^2 c.}_{F} \qquad (9.120)$$

Noticing that the (unknown) field amplitude E_0 cancels out and doing the rest of the algebra, we get

$$\frac{P_{\mathrm{dipole}}}{F} = \underbrace{\frac{8\pi}{3}\left(\frac{e^2}{4\pi\epsilon_0}\frac{1}{m_e c^2}\right)^2}_{\mathrm{area}\ \sigma} \times \left(\frac{\omega}{\omega_0}\right)^4. \qquad (9.121)$$

The left side, P_{dipole}/F, is power divided by power per area. Thus, P_{dipole}/F has dimensions of area. It represents the area of the beam whose energy flux gets removed and scattered in all directions. The area is therefore called the scattering cross section σ (a concept introduced in Section 6.4.5, when we estimated the mean free path of air molecules).

On the right side, the frequency ratio ω/ω_0 is dimensionless, so the prefactor must also be an area. It is called the Thomson cross section σ_{T}:

$$\sigma_{\mathrm{T}} \equiv \frac{8\pi}{3}\left(\frac{e^2}{4\pi\epsilon_0}\frac{1}{m_e c^2}\right)^2 \approx 7\times 10^{-29}\ \mathrm{m}^2. \qquad (9.122)$$

Because σ_{T} is an area, the factor inside the parentheses must be a length. Indeed, it is the classical electron radius r_0 of Problems 5.37 and 5.44(a). It is comparable to the proton radius of 10^{-15} meters:

$$r_0 \equiv \frac{e^2}{4\pi\epsilon_0}\frac{1}{m_e c^2} \approx 2.8\times 10^{-15}\ \mathrm{m}. \qquad (9.123)$$

It approximately answers the question: For the electron to acquire its mass from its electrostatic energy, how large should the electron be? This electron's cross-sectional area is, roughly, the Thomson cross section. In terms of the Thomson cross section, our scattering cross section σ is

$$\sigma = \sigma_{\mathrm{T}}\left(\frac{\omega}{\omega_0}\right)^4. \qquad (9.124)$$

Each scattering electron converts this much area of the beam into scattered radiation. As a rough estimate, each air molecule contributes one scattering electron (the inner electrons, which are tightly bound to the nucleus, have a large ω_0 and a tiny scattering cross section). As we found in Section 6.4.5, the mean free path λ_{mfp} and scattering cross section σ are related by

$$n\sigma\lambda_{\mathrm{mfp}} \sim 1, \tag{9.125}$$

where n is the number density of scattering electrons. The mean free path is how far the beam travels before a significant fraction of its energy (at that frequency) is scattered in all directions and is no longer part of the beam.

With one scattering electron per air molecule, n is the number density of air molecules. This number density, like the mass density, varies with the height above sea level. To simplify the analysis, we will use a lumped atmosphere. It has a constant temperature, pressure, and density from sea level to the atmosphere's scale height H, which we estimated in Section 5.4.1 using dimensional analysis (and you estimated using lumping in Problem 6.36). At H, the atmosphere ends, and the density abruptly goes to zero.

In this lumped atmosphere, 1 mole of air molecules at any height occupies approximately 22 liters. The resulting number density is approximately 3×10^{25} molecules per cubic meter:

$$n = \frac{1\,\mathrm{mol}}{22\,\ell} \times \frac{6\times10^{23}}{1\,\mathrm{mol}} \times \frac{10^3\,\ell}{1\,\mathrm{m^3}} \approx 3\times10^{25}\,\mathrm{m^{-3}}. \tag{9.126}$$

Using this n and $\sigma = \sigma_{\mathrm{T}}(\omega/\omega_0)^4$, the mean free path becomes

$$\lambda_{\mathrm{mfp}} \sim \frac{1}{n\sigma} \approx \underbrace{\frac{1}{3\times10^{25}\,\mathrm{m^{-3}}}}_{n} \times \underbrace{\frac{1}{7\times10^{-29}\,\mathrm{m^2}}}_{\sigma_{\mathrm{T}}} \times \left(\frac{\omega_0}{\omega}\right)^4 \tag{9.127}$$

$$\approx \frac{1}{2}\,\mathrm{km} \times \left(\frac{\omega_0}{\omega}\right)^4.$$

Unlike the scattering cross section σ, which grows rapidly with ω, the mean free path falls rapidly with ω: High frequencies scatter out of the beam rapidly and travel shorter distances before getting significantly attenuated.

To estimate the frequency ratio ω_0/ω, let's estimate the equivalent energy ratio $\hbar\omega_0/\hbar\omega$. The numerator $\hbar\omega_0$ is the bond energy. Because air is mostly N_2, and the nitrogen–nitrogen triple bond is much stronger than a typical chemical bond (roughly 4 electron volts), the natural frequency ω_0 corresponds to an energy $\hbar\omega_0$ of about 10 electron volts.

The denominator $\hbar\omega$ depends on the color of the light. As a lumping approximation, let's divide light into two colors: red and, to represent nonred light, blue–green light. A blue–green photon has an energy $\hbar\omega$ of approximately 2.5 electron volts, so $\hbar\omega_0/\hbar\omega \approx 4$ and $(\omega_0/\omega)^4 \approx 200$. Because the mean free path is

$$\lambda_{\mathrm{mfp}} \approx \frac{1}{2}\,\mathrm{km} \times \left(\frac{\omega_0}{\omega}\right)^4, \tag{9.128}$$

for blue–green light $\lambda_{\mathrm{mfp}} \sim 100$ kilometers. After a distance comparable to 100 kilometers, a significant fraction of the nonred light has been removed (and scattered in all directions).

At midday, when the Sun is overhead, the travel distance is the thickness of the atmosphere H, roughly 8 kilometers. This distance is much shorter than the mean free path, so very little light (of any color) is scattered out of the sunbeam, and the Sun looks white as it would from space. (Fortunately, our theory doesn't predict that the midday Sun looks red—but do not test this analysis by looking directly at the Sun!) As the Sun descends in the sky, sunlight travels ever farther in the atmosphere.

▶ *At sunset, how far does sunlight travel in the atmosphere?*

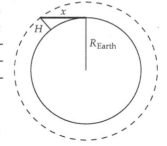

This length is the horizon distance: Standing as high as the atmosphere ($H \approx 8$ kilometers), the horizon distance x is the distance that sunlight travels through the atmosphere at sunset. It is the geometric mean of the atmosphere height H and of the Earth's diameter $2R_{\mathrm{Earth}}$ (Problem 2.9), and is approximately 300 kilometers:

$$x = \sqrt{H \times 2R_{\mathrm{Earth}}} \approx \sqrt{8\,\mathrm{km} \times 2 \times 6000\,\mathrm{km}} \tag{9.129}$$
$$\approx 300\,\mathrm{km}.$$

This distance is a few mean free paths. Each mean free path produces a significant reduction in intensity (more precisely, a factor-of-e reduction). Therefore, at sunset, most of the nonred light is gone.

However, for red light the story is different. Because a red-light photon has an energy of approximately 1.8 electron volts, in contrast to the 2.5 electron volts for a blue-green photon, its mean free path is a factor of $(2.5/1.8)^4 \approx 4$ longer than 100 kilometers for a blue–green photon. The trip of 300 kilometers in the atmosphere scatters out a decent fraction of red light, but much less than the corresponding fraction of blue–green light. The moral of the story is visual: Together the springs in the air molecules and the steep dependence of dipole radiation on frequency produce a beautiful red sunset.

9.5 **Summary and further problems**

Many physical processes contain a minimum-energy state where small deviations from the minimum require an energy proportional to the square of the deviation. This behavior is the essential characteristic of a spring. A spring is therefore not only a physical object but a transferable abstraction. This abstraction has helped us understand chemical bonds, sound speeds, and acoustic, electromagnetic, and gravitational radiation—and from there the colors of the sky and sunset.

Problem 9.10   Two-dimensional acoustics

An acoustic line source is an infinitely long tube that can expand and contract. In terms of the source strength per length, find the power radiated per length.

Problem 9.11   Decay of a lightly damped oscillation

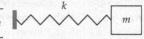

In an undamped spring–mass system, the motion is described by $x = x_0 \cos \omega_0 t$, where x_0 is the amplitude and ω_0, the natural frequency, is $\sqrt{k/m}$. In this problem, you use easy cases and lumping to find the effect of a small amount of (linear) damping. The damping force will be $F = -\gamma v$, where γ is the damping coefficient and v is the velocity of the mass.

a. In terms of ω_0 and x_0, estimate the typical speed and damping force, and then the typical energy lost to damping in 1 radian of oscillation (a time $1/\omega_0$).

b. Express the dimensionless measure of energy loss

$$\frac{\Delta E}{E} = \frac{\text{energy lost per radian of oscillation}}{\text{oscillation energy}} \tag{9.130}$$

by finding the scaling exponent n in

$$\frac{\Delta E}{E} \sim Q^n \tag{9.131}$$

where Q is the quality factor from Problem 5.53. In terms of Q, what does "a small amount of damping" mean?

c. Find the rate constant C, built from Q and ω_0, in the time-averaged oscillation energy:

$$E \approx E_0 e^{Ct}, \tag{9.132}$$

d. Using the scaling between amplitude and energy, represent how the mass's position x varies with time by finding the rate constant C' in the formula

$$x(t) \approx x_0 \cos \omega_0 t \times e^{C't}, \tag{9.133}$$

Sketch $x(t)$.

Problem 9.12   Driving a spring at high frequency

Sunlight driving an electron in nitrogen is the easy-case regime $\omega \ll \omega_0$. The opposite regime is $\omega \gg \omega_0$. This regime includes metals, where the electrons are free (the spring constant is zero, so ω_0 is zero). Estimate, using characteristic or typical values, the amplitude x_0 of a spring–mass system driven by a force

$$F = F_0 \cos \omega t, \tag{9.134}$$

where $\omega \gg \omega_0$. In particular, find the scaling exponent n in the transfer function $x_0/F_0 \propto \omega^n$.

Problem 9.13   High-frequency scattering

Use the result of Problem 9.12 to show that, in the $\omega \gg \omega_0$ regime (for example, for a free electron), the scattering cross section P_{dipole}/F is independent of ω and is the Thomson cross section σ_{T}.

Problem 9.14   Fiber-optic cable

A fiber-optic cable, used for transmitting telephone calls and other digital data, is a thin glass fiber that carries electromagnetic radiation. High data transmission rates require a high radiation frequency ω. However, scattering losses are proportional to ω^4, so high-frequency signals attenuate in a short distance. As a compromise, glass fibers carry "near-infrared" radiation (roughly 1-micrometer in wavelength). Estimate the mean free path of this radiation by comparing the density of glass to the density of air.

Problem 9.15   Resonance

In Problem 9.12, you analyzed the case of driving a spring at a frequency ω much higher than its natural frequency ω_0. The discussion of blue skies in Section 9.4.1 required the opposite regime, $\omega \ll \omega_0$. In this problem, you analyze the middle regime, which is called resonance. It is a lightly damped spring–mass system driven at its natural frequency.

Assuming a driving force $F_0 \cos \omega t$, estimate the energy input per radian of oscillation, in terms of F_0 and the amplitude x_0. Using Problem 9.11(b), estimate the energy lost per radian in terms of the amplitude x_0, the natural frequency ω_0, the quality factor Q, and the mass m.

By equating the energy loss and the energy input, which is the condition for a steady amplitude, find the quotient x_0/F_0. This quotient is the gain G of the system at the resonance frequency $\omega = \omega_0$. Find the scaling exponent n in $G \propto Q^n$.

Problem 9.16   Ball resting on the ground as a spring system

A ball resting on the ground can be thought of as a spring system. The larger the compression δ, the larger the restoring force F. (Imagine setting weights on top of the ball so that the extra weight plus the ball's weight is F.) Find the scaling exponent n in $F \propto \delta^n$. Is this system an ideal spring (for which $n = 1$)?

Problem 9.17   Inharmonicity of piano strings

An ideal piano string is under tension and has vibration frequencies given by

$$f_n = \frac{n}{2L}\sqrt{\frac{T}{\rho A}}, \tag{9.135}$$

where A is its cross-sectional area, T is the tension, L is the string length, and $n = 1, 2, 3, \dots$. (In Section 9.2.2, we estimated f_1.) Assume that the string's cross section is a square of side length b. The piano sounds pleasant when harmonics match: for example, when the second harmonic (f_2) of the middle C string has the same frequency as the fundamental (f_1) of the C string one octave higher.

However, the stiffness of the string (its resistance to bending) alters these frequencies slightly. Estimate the dimensionless ratio

$$\frac{\text{potential energy from stiffness}}{\text{potential energy from stretching (tension)}}. \tag{9.136}$$

This ratio is also roughly the fractional change in frequency due to stiffness. Write this ratio in terms of the mode number n, the string side length (or diameter) b, the string length L, and the tension-induced strain ϵ.

Problem 9.18   Buckling

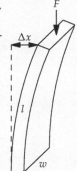

In this problem you estimate the force required to buckle a strut, such as a leg bone landing on the ground. The strut has Young's modulus Y, thickness h, width w, and length l. The force F has deflected the strut by Δx, producing a torque $F\Delta x$. Find the restoring torque and the approximate condition on F for $F\Delta x$ to exceed the restoring torque—whereupon the strut buckles.

Problem 9.19   Buckling versus tension

A strut can withstand a tension force

$$F_T \sim \epsilon_y Y \times hw, \tag{9.137}$$

where ϵ_y is the yield strain. The force F_T is just the yield stress $\epsilon_y Y$ times the cross-sectional area hw. The yield strain typically ranges from 10^{-3} for brittle materials like rock to 10^{-2} for piano-wire steel.

Use the result of Problem 9.18 to show that

$$\frac{\text{force that a strut can withstand in tension}}{\text{force that a strut can withstand against buckling}} \sim \frac{l^2}{h^2}\epsilon_y, \tag{9.138}$$

and estimate this ratio for a bicycle spoke.

Problem 9.20   Buckling of leg bone

How much margin of safety, if any, does a typical human leg bone ($Y \sim 10^{10}$ pascals) have against buckling, where the buckling force is the person's weight?

Problem 9.21   Inharmonicity of a typical piano

Estimate the inharmonicity in a typical upright piano's middle-C note. (Its parameters are given in Section 9.2.2.) In particular, estimate the frequency shift of its fourth harmonic ($n = 4$).

Problem 9.22   Cylinder resting on the ground

For a solid cylinder of radius R resting on the ground (for example, a train wheel), the contact area is a rectangle. Find the scaling exponent β in

$$\frac{x}{R} = \left(\frac{\rho g R}{Y} \right)^{\beta} ,$$ (9.139)

where x is the width of the contact strip. Then find the scaling exponent γ in $F \propto \delta^{\gamma}$, where F is the contact force and δ is the tip compression. (In Problem 9.16, you computed the analogous scaling exponent for a sphere.)

Bon voyage:
Long-lasting learning

The world is complex! But our nine reasoning tools help us master and enjoy the complexity. Spanning fields of knowledge, the tools connect disparate facts and ideas and promote long-lasting learning.

An analogy for the value of connected knowledge is an infinite two-dimensional lattice of dots: a percolation lattice [21]. Every dot marks a piece of knowledge—a fact or an idea. Now add bonds between neighboring pieces of knowledge, with a probability p_{bond} for each bond. The following figures show examples of finite lattices starting at $p_{bond} = 0.4$. Marked in bold is the largest cluster—the largest connected set of dots. As p_{bond} increases, this cluster unifies an ever-larger fraction of the lattice of knowledge.

$p_{bond} = 0.40$

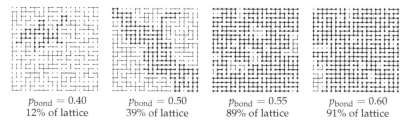

| $p_{bond} = 0.40$ | $p_{bond} = 0.50$ | $p_{bond} = 0.55$ | $p_{bond} = 0.60$ |
| 12% of lattice | 39% of lattice | 89% of lattice | 91% of lattice |

An infinite lattice might hold many infinite clusters, and the measure analogous to the size of the largest cluster is the fraction of dots belonging to an infinite cluster. This fraction f_∞ is, like the number of infinite clusters, zero until p_{bond} reaches the critical probability 0.5. Then it rises above zero and, as p_{bond} rises, eventually reaches 1.

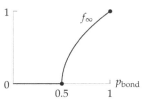

For long-lasting learning, the pieces of knowledge should support each other through their connections. For when we remember a fact or use an idea, we activate connected facts and ideas and solidify them in our minds.

Our knowledge lives best in infinite, self-supporting clusters. But if we learn facts and ideas in isolation, we make dots without bonds. Then p_{bond} falls and, with it, the membership in infinite clusters. If p_{bond} falls too much, the infinite clusters simply vanish.

So, for long-lasting learning and understanding, make bonds; connect each new fact and idea to what you already know. This way of thinking will help you learn in one year what took me two or twenty. Use your reasoning tools to weave a richly connected, durable tapestry of knowledge. Bon voyage as you learn and discover new ideas and their fascinating connections!

Only connect! That was the whole of her sermon... Live in fragments no longer.
—E. M. Forster [16]

Bibliography

[1] Kenneth John Atchity. *A Writer's Time: Making the Time to Write*. W. W. Norton & Company, New York, revised and expanded edition, 1995.

[2] Benjamin S. Bloom. The 2 sigma problem: The search for methods of group instruction as effective as one-to-one tutoring. *Educational Researcher*, 13(6):4–16, 1984.

[3] Carl H. Brans and Robert H. Dicke. Mach's principle and a relativistic theory of gravitation. *Physical Review*, 124:925-935, 1961.

[4] Edgar Buckingham. On physically similar systems. *Physical Review*, 4(4):345–376, 1914.

[5] Adam S. Burrows and Jeremiah P. Ostriker. Astronomical reach of fundamental physics. *Proceedings of the National Academy of Sciences of the USA*, 111(7):2409-16, 2014.

[6] Robert A. Caro. *The Power Broker: Robert Moses and the Fall of New York*. Vintage Books, New York, 1975.

[7] Thomas P. Carpenter, Mary M. Lindquist, Westina Matthews and Edward A. Silver. Results of the third NAEP assessment: Secondary school. *Mathematics Teacher*, 76:652–659, 1983.

[8] Michael A Day. The no-slip condition of fluid dynamics. *Erkenntnis*, 33(3):285–296, 1990.

[9] Stanislas Dehaene. *The Number Sense: How the Mind Creates Mathematics*. Oxford University Press, New York, revised and updated edition, 2011.

[10] Persi Diaconis and Frederick Mosteller. Methods for studying coincidences. *Journal of the American Statistical Association*, 84(408):853–861, 1989.

[11] Peter G. Doyle and Laurie Snell. *Random Walks and Electric Networks*. Mathematical Association of America, Washington, DC, 1984.

[12] Arthur Engel. *Problem-Solving Strategies*. Springer, New York, 1998.

[13] William Feller. *An Introduction to Probability Theory and Its Applications*, volume 1. Wiley, New York, 3rd edition, 1968.

[14] Richard P. Feynman, Robert B. Leighton and Matthew L. Sands. *The Feynman Lectures on Physics*. Addison-Wesley, Reading, MA, 1963. A "New Millenium" edition of these famous lectures, with corrections accumulated over the years, was published in 2011 by Basic Books.

[15] Neville H. Fletcher and Thomas D. Rossing. *The Physics of Musical Instruments*. Springer, New York, 2nd edition, 1988.

360

[16] Edward M. Forster. *Howard's End*. A. A. Knopf, New York, 1921.

[17] Mike Gancarz. *The UNIX Philosophy*. Digital Press, Boston, 1995.

[18] Mike Gancarz. *Linux and the Unix Philosophy*. Digital Press, Boston, 2003.

[19] Robert E. Gill, T. Lee Tibbitts, David C. Douglas, Colleen M. Handel, Daniel M. Mulcahy, Jon C. Gottschalck, Nils Warnock, Brian J. McCaffery, Philip F. Battley and Theunis Piersma. Extreme endurance flights by landbirds crossing the Pacific Ocean: Ecological corridor rather than barrier? *Proceedings of the Royal Society B: Biological Sciences*, 276(1656):447-457, 2009.

[20] Simon Gindikin. *Tales of Mathematicians and Physicists*. Springer, New York, 2007.

[21] Geoffrey Grimmett. *Percolation*. Springer, Berlin, 2nd edition, 1999.

[22] John Harte. *Consider a Spherical Cow: A Course in Environmental Problem Solving*. University Science Books, Mill Valley, CA, 1988.

[23] Sighard F. Hoerner. *Fluid-Dynamic Drag: Practical Information on Aerodynamic Drag and Hydrodynamic Resistance*. Hoerner Fluid Dynamics, Bakersfield, CA, 1965.

[24] Williams James. *The Principles of Psychology*, volume 2. Henry Holt, New York, 1890.

[25] Edwin T. Jaynes. A backward look into the future. In W. T. Grandy Jr. and P. W. Milonni, editors, *Physics and Probability: Essays in Honor of Edwin T. Jaynes*. Cambridge University Press, Cambridge, UK, 1993.

[26] Edwin T. Jaynes. *Probability Theory: The Logic of Science*. Cambridge University Press, Cambridge, UK, 2003.

[27] Thomas B. Greenslade Jr.. Atwood's machine. *The Physics Teacher*, 23(1):24–28, 1985.

[28] Anatoly A. Karatsuba. The complexity of computations. *Proceedings of the Steklov Institute of Mathematics*, 211:169–183, 1995.

[29] Anatoly A. Karatsuba and Yuri Ofman. Multiplication of many-digital numbers by automatic computers. *Doklady Akad. Nauk SSSR*, 145:293–294, 1962. English translation in *Physics-Doklady* 7:595–596 (1963).

[30] Doug King. Design masterclass 2: Thermal response. *CIBSE Journal*, pages 47–49, August 2010.

[31] Rodger Kram, Antoinette Domingo and Daniel P. Ferris. Effect of reduced gravity on the preferred walk-run transition speed. *Journal of Experimental Biology*, 200(4):821–826, 1997.

[32] Sanjoy Mahajan. *Order of Magnitude Physics: A Textbook with Applications to the Retinal Rod and to the Density of Prime Numbers*. PhD thesis, California Institute of Technology, 1998.

[33] Sanjoy Mahajan. *Street-Fighting Mathematics: The Art of Educated Guessing and Opportunistic Problem Solving*. MIT Press, Cambridge, MA, 2010.

[34] Ned Mayo. Ocean waves—their energy and power. *The Physics Teacher*, 35(6):352-356, 1997.

[35] Karen McComb, Craig Packer and Anne Pusey. Roaring and numerical assessment in contests between groups of female lions, *Panthera leo*. *Animal Behaviour*, 47(2):379–387, 1994.

[36] George Pólya. Über eine Aufgabe der Wahrscheinlichkeitsrechnung betreffend die Irrfahrt im Strassennetz. *Mathematische Annalen*, 84(1):149–160, 1921.

[37] George Pólya. Let us teach guessing: A demonstration with George Pólya [video-recording]. Mathematical Association of America, Washington, DC, 1966

[38] George Pólya. *How to Solve It: A New Aspect of Mathematical Method*. Princeton University Press, Princeton, NJ, 2004.

[39] Edward M. Purcell. Life at low Reynolds number. *American Journal of Physics*, 45:3–11, 1977.

[40] Kenneth A. Ross and Donald E. Knuth. A programming and problem solving seminar. Technical Report, Stanford University, Stanford, CA, 1989. STAN-CS-89-1269.

[41] Knut Schmid-Nielsen. *Scaling: Why Animal Size is So Important*. Cambridge University Press, Cambridge, UK, 1984.

[42] Gilbert Strang. *Linear Algebra and its Applications*. Thomson, Belmont, CA, 2006.

[43] David Tabor. *Gases, Liquids and Solids and Other States of Matter*. Cambridge University Press, Cambridge, UK, 3rd edition, 1990.

[44] Geoffrey I. Taylor. The formation of a blast wave by a very intense explosion. II. The atomic explosion of 1945. *Proceedings of the Royal Society of London. Series A, Mathematical and Physical*, 201(1065):175–186, 1950.

[45] John R. Taylor. *Classical Mechanics*. University Science Books, Sausalito, CA, 2005.

[46] David J. Tritton. *Physical Fluid Dynamics*. Oxford University Press, Oxford, UK, 1988.

[47] Lawrence Weinstein. *Guesstimation 2.0: Solving Today's Problems on the Back of a Napkin*. Princeton University Press, Princeton, NJ, 2012.

[48] Lawrence Weinstein and John A. Adam. *Guesstimation: Solving the World's Problems on the Back of a Cocktail Napkin*. Princeton University Press, Princeton, NJ, 2009.

[49] Kurt Wiesenfeld. Resource letter: ScL-1: Scaling laws. *American Journal of Physics*, 69(9):938-942, 2001.

[50] Michael M. Woolfson. *Everyday Probability and Statistics: Health, Elections, Gambling and War*. Imperial College Press, London, 2nd edition, 2012.

Index

An italic page number refers to a problem on that page.

This book was typeset entirely with free software and fonts. The text is set in Palatino, designed by Hermann Zapf and available as TeX Gyre Pagella. The headings are set in Latin Modern Sans, based on Computer Modern Sans, designed by Donald Knuth.

The source files were created with GNU Emacs and managed with the Mercurial revision-control system. The figure source files were compiled with MetaPost 1.999 and Asymptote 2.44. The TeX source was compiled to PDF using ConTeXt 2014.05.17 and LuaTeX 0.79.1. The compilations were managed with GNU Make and took 3 minutes on a 2015-vintage Thinkpad T450s laptop. All software was running on Debian GNU/Linux.

A heartfelt thank you to all who contribute to the software commons!